CABLE COMMUNICATION

Second Edition

Thomas F. Baldwin
Michigan State University

D. Stevens McVoy
Coaxial Communications

PRENTICE HALL
Englewood Cliffs, New Jersey 07632

Library of Congress Cataloging-in-Publication Data

Baldwin, Thomas F.
    Cable communication / Thomas F. Baldwin, D. Stevens
  McVoy.
    448 p. 15.2 × 22.9 cm.
    Includes bibliographical references and index.
    ISBN 0–13–110263–X
    1. Cable television. I. McVoy, D. Stevens. II. Title.
HE8700.7.B35 1988   87–16109
384.55′56—dc 19   CIP

Editorial/production supervision and
  interior design: Fred Dahl
Manufacturing buyer: Edward O'Dougherty

© 1988 by Prentice-Hall, Inc.
A Division of Simon & Schuster
Englewood Cliffs, New Jersey 07632

Printed in the United States of America

10 9 8 7 6 5 4 3 2 1

ISBN   0-13-110263-X

Prentice-Hall International (UK) Limited, *London*
Prentice-Hall of Australia Pty. Limited, *Sydney*
Prentice-Hall Canada Inc., *Toronto*
Prentice-Hall Hispanoamericana, S.A., *Mexico*
Prentice-Hall of India Private Limited, *New Delhi*
Prentice-Hall of Japan, Inc., *Tokyo*
Prentice-Hall of Southest Asia Pte. Ltd., *Singapore*
Editora Prentice-Hall do Brasil, Ltda., *Rio de Janeiro*

To Jan Baldwin and Sue McVoy

Contents

PART III   PUBLIC POLICY

PART IV   ORGANIZATION AND OPERATIONS

Preface

The second edition of *Cable Communication* updates all of the original material and adds two chapters. Major changes were required in the chapters dealing with public policy to reflect the 1984 Cable Act. A new section discusses cable's emergence as a First Amendment speaker. Greater emphasis is now placed on renewal of franchises since so much of the U.S. will be in that process in the next few years. With advertising sales developing rapidly as a function in the cable industry, a separate chapter treats the unique character of cable as an advertising medium. Another new chapter, by Joseph Straubhaar, describes cable development outside the United States.

Appendices include the Cable Act of 1984, FCC Rules for local origination, sample access rules, local origination rules, descriptive information on the basic and pay satellite networks, sample advertising production rates, a set of typical operating cost figures for cable systems, procedures for assessing cable-related communication needs and cable audience survey methods.

A great many people made substantial contributions to this book. Georgella Muirhead, public information office for the City of Southfield, Michigan, supplied information about the operation of government channels. Randy VanDalsen, former national coordinator of local origination programming for United Cable, and Robert DiMatteo, *CableVision* magazine, provided material used in the public access and community channels sections.

Barry Litman, in the Department of Telecommunication, Michigan State University, read the original pay cable chapter, making a number of suggestions. Dave Hanson, HBO Chicago, was very helpful in supplying information. Involved in the Michigan State University, Rockford, Two-Way Cable Project were James Cragan, former Rockford, Illinois, Fire Chief; James Wright, then with Rockford Cablevision; and Martin Block, John Eulenberg, Bradley Greenberg, and Tom Muth. This project provided technical and applications knowledge reported in Chapters 5 and 9. Geoffrey Gates, Cox Cable Communications, read the original Chapters 5 and 9, making numerous useful suggestions.

The National Science Foundation supported work reported in Chapters 5, 9, and 18. Charles Brownstein was the program manager.

Robert Yadon, now with Ball State University, read several sections on business organization and made suggestions that were incorporated. Gil Hernandes, Brian McNamara, formerly of Coaxial Communications, and Frank Prosen, Continental Cablevision, contributed parts of that chapter on business organization. Glenn Friedly, Horizon Cablevision, helped write the section on cable finance, reflecting new business conditions and tax laws. Genelle Armstrong, Director of Customer Service, and Harry Cushing, Director of Field Operations, Coaxial Communications, provided the basic information for the section on customer service. Harry Cushing also reviewed the original chapter on distribution plant design and construction. Doug Grace, Chief Engineer for Coaxial Communications, reviewed the original chapter on headends.

Scott Westerman, Regional Marketing Manager, Continental Cablevision, made numerous contributions to the chapter on marketing. Carol Mackey of AT&T and Ronald Paugh of Ashland College also made contributions to the marketing chapter. Kensinger Jones of Michigan State University, Martin Block of Northwestern University, David Gettys of Coaxial Communications, and Shirley Szabadi of HBO, Los Angeles, contributed to the development of the advertising chapter.

Bruce Franca, of the FCC, responsible for developing the FCC response to the 1984 Cable Act, read all of the public policy sections and made very helpful comments. Robert Whitehead, Bobby Baker, and Rick Kalb of the FCC Cable Branch were also helpful. Wesley Heppler, a communications attorney with Cole, Raywid and Braverman, read parts of the public policy chapters making useful contributions. Todd Simon of Michigan State made numerous suggestions on the interpretation of cable status under the constitution. Sharon Briley of the FCC and Jim Ewalt of NCTA were helpful in the section on state government.

Jean-Luc Renaud, Megumi Komiya, and Charles Steinfield, all of Michigan State University, made contributions to Chapter 17 on cable development outside the United States.

We drew on the work of Carrie Heeter and Bradley Greenberg, both of Michigan State, in the chapter on audiences.

Several people were most helpful in searching out photographs and illustrations: Jessica Baron, Warner Amex, Cincinnati; Sally Cahur, HBO; Linda Holland, Tocom; Shirley Leslie, FCC; John Feight, Scientific Atlanta; Leo Murray, Warner Amex; Harry Cushing, Coaxial Communications; David Anderson, Cox; John Reinhart, Continental; Whit Sibley, X-Press; Sandy Neuzil, Electronic Program Guide; Jim DeBold, Cable Television Network of New Jersey; Lawrence Pike, Silent Network; Caroline Bock, USA Network;

Barbara Shulman, MTV Networks, Inc.; J. I. Taylor, Zenith Electronics Corporation; Marilyn Bellock, CTSS Cable Connect; Dennis Melton, Channelmaster; Alan Taylor, Channelmatic; Molly Seagrave, HBO; Rob Maynor, Disney; Kitsie Bassett, CNN; Susan Swain, C-SPAN; Terri Luke, A. C. Nielsen; Tola Murphy-Baran, Showtime; Kazie Metzger, Group W.

John Duhring recognized the need for this book and was responsible for its original publication by Prentice-Hall.

Reviews of the manuscript for the second edition, by Dan Agostino of Indiana University, Morleen Getz Rouse of the University of Cincinnati, and Manjunath Pendakur of Northwestern University, were extremely helpful.

Ann Alchin handled much of the manuscript typing in East Lansing; Phyllis Podkin in Columbus. Peggy Wong, in Hong Kong, worked on the index. Again, we acknowledge the patience of our families with this continuing project.

To all these people, we are very grateful.

CHAPTER 1

Introduction

Channel lineup, Portland, Oregon

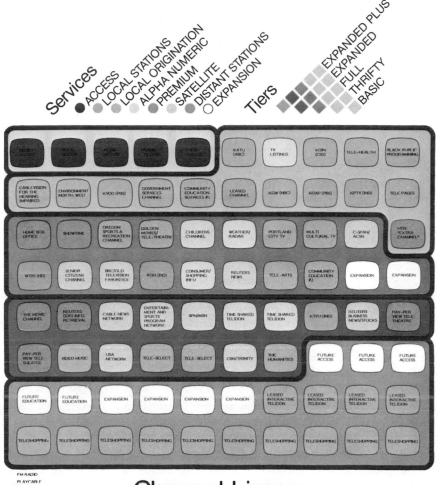

Channel Lineup

Starting as a means of capturing broadcast television signals for people at too great a distance from a transmitter or blocked by mountains, cable has grown to be a proliferating television delivery system in cities as well as remote areas. It owes its more recent growth to what economists have called a *consumer under-investment* in television. According to Noll, Peck, and McGowan, "The available evidence from both STV [subscription TV] and cable experience suggests the existence of a considerable unfulfilled demand for television programming, both of the conventional type and a few categories not well represented in the present programs logs."[1] The extent of the unfulfilled demand was not fully appreciated until the late 1970s when satellite-delivered movies, superstations, and other satellite cable networks came into being and cable began selling a wide range of nonbroadcast services. The discovery of this new appetite for television went far beyond the broadcast retransmission business of the original community antenna television (CATV) operators.

Now cable television is everywhere in the United States with the exception of very low-density housing areas where it is not practical and some major cities that are now being franchised and built. Elsewhere in the world, cable is also developing or under consideration.

CABLE PROMISE

Early in its history, cable captured the imagination of communication scholars, television critics, communication policymakers, and others who saw more promise for the medium than the products of limited-channel commercial television. In this section we review the traditional criticisms of the broadcast system and suggest various promises of cable purported to respond to each.

Diversity

Technical factors require separation of broadcast television channel assignments. Adjacent channels (such as channels 7 and 8), with some exceptions, cannot be assigned to the same geographic area.[2] They would interfere with each other. Two stations on the *same* channel must be separated by 150 miles or more, so that their signals don't overlap and make a muddle of the signals for people in the middle. It would have been possible for the FCC (Federal Communications Commission) to have created a system of high-powered regional stations so that every household would have six or more stations available, but this would have sacrificed *local* service, which was a crucial goal of the FCC in creating the table that assigned frequencies to cities. As the table was worked out, most places in the United States had three or fewer stations. This effectively limited the number of networks to three.

In programming television networks, it made better economic sense for each of the networks to aim for the mass audience with similar tastes, that is, to create the "lowest-common-denominator" programs that would attract a one-third share of the majority.[3] Following this strategy, the networks, and also individual stations programming their own option time, imitated each other's successes, and all television stations were pretty much alike. People who didn't conform to the interests of the mass audience were under-served. Critics and intellectuals railed at this development, but the economic incentives of the commercial television system, under the table of assignments, all favored the system as it existed.

Cable offered an answer—*unlimited* channels: 12, 35, 54, even more. With this abundance, new networks could arise, and, since the commercial broadcast networks served the mass audience well, at least some of the new networks would *have* to be more specialized. With signals beamed from communication satellites to cable system earth stations, this hope for diversity has actually been realized to a degree.

New Opportunities

Another complaint against the broadcast television system arising from the limited-broadcast-channel assignments was the monopoly of communication and market power it gave to a few corporations. The three commercial broadcast networks controlled prime time and much of the rest of the day. This presented a tight market to creative talent. Very few could break in. There was no room to experiment with new ideas. Some felt that, in news and public affairs, television was dominated by a few *white men* in New York and Washington, and in entertainment by *white men* in New York and Hollywood.

Cable could loosen the hold of the networks and their affiliates, as well as open television to new talent and fresh ideas. Certainly cable has provided new options in prime time as the satellite cable networks have emerged. As cable reaches a higher proportion of U.S. homes more new entertainment material will be produced originally for cable, thereby increasing the market for talent and ideas. Now there is a Cable News Network (CNN) with headquarters, not in New York, but in Atlanta, Georgia.

More News

Many have said there is no breadth or depth to broadcast news. Broadcast television is only a headline news service that is not always offered at convenient viewing times. Commercial television stations have expanded news time to what they believe to be the tolerable limits, economically. Only for crisis news events can entertainment programming be sacrificed to news broadcasts.

Cable can devote whole channels to news. A cable system may have CNN and CNN Headline 24-hour news services, local information channels, and two or three full-time text news channels. Cable can present *raw* (unedited) news—gavel-to-gavel coverage of the U.S. House of Representatives, city council meetings, school boards, trials, special events, and so on.

Many neighborhoods and communities within metropolitan areas are too small to win the attention of the big media—newspapers, broadcast television, and radio—that must cover the entire metro market. Cable can offer community news and information in either full audiovisual or alphanumeric text format.

Access

Critics of broadcast television have lamented the fact that, under the constraints of spectrum scarcity, not everyone can own a station. There is no access to the airwaves comparable to the freedom to print a newspaper or a handbill. Efforts to impose some elements of free expression on broadcast television produced the FCC's Fairness Doctrine and other federal rules that encroached on the freedom of the broadcaster and were not an entirely satisfactory solution.

Cable can provide a community soapbox, giving over one or more channels to the public. Many cable franchises require a public access channel.

Less Commercial Intrusion

Some people are offended by television commercials, although Americans are generally tolerant. The critics say that commercials interrupt program flow, influence television content, are tasteless, and invade viewer privacy.

The cable subscriber can experience commercial-free television on several cable channels in addition to PBS.

Education and Government

Educators and public service providers note that in most countries TV first serves public communication and education needs and then commercial interests, if at all. In the United States it is the other way around. Commercial broadcasting dominated the system and took the best channels first. Education was second, and government public service channels operated in only a handful of cities.

Cable has room for government, education, and library channels. Some cities and cable systems are proud of the innovative uses and accomplishments of these channels.

Interactive Television

Finally, almost everybody is somewhat uneasy about hours of passive television viewing. It doesn't seem healthful for kids, and adults feel guilty about their own viewing.

Interactive two-way cable can offer a modicum of involvement to the television user. Through polling and instant feedback, viewers can have some sense of the rest of the audience. Television can be used as a reference service where the user may order text and graphic information to serve a variety of individual needs.

HISTORY

The origins of cable are humble. There was no vision of its current services and impacts. When broadcast television became a reality for many areas of the country in 1948, people in remote or shielded areas felt a sense of deprivation. Appliance stores and radio repair shops in these areas were denied the booming new business in television. The most imaginative of the appliance dealers and repair persons began to look for a way into the market. Several of them laid claim to the original community antenna television (CATV) system.

One is Robert J. Tarlton of Lansford, Pennsylvania, a radio sales and service person. Lansford was only 65 miles from Philadelphia, close enough to receive weak television signals, but cut off by the Allegheny Mountains. A few venturesome people bought television sets. The reception was terrible. Tarlton went to the top of the mountain in 1949 and tried to set up individual antennas for the set owners. It worked, but it would have resulted in a mountain-top forest of antennas and rivers of cable coming down the hill with tributaries all over town.

Tarlton thought of a better way. He found some friends to invest with him in a company called Panther Valley Television. They built a master antenna at the mountain summit, amplified the weak signals from Philadelphia, and distributed them house-to-house by coaxial cable hung on poles. Panther Valley charged an initial installation fee of $125 and $3 per month. The system brought in the three Philadelphia television stations clearly, and Tarlton began selling television sets.[4]

At the same time, in Astoria, Oregon, Ed Parsons at KAST radio was experimenting with antennas to get television from Seattle for his wife. He ran the wire from his home to a hotel lobby across the street and to a nearby music store. Even earlier, in 1948, John Walson, a power and light maintenance employee and appliance dealer in Mahanoy City, Pennsylvania, claimed to provide a master antenna-cable system. However, all his records were wiped out in a fire.

From 1949, the number of cable systems grew slowly but steadily. By 1961 there were 700 community antenna TV systems. Growth accelerated so that in 1971 there were 2,750 systems serving nearly six million homes.[5] During this period, cable was first providing television to homes that were entirely out of rooftop antenna range of any television stations. Later, cable came to be a business of filling out the complement of television services in communities that had less than the three commercial networks. Through much of this period, cable was viewed as ancillary to the broadcast service, simply a community antenna. There was no intent to bring in anything except the nearest television stations, although some CATV systems reached out a fair distance to get those stations with the help of microwave relay stations.

In the 1970s, the concept of originating programs at a cable system was established. The programs were fed directly into the cable. Cable systems could also receive television signals from satellites. This meant that cable systems became more than an extension of broadcasting stations. They became a programming *source*. The cable systems of the 1970s were capable of two-way communication. This meant that their amplifiers were constructed to accommodate signals *from* the receiving household as well as *to* the household if the demand ever arose for such a service. Cable became a true *multichannel* medium with basic and premium services from satellites making cable attractive to urban areas in addition to its traditional markets.

In the latter half of the 1980s the cable industry can be expected to solidify its position as a supplier and distributor of multichannel programming and mature in the administrative details of system operation. The industry will improve its stature as a national and local advertising medium. These prospects will be discussed in the body of the text.

In 1990 cable will pass about 80 percent of all U.S. households and be connected to more than 50 percent of all television households. Industry revenues will be about $16 billion, twice the 1984 revenues.[6]

BOOK ORGANIZATION

This book is in five parts. Part I is the technology of cable, tracing a cable system from the control center, or "headend," through the distribution plant to the home. A final chapter presents the special technology of two-way cable.

Part II describes cable services, present and future, including over-the-air television and radio; automated channels; public, government, and education access channels; community channels; satellite-delivered programming; two-way services; and audiences.

Part III is on public policy—the Congress, FCC, copyright law, state regulation, and local franchising.

Part IV is the business of cable. It includes ownership patterns, finance, accounting, audiences, marketing, advertising, customer relations, maintenance, engineering, and personnel. This part also contains a section on professional resources.

Part V discusses multichannel communication technologies as they relate to each other and to the general media environment in the U.S. and elsewhere in the world.

NOTES

[1]Roger G. Noll, Merton J. Peck, and John J. McGowan, *Economic Aspects of Television Regulation* (Washington, D.C.: The Brookings Institution, 1973), p. 32.

[2]This is true except for channels 6 and 7, which have the entire FM band separating them, and channels 4 and 5 which are separated by an air navigation band.

[3]For a discussion of the economics of programming, see Bruce M. Owen, Jack H. Beebe, and Willard G. Manning, Jr., *Television Economics* (Lexington, Mass.: Lexington Books, 1974), Chapter 3.

[4]Ralph Lee Smith, *The Wired Nation* (New York: Harper Colophon Books, 1972), p. 3.

[5]Sloan Commission on Cable Communication, *On the Cable* (New York: McGraw-Hill, 1971), p. 31.

[6]Peter D. Shapiro and Donald T. Schlosser, "Prosperity for Cable TV: Outlook 1985–1990," Arthur D. Little, Inc., p. 46.

CHAPTER **2**

Headend

Base of antenna tower and satellite receive stations at the headend, Cincinnati, Ohio.

OVERVIEW

The technology of cable television is very important for several reasons. Providing a cable service requires a significant capital investment in construction and the equipment, which captures, creates, processes, and distributes the signals used by the subscribers. The equipment sets the limits on how many channels are available. It determines the ease of use in the home, the compatibility with television receiving equipment and videocassette recorders, and the quality of the picture. Chapters 2 through 5 describe how cable equipment works and the relationship of the cable technology to the economics and operation of a cable system.

A *cable television system* is in essence a method of distributing television, radio, and data signals from a central originating location to residences and businesses by way of coaxial cable. It consists of three main elements. First is the *headend*, which is the point at which all the program sources are received, assembled, and processed for transmission by the distribution network. Second is the *distribution network*, consisting of coaxial cable leaving the headend on power or telephone company poles or, in some cases, buried underground, and going down each street within the community serviced. Third is the *subscriber drop*, which consists of the coaxial cable going from the street into the individual subscriber's home, and the related equipment required to connect the cable to the subscriber's television receiver and other devices.

The operation of a cable television system can be compared with that of a municipal water system as shown in Figure 2–1. In a water system, the supply of water comes from either a well field or a reservoir. In cable television, the headend is the source for the signals to be transmitted to the subscribers.

Water mains leave the well field to carry the water into individual areas of the community to be served. In the cable system, the trunk cable serves a similar function for transmission of the television signals.

Along the way, friction in the water pipes slows down the flow of water, and pumping stations are required to bring the water flow up to the required rate. In a cable television system, electronic signals diminish as they travel through the cable, and amplifiers are required periodically to boost the signal back up to an acceptable level.

Within individual neighborhoods, a water system branches out, with smaller-diameter pipes along each street. Similarly, in a cable system, smaller-diameter cables are used to transport TV signals down individual streets. Each water system customer is connected to the pipe at the street by a junction box located near the street, typically next to the sidewalk. A portion of the water is tapped out of the pipe and is sent into the customer's home. In a cable television system, the multitap serves a similar function, with electronic signals tapped from the distribution cable and fed over a still smaller-diameter cable into the subscriber's home for connection to television and radio receivers.

FIGURE 2-1. Cable system compared to a water system.

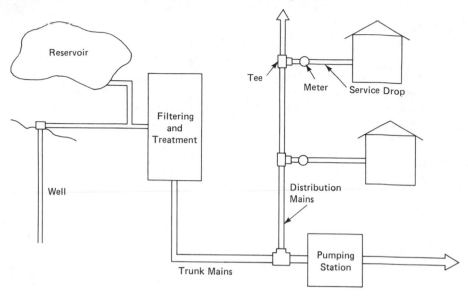

Water System

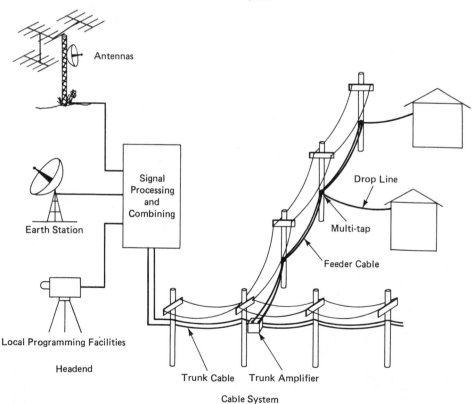

Cable System

THE CABLE TELEVISION HEADEND

As mentioned, the headend is the originating point for all the services carried on a cable television system. Over the years, it has grown in complexity from perhaps ten pieces of electronic equipment in a small utility building to an elaborate control center consisting of thousands of pieces of equipment, including satellite-receiving equipment, towers, antennas, computers, TV production studios, test instruments, and radio and television receivers. The production studios may be located at the headend or in some more convenient site and connected to the headend.

Over-the-Air Channels

Early cable television systems served only one purpose: to pick up distant television channels that could not be received by residents of a community and to deliver those channels to subscribers. Typically, a cable television headend would be located on the top of a mountain near the community, or a very tall tower (300 to 1,000 feet) was built. Antennas were installed to bring in stations from distant cities, and a small building was constructed to contain *signal processors* (devices that "clean up" the pictures from the desired television channel). Since these channels came from some distance away, they would often vary greatly in signal strength. One of the signal processor's purposes is to level out the strength of these signals so that they are more uniform. Signal processors also filter out interference from adjacent channels, and sometimes convert the incoming channel to a different frequency for carriage on the cable system. After each of the signals has been processed, all are joined together in a combining network and fed into the cable distribution plant.

Later, cable television operators began originating some of their own services, most commonly a time and weather channel. At the headend, a television camera was mounted on a motorized platform that would rotate back and forth televising a thermometer, a barometer, a windspeed indicator, and a clock. To put this television picture on the cable system, a *modulator* is required. A modulator is a miniature television transmitter that generates a signal on a television channel. The output of this modulator is combined with the other channels in the headend *combiner* for transmission over the cable distribution plant. (See Figure 2–2.)

Microwave

Another source of signals for the cable television headend is *microwave* transmission. Sometimes, when a cable operator wishes to provide its subscribers with a distant television station, it is impossible to receive that station with an antenna located on a tower. In this case, the cable television operator has two options. First, if there is an existing *common carrier* microwave net-

FIGURE 2-2. Simplified diagram of cable headend.

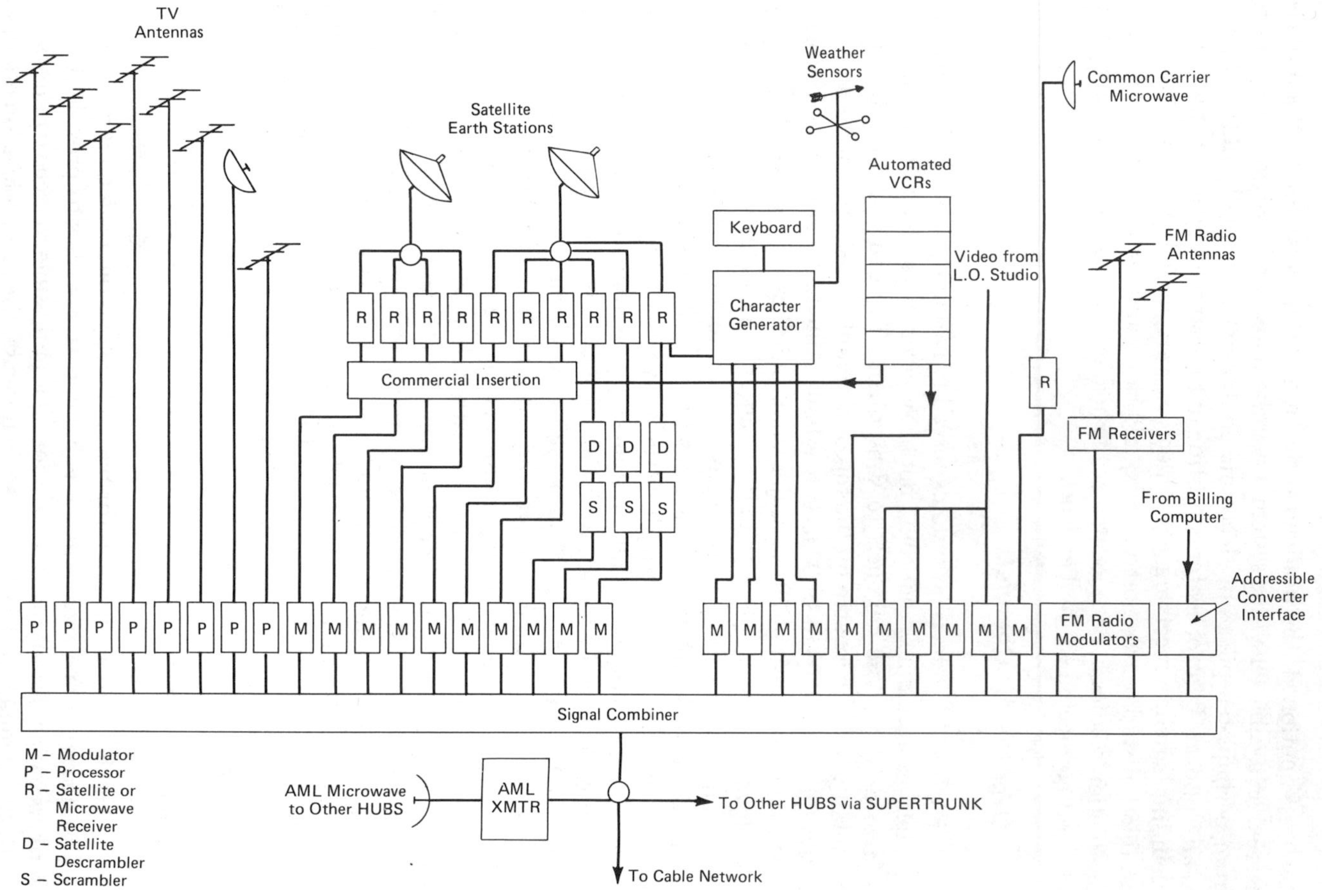

M – Modulator
P – Processor
R – Satellite or
    Microwave
    Receiver
D – Satellite
    Descrambler
S – Scrambler

work in the area, the operator might contract with that common carrier for delivery of the desired channels to the headend. Common carriers are licensed by the FCC (Federal Communications Commission) to deliver television channels to a number of cable television systems on a *tariff* (fee schedule) established by the company and approved by the FCC. In this case, the common carrier company provides a microwave-receiving antenna on the cable company's tower and the necessary receiving equipment in the headend. The cable operator provides only the modulators to put the desired channels on the cable system.

The alternative is for the cable TV company to build its own *community antenna relay service* (CARS) microwave system. The cable operator purchases or leases land close enough to the desired television station to receive a good picture, and installs a small building and tower with the necessary microwave-transmitting equipment to relay the signal to the headend. In some cases intermediate relay points are required.

A microwave signal can travel approximately 25 miles before it must be retransmitted. Microwave is also being used for interconnection of portions of large cable systems and for exchange of programming between systems.

Satellite Earth Stations

During the mid-1960s, the technology became available for distribution of telephone, data, and video signals via communications satellite. These satellites are located in a *geosynchronous orbit* 22,300 miles above the earth at the equator. At this altitude, they make one revolution per day around the earth so that they appear to be standing still in relationship to the ground below.

Communications satellites are broadcast relay stations. They pick up signals that are beamed from *uplink* stations on the ground and retransmit them back down toward the earth to a *receiving station*, often called an *earth station*, a *TV receive only* (TVRO), *downlink*, or simply a *dish*.

Communications satellites use *transponders* to receive the signals from the ground and relay them back to the earth stations. Although there are many types of communications satellites, the ones used for cable purposes are capable of transmitting 24 TV channels, together with a number of *subcarrier* audio and data channels within each TV channel.

Given the great distances of satellites from earth and the state of their art in design, the signals that arrive at an earth station are extremely weak. For this reason, large-diameter receiving antennas are required, together with sophisticated preamplifiers and receivers. When communications satellites were first used for cable service, cable operators were required to install antennas that were 10 meters in diameter. These earth stations cost in excess of $100,000. Later, the FCC authorized the use of smaller earth stations. Typically, satellite earth receiving stations of 4 to 5 meters in diameter are used by the cable industry.

In some areas of the country, and particularly in Hawaii and Alaska, the signal from the satellites is weaker than in other areas, and larger-diameter antennas are required. For instance, in Alaska a 10-meter earth station is almost always necessary. In the southern portion of Florida, a 7-meter antenna is usually required.

At present, all programming for cable use is in the *C band*, frequencies at about 4,000 MHz (MegaHertz, see Figure 3–1). However, satellites now in orbit use the *Ku band* (12,000 MHz). Ku-band satellites have the advantage of requiring much smaller earth stations (one to two meters in diameter), but the receiving equipment is more expensive than C-band equipment, and Ku-band transmissions are more subject to interference from heavy rainfall.

Before an earth station is installed, an engineering study must be performed to determine that the antenna size is sufficient to produce a good quality picture and that nearby microwave transmitters, which share the same frequencies, will not interfere with the reception of the satellite signals. This work, called a *frequency coordination study*, is performed using a computerized listing of all the existing microwave transmission facilities in the United States. A cable operator usually relies on a consulting service to provide this study.

There is no advantage in putting in an earth station larger than the one necessary for good quality reception. If a 5-meter station is called for by the engineering study, installation of a 7-meter station will not provide better picture quality; however, installation of a 4-meter station will most likely produce inferior results.

Three major types of earth stations are in use today. The first and most common is the *parabolic reflector*. This type of antenna looks like a giant dish and is mounted on a concrete slab and aimed toward a point above the equator where the satellite is located. (See the photograph on the chapter title page.) Geosynchronous satellites are located at specified positions above the equator at least two degrees apart and each transmitting on the same frequencies. By aiming an earth station at one satellite, and aligning it carefully, transmission from other satellites will not interfere with the desired signal.

Recently, with the split of most cable programming between two primary satellites which are located a few degrees apart, multisatellite feeds have become common. The same parabolic reflector concentrates signals from two different satellites into two different collecting points, or *feeds*. In this way two, or even three satellites, can be received by a single dish.

The second type of earth station is a *multibeam antenna*. These earth stations are similar to the parabolic reflector except that they are oval. They are designed to receive signals from several satellites at the same time over a wide angle. These antennas are used when reception of signals from three or more satellites is required, and when the satellites are not located near each other over the equator.

The third type of earth station is called a *conical horn* and looks like an immense horn of plenty. The conical horn is substantially more expensive than the parabolic reflector, but it is sometimes necessary in areas with interference from nearby terrestrial microwave sources.

Each earth satellite is capable of transmitting 24 channels, and a single earth station is capable of receiving all of them. At the antenna feed, *low noise amplifiers* are used to boost the extremely weak satellite signals. A separate receiver and modulator is necessary for each channel received from the satellite.

At first, only a few channels of satellite programming were offered, and there was no problem getting them all on a single satellite—the RCA Satcom I. (See picture on the title page of Chapter 7.) However, as new services became available, the RCA satellite filled up, and program suppliers were forced to go to additional satellites. At present, most cable programming is on two adjacent satellites, the RCA Satcom III-R and Hughes Galaxy I. Most cable systems have only one earth station. Many, however, have retrofitted their stations with multisatellite feeds to pick up both satellites, or they have installed a second or even a third earth station to take advantage of all the programming that is available.

Recently, cable programmers have begun scrambling their satellite feeds to control unauthorized use and to position themselves to charge the growing *backyard dish* (home satellite earth station) population for their services. The scrambling techniques used are relatively secure, making it unlikely that bootleg descramblers will allow unauthorized reception.

Descramblers are installed after the earth station receivers, and one is required for each scrambled channel. In the future, it is likely that all cable services will be scrambled and that the same scrambling technology will be used by all programmers to allow home earth station users to receive all cable programming with a single descrambling unit.

Headed Buildings

A headend with 20 to 35 channels, with reception of 20 or so satellite channels, can be housed in a building with no more than 200 square feet of floor space. Today's equipment is very reliable and can be operated unattended. Proper heating and cooling is required to keep the equipment around 60 to 90 degrees Farenheit, and a standby power generator, which starts automatically in case of power failure, is usually installed. Many cable headends in the U.S. today are unattended.

Automated Channels

With the advent of cable systems in major markets, the complexity of cable headends has increased dramatically. In addition to all the signal sources noted, cable systems have started originating a large number of TV channels. Most of them are *automated services;* that is, they are alphanumeric channels, such as news, weather, sports, program listings, stock market quotations, and community bulletin boards. A device called a *character generator* is required to provide these channels. A character generator is a special-purpose computer that accepts input from a keyboard or other source and formats the information in a pleasing way to be displayed on a television screen. Initially, character generators were used by cable systems to provide national and international news service from the existing news wire services, such as the Associated Press or Reuters. The cable operator leased a telephone line from the nearest Reuters or AP center and connected it to a character generator. The telephone line transmits digital information to the character generator at the cable headend, where it is formatted and displayed on a TV channel. Today the digital information is transmitted over a satellite subcarrier rather than a telephone line.

As more channels of alphanumeric information were required, cable operators began installing complex *multichannel character generators*. These systems have the capability of programming on 10 or more different television channels and of storing thousands of *pages* of information. A typical multichannel character generator system has some channels programmed by a keyboard (the program schedule, community bulletin board, and the like), some channels programmed by satellite subcarrier data channels (AP, Reuters news), and some channels programmed by automatic sensors (weather).

Many cable systems offer a weather radar channel. The National Weather Service radar centers have devices that create a digital data stream from its weather radar display. The information from this radar picture is transmitted over a telephone line to the cable system headend, where it is reformatted to produce a video display of the weather radar scan.

Headend Automation

As newer cable systems have become more sophisticated, headends have changed from unmanned sheds located on the outskirts of town to large operation centers manned 24 hours a day. (See Figure 2–3.)

Many cable systems now sell commercial advertising time on their satellite-delivered channels. Most of the advertiser-supported channels have local availabilities for insertion of 30- or 60-second commercials, and many channels are programmed with cue tones to control *automated insertion equipment* at the cable headend.

A *cue tone* is a series of audio tones, similar to those produced by a touch-tone telephone, that are transmitted on a satellite channel a few seconds before a local commercial availability is coming up. At the cable headend, the automation equipment recognizes cue tones and plays back the appropriate video-taped commercial at the proper time on the proper channel. Automated insertion systems generally are controlled by a microcomputer, which operates several videocassette players (one is required for each channel to be automated).

FIGURE 2-3. Interior of the headend, Cincinnati, Ohio. Racks contain modulators for each channel.

Automation equipment for large cable systems must control many channels and is complex and expensive.

In addition to commercial insertion, modern cable systems have other switching requirements that are best handled by computer automation. Often, two programming sources share a single channel, or a channel is programmed entirely by the operator using a combination of programming sources, such as videotape, microwave, over-the-air broadcast, and satellite. However, because of the complexity of the switching operation, and because of last-minute changes in program schedules, automation is not, in itself, sufficient. Many major market cable systems today provide around-the-clock monitoring of program quality and switching to assure that subscribers are receiving the programs and quality that they expect. With the enactment of FCC and copyright rules on network nonduplication and and sports blackouts, systems were required to start switching program sources. For instance, if a distant network affiliate television channel was a program source on a cable channel, the cable operator might be required to black out the network part of the programming. Rather than leaving that channel time blank, many cable operators would opt to insert programming from another television station in its place. To do this, switching equipment is required. In the early days of cable, the switching was often done manually or with crude rotating drums. As the drum rotated, switches would be activated to enable or disable different program sources.

Usually this monitoring is done by video operators who constantly check high-quality television monitors. The operators looks sequentially at each of the signals carried on the cable system to make certain that the picture quality is good. The operators also check to make certain that the proper program is being carried on the proper channel at all times. Often cable systems have alternative sources for programming. For instance, if a program is being carried on the system and the picture quality is poor from one station, the video operator may switch to a second broadcast station signal. Also, the operator may have spare equipment to be used in case one of the channels fails. Video operators also handle the insertion of videotapes into the automated playback machines, and monitor playback quality.

Addressability

Many cable systems now use one-way addressability (see Chapter 4, "Home Drop"). Addressability requires that a computer be located at the headend to store the address codes for each subscriber converter, together with the services that the subscriber is authorized to receive. This computer is sometimes linked with the billing computer, so that changes in subscriber status (which pay television services and programming tiers are subscribed to) need only be entered once. The addressing computer is connected to the cable system using a transmitter or modulator, which converts the digital codes from the computer to a frequency for transmission on the cable system to the subscriber's converter.

Cable Audio Services

Almost from the beginning cable systems have carried audio services, utilizing the FM radio spectrum of 88 to 108 MHz (see Figure 3–1). Originally, the only services provided were rebroadcasts of local and distant FM radio stations, brought in using methods identical to those used for video signals. Today, however, there are a growing number of audio services being offered over cable in addition to rebroadcast of over-the-air stations.

Each video channel on a satellite is capable of eight or more audio subcarriers, which are being used to deliver a number of audio programming sources to cable systems, including distant FM radio broadcast stations, stereo audio for pay TV and other satellite video services, and packaged multichannel audio services to be sold as a premium tier to consumers (see Chapter 6).

At the headend, radio signals are processed similarly to video signals. For over-the-air FM stations, signal processors convert stations to a different frequency and stabilize the signal levels. For locally generated radio signals, a modulator is used to put the audio signal on an FM radio frequency. For signals carried on satellite subcarriers, subcarrier receivers extract the audio from the video channels. Modulators are then used to produce the FM radio frequency.

Up to 40 audio channels can be carried on a cable system in the FM band. However, many engineers are concerned about the effects of carrying a large number of FM stations, especially in 50-channel systems. They fear that degradation of video services could occur. In addition, it is difficult to provide undergraded FM audio service, since the signal levels used for radio transmission on cable cause reception to be noisy, a liability that will become more critical as FM receivers and home stereo equipment improve.

Recent cable audio technology will deliver much higher sound quality. Using either analog or digital techniques, these systems deliver sound quality equal to the compact disk audio players, but reduce the number of channels that can be carried in the FM radio band. (See Chapter 5, "The Technological Future," for more on cable audio.)

Stereo Sound

As already described, many satellite programming sources transmit their audio in stereo, and cable systems carry these stereo signals in the FM radio band. To get stereo sound, the customer must tune an FM radio to the appropriate frequency for the channel being viewed.

Recently, television stations have begun broadcasting in stereo, using subcarriers within the standard television channel. This technique, called *BTSC (Broadcast Television Standards Committee)* stereo or *MTS (Multichannel Television Sound)*, requires a special television receiver or an external tuner and decoder to deliver stereo sound. A third channel, which can be used for bilingual audio, is also available. As the number of television sets that can receive stereo sound increases, cable operators will begin programming in BTSC stereo. For over-the-air signals, standard headend processors will pass the stereo signal with little or no modification. For satellite-delivered channels, a stereo sound encoder is required after the satellite receiver and before the television modulator to create the stereo sound subcarriers.

Local Origination Facilities

Most medium-sized and large cable television systems offer locally produced programming on at least one television channel. Except for smaller markets, most cable systems have a small studio and control center consisting of two or three low-cost cameras and perhaps a portable videotape-camera combination. Such a studio might cost $75,000 to $100,000 and might be staffed with two full-time and three or four part-time employees. Maintenance is usually handled by the chief engineer of the cable system or perhaps a part-time video engineer.

With the explosion of the franchise competition in 1979 to 1982, cable operators began promising more elaborate local origination facilities. In urban systems, capital investments of $1-2 million for local origination equipment was not unusual, providing several origination studios, equipped with broadcast-quality camera and videotape machines. Expensive mobile production vans were available. Portable microwave equipment is included to allow the production of live programming from anywhere within the community. Major city franchise applicants often promised 50 or more employees for local program production, with operating budgets of several million dollars per year.

As operators experienced financial difficulties in these markets, they renegotiated franchise requirements and, in most cases, drastically cut back on the amount of local programming they produced. Many of the elaborate local origination studios that were promised were never built, and staff for local programming has been cut to a fraction of the original proposals.

Most urban systems now have local production facilities which, although more modest than originally planned, are of high quality and are capable of producing excellent programming. A typical urban system has three or four medium-priced cameras, a mobile production van, and several camcorders.

HUB INTERCONNECTION

Cable television systems are limited in size to a radius of about five miles from a headend location. A detailed explanation of these limitations is provided in Chapter 3. To service large metropolitan areas, a single headend is not sufficient. However, it is desirable to centralize the channel origination, processing, switching, and monitoring at one location. To serve large metropolitan areas, a *hub* system is typically used. (See Figure 2–4.) The area around the main headend is served by trunk lines leaving the headend. Other areas of the community are served by remote hubs, each of which serves an area with a radius of about five miles around the hub. To deliver the television programming from the main headend to each of the hubs, two techniques are employed.

The first is microwave. A complex microwave transmitter is installed at the main headend feeding several antennas, one directed at each of the remote hubs. At each hub, a relatively inexpensive microwave-receiving system is installed, which converts these microwave channels to regulate television channels for carriage on the cable system within that hub.

Because new systems are capable of large numbers of channels (50 to 100), a special microwave technique called *AML (amplitude modulated link)*

FIGURE 2-4. Hub system.

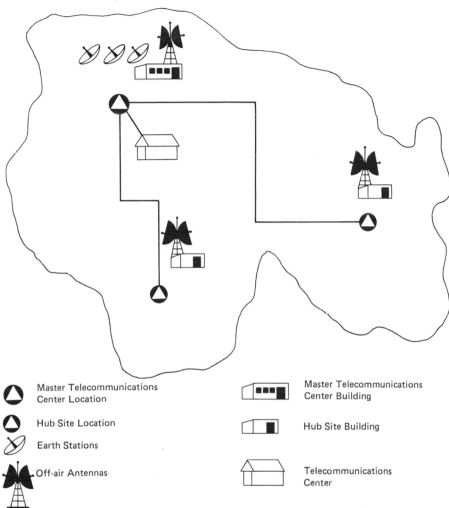

	Master Telecommunications Center Location		Master Telecommunications Center Building
	Hub Site Location		Hub Site Building
	Earth Stations		
	Off-air Antennas		Telecommunications Center

is used. AML systems have proven reliable and capable of delivering good-quality television signals in many locations. However, they have two major drawbacks. The first is that a failure of the transmitter system at the main headend will cause loss of service to all the hubs and therefore all subscribers in the cable system. Redundant equipment adds substantially to the total cost. Second, and more important, AML systems are not reliable in areas with frequent and severe thunderstorms. Rain attenuates the microwave signals very dramatically, and during heavy periods of rainfall, AML microwave signals deteriorate. AML hub interconnection, therefore, is not a wise solution in some locations.

A second interconnection method is the use of a *supertrunk.* A large diameter cable interconnects the main headend with each of the hubs. This cable is equipped with high-quality cable amplifiers. In some cases, FM transmission of television signals is utilized over the supertrunk, reducing the signal degradation. Another approach is to use *feedforward* amplifiers, which are expensive but have substantially less distortion than conventional cable amplifiers.

Supertrunk interconnection of hubs also has two disadvantages. First, signal degradation may be greater over a supertrunk interconnection than over AML microwave in certain cases. Careful design techniques, however, can keep this degradation to an insigificant level. Second, a supertrunk is susceptible to outages caused by failure of amplifiers or by physical damage to the cable plant. For instance, an automobile accident that involves knocking down a utility pole could put a supertrunk out of operation for hours. Recently, fiber optic cables have been used for supertrunks. (See Chapter 5.)

Reliability in modern cable television systems is of extreme importance, especially when home security or data transmission services are offered. Interconnection of the main headend with hubs, therefore, must be designed so as to provide almost flawless reliability. Redundant paths from the main headend to the hub provide one solution. For instance, interconnection might be done using both AML and supertrunk, or two supertrunks utilizing different physical routes might be employed. Either of these approaches is costly, and redundant interconnection is done in only a few of the very largest systems.

REGIONAL INTERCONNECTS

Many cable systems are now interconnected to allow common programming and advertising. Nearby systems can be interconnected in the same ways that hubs are connected, but systems that are some distance apart must use microwave. In this case, FM microwave is used in place of AML. FM transmission, which is the method used for CARS and common carrier microwave, is highly reliable and introduces little picture degradation. However, a great deal of spectrum space is utilized for FM and the cost per channel is high. Generally only a few channels are transmitted over a regional interconnect. Figure 2–5 shows a large statewide interconnect using microwave and cable.

HRC AND IRC HEADENDS

Some new cable systems are designed using two types of headends: *HRC (harmonically related carrier)* or *IRC (interval related carrier).* In Chapter 3, signal degradation due to distortion in cable equipment is described. These distortions can cause poor picture quality, and their severity increases as the

FIGURE 2–5. Cable Television Network of New Jersey covers Cape May to Mahwah by common carrier, microwave, wire, and AML.

number of channels carried on the system is increased. HRC and IRC headends provide one way of minimizing the problem.

HRC and IRC headends tie television channels to a common frequency reference, which causes the distortions that are created in amplifiers to cancel each other out or to occur at places where they are less visible to cable subscribers. HRC and IRC techniques will allow a cable operator to increase the system size without adding more headends or hubs, or to increase the performance of a cable system of a given size.

Although improvements can be made in picture quality using HRC and IRC headends, the distortions are still present, only hidden, and future services, such as teletext, may cause these distortions to become visible again. Also, certain other problems, such as signal ingress (see Chapter 5), are made more severe using these headends.

For these reasons a prudent cable operator will use HRC and IRC headends as a method of enhancing system performance rather than extending system size or reducing the number of hubs.

CHAPTER **3**

Distribution Plant

Section of a strand map, Coaxial Communications, Temple Terrace, Florida.

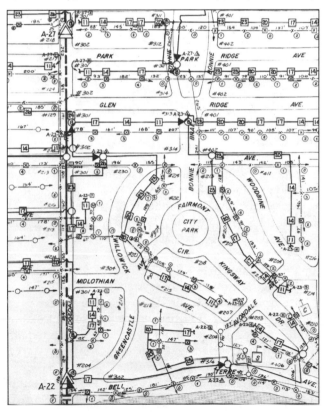

THE NATURE AND FUNCTION
OF THE DISTRIBUTION PLANT

In contrast to the cable headend, the complexity of cable television distribution systems has changed very little from the early days of cable. However, many improvements have been made in reliability, performance, and channel capacity.

The Cable Spectrum

To understand cable television systems, a knowledge of the television spectrum is required. (See Figure 3–1.) Each television channel occupies 6 MHz of space in the electromagnetic spectrum. Over-the-air TV broadcasts take place in three regions of the electromagnetic spectrum: channels 2 through 6 are broadcast at 54 to 88 MHz, and channels 7 through 13 are at 174 to 216 MHz. The third band, the UHF channels 14 through 83, are in the range of 470 to 890 MHz.

AM radio broadcasting takes place at about .5 to 1.5 MHz and FM radio at 88 to 108 MHz. Other frequencies are used for a variety of broadcast services, such as business, governmental, aircraft, military, and international broadcasting and communication.

Frequencies above 2,000 MHz are referred to as *microwave frequencies.* They are generally used for point-to-point fixed service, such as relay of telephone calls from one city to another, and relay of television programming. Communications satellites use microwave frequencies in the C band (3,700–4,200 MHz) and the Ku band (around 12,000 MHz) for downlinks.

Channel Capacity

The very earliest cable distribution systems consisted of coaxial cable about half an inch in diameter. Coaxial cable has a copper *center conductor,* which carries the TV signals. The conductor is surrounded by a plastic insulator, which in turn is surrounded by a braided copper shield, which prevents cable signals from leaking *out of* the cable and over-the-air signals from leaking *into* it. A plastic jacket is placed around the cable for protection from moisture and physical damage. In the CATV system, the amplifiers, which maintain signal strength throughout the system, were located in utility boxes on poles. A single amplifier was required for each channel carried on the cable system, and for this reason, three to five channels were all that could be carried.

These amplifiers used electron tubes, which have a limited life and which deteriorate in performance with age. Maintenance of this type of cable television system was extremely costly and difficult.

In the next version of cable TV systems, single-channel amplifiers were replaced with *broadband* amplifiers. These amplifiers were capable of simul-

FIGURE 3-1. Television spectrum, over-the-air and cable.

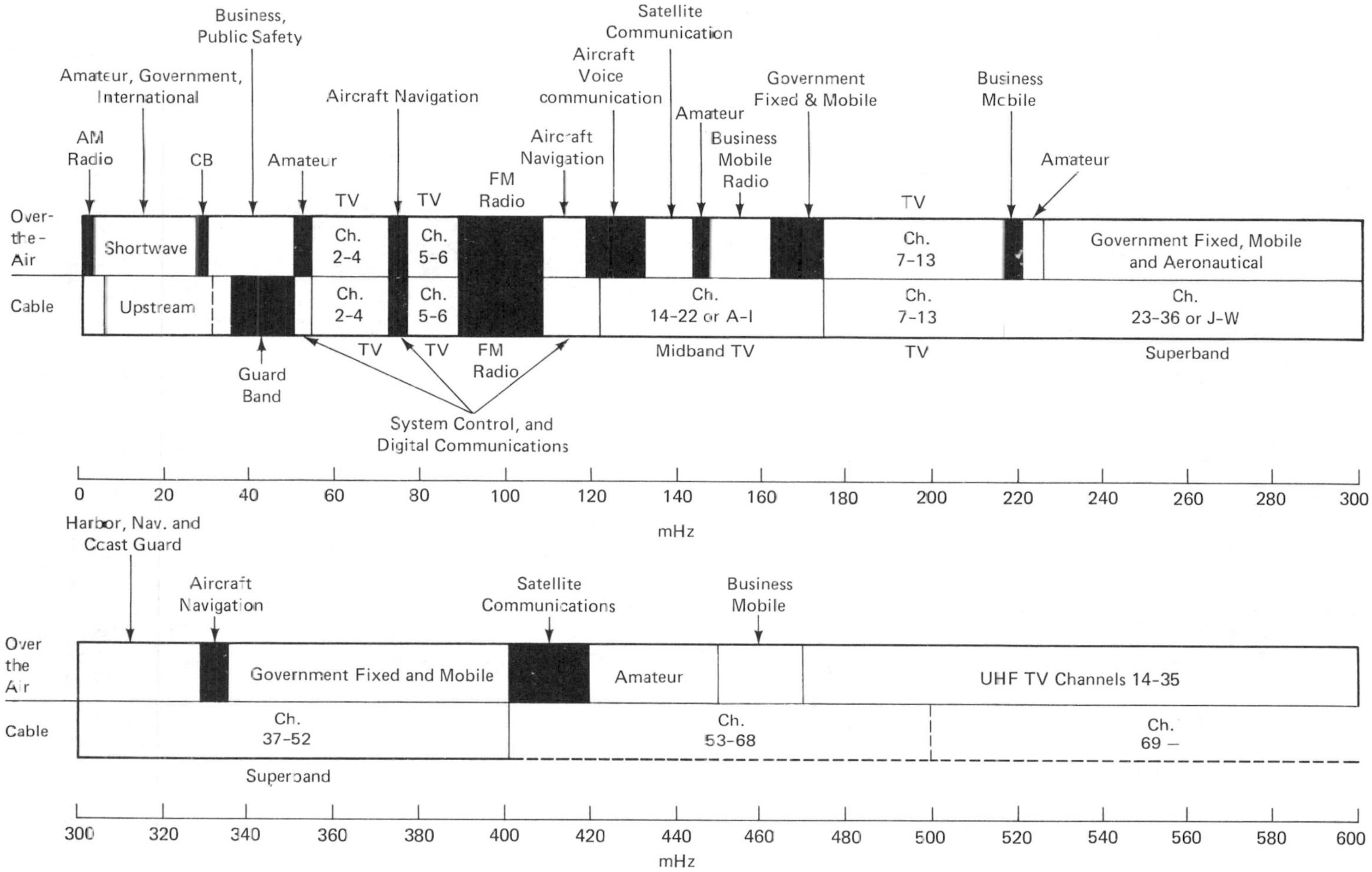

taneously carrying 54 to 108 MHz (5 channels) and later 54 to 108 and 174 to 216 MHz (12 channels). At first, these amplifiers also used tubes and had the same maintenance and reliability problems.

Later, in the mid-1960s, transistorized amplifiers began to appear. Although promising greater reliability than electron tube amplifiers, they were less reliable for the first several years after their introduction. Transistors are much more susceptible to damage by lightning and surges of electrical current than are electron tubes.

During the early 1960s, the braided copper cable was replaced by a solid sheathed aluminum cable with a center conductor of copper or copper-covered aluminum. With minor variation, this is the type of cable in use today. (See Figure 3–2.)

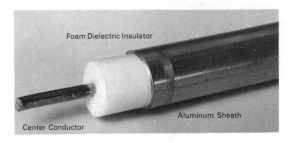

FIGURE 3-2. Coaxial cable.

In the early 1970s, the first hybrid cable amplifiers were developed. An outgrowth of integrated circuit development, a *hybrid* is a packaged device that contains integrated circuits and individual electronic components. An *integrated circuit* is a silicon chip, the surface of which is processed (*printed*) to create transistors and other devices that are connected to make an electronic circuit. The hybrid amplifiers, using integrated circuits, offer much higher reliability and performance than do the older transistor amplifiers.

About the same time, cable system amplifiers became capable of carrying 20 TV channels with reasonable performance. To increase a cable system's capacity from 12 to 20 channels, the space between channel 6 and channel 7 (121 to 174 MHz) is utilized. These frequencies are used by over-the-air broadcasters for other purposes such as aircraft communication, police and fire department transmissions, amateur radio, and military communications. However, since the cable system distributes signals only within a wire and not over the air, it can share these same frequencies without interference.

A normal television receiver cannot tune these *midband* channels (commonly labeled channels 14–22 in cable systems). To receive them, a converter is required. (Converters are described in Chapter 4.)

Going from 12 to 20 channels was not as simple as providing amplifiers that covered the frequency range of 54 to 216 MHz, because amplifiers are not perfect and introduce distortion to television pictures. All cable TV systems, no matter how well designed, contain these distortions, but in a well-designed and well-maintained system, the distortions are small enough that they are not visible to cable TV subscribers. As the number of channels is increased on a cable system, the severity of the distortion increases substantially. So going from 12 to 20 channels requires amplifiers with a great deal higher performance to produce good-quality television pictures. In the late 1960s and early 1970s, the distortion problem was the main limitation on the number of channels that could be carried on a cable system, and in fact, today, it is a major problem with the 50- to 80-channel systems that are being constructed.

In the early 1970s, channel capacity was increased to around 30 channels, and by the mid-1970s, 35 channels could be carried with excellent performance. To go from 20 channels to 35 channels required amplifiers to cover the frequency range of 54 to 300 MHz, with the extra channels being carried above channel 13 in the *superband* (216 MHz and up). These channels are commonly labeled channels 23–36. The most recent cable TV systems carry 54 to 80 channels. This is accomplished by increasing the amplifier's frequency spectrum up to 400–550 MHz.

There is much confusion about the actual channel capacity of these systems. Systems that carry from 54 to 400 MHz are variously described by different equipment manufacturers and cable operators as being 50-, 52-, or 54-channel systems. Frequencies below 54 MHz are not used for cable video services. The actual channel capacity depends on FCC regulations concerning use of frequencies in the aeronautical navigational and communications spectrum. Many cable systems are prohibited from using channels 14–16 in the midband, and some cannot use other channels in the midband or superband.

Signal Levels and Quality

The cable industry uses the *decibel* (dB) as a way of expressing the differences in signal level. The decibel is convenient because signal losses and gains can be calculated using simple addition and subtraction of dBs.[1]

Cable Attenuation

Signal strength diminishes as the signals pass through coaxial cable. This loss is referred to as *attenuation.* Cable attenuation increases at higher frequencies, with attenuation approximately doubling every time frequency is quadrupled. In addition, cable attenuation changes with temperature. Generally, the larger the cable's diameter, the lower its attenuation. Table 3–1 shows the attenuation of commonly used coaxial cables.

TABLE 3-1. Attenuation of Commonly Used Coaxial Cables

Type of Cable	Use	Attenuation (dB per 100 feet)	
		@ 50 MHz	@ 400 MHz
three-fourth-inch diameter	Trunk	.3	.9
one-half-inch diameter	Feeder	.5	1.3
RG-59	Drop	1.5	4.0

Amplifier Noise and Distortion

Amplifiers are located when the signal level is attenuated to the point where *noise* (snow) would be visible in the television picture. Amplifiers boost (that is, add *gain* to) the signal by an amount equal to the attenuation of the previous section of cable. Therefore, the amount of signal coming out of each amplifier will be equal throughout the entire cable system.

If a cable amplifier were perfect, that is, created no distortion, then one large amplifier with a gain of, say, 500 dB could be placed at the cable head-end, which would produce a signal strong enough to compensate for the loss of all the cable in the entire cable system. No amplifiers would be required in the cable plant. However, two factors limit the practical gain of cable amplifiers.

The maximum signal level that an amplifier can put out is limited by the *distortion* that the amplifier creates. Just as when a stereo system is turned up to a high volume the sound becomes distorted, signals passed through cable amplifiers are distorted. There are many forms of distortion. However, the types of distortion that most affect cable systems are: (1) *cross modulation*, when the video content of one channel modulates another channel, causing a weak, visible picture in the background; (2) *second order distortion*, when a channel creates interference on exactly twice its frequency, thereby distorting the channel on that frequency; and (3) *third order distortion*, when three channels mix with each other to create a multitude of interfering signals.

The minimum signal level that can be fed to an amplifier is limited by the electronic *noise* that amplifiers generate. Noise shows up as snow in pictures and as hiss in sound. Signal levels must always be enough above the noise generated by amplifiers to make the noise insignificant.[2]

Cable amplifiers are specified to have certain performance in each category of distortion. The *maximum* level that an amplifier can put out is determined by the specification for that amplifier.[3]

To choose the amplifier signal levels to be used in a cable system, one more factor must be considered. Both *noise* and *distortion* increase with the number of amplifiers that are strung together in *cascade*. For example, noise increases 3 dB each time the number of amplifiers is doubled. Cross modulation increases 6 dB for each doubling of the number of amplifiers.[4]

Therefore, the maximum cascade of a cable system is limited. Over the years, the cable industry has standardized amplifier design around a 20–22 dB gain. With these amplifiers, a maximum cascade of 64 amplifiers is theoretically possible. In practice, cascades normally are limited to around 32 amplifiers. With about three amplifiers per mile, the maximum trunk length is limited to about ten miles. However, to provide a safety margin and to allow for future growth, most well-designed systems will have trunks of no more than about five miles in length.

THE TRUNK

A cable TV distribution plant consists of two major elements: the trunk system and the feeder network.[5] The trunk system (shown in Figure 3–3) consists of a large-diameter cable (three-quarters to one-inch-diameter) leaving the headend or hub and going through a community, splitting at various points, and finally terminating at the extreme of the service area. The trunk's only purpose is to deliver television signals to individual neighborhoods of the community. No subscribers are served directly from the trunk; instead, *bridger*

FIGURE 3-3. Trunk system.

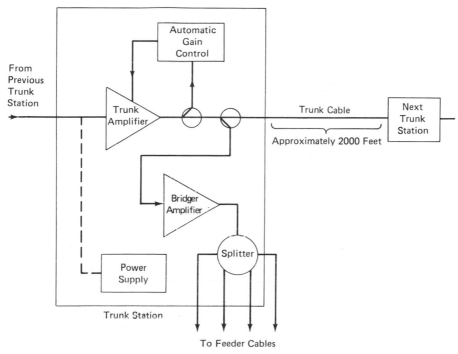

amplifiers are located periodically along the trunk at intervals of approximately 0.35–0.50 miles to take signals from the trunk and feed them to the feeder network. There is about one mile of trunk cable for every three to four miles of feeder cable in a cable system.[6]

The Trunk Station

Trunk amplifier stations consist of a cast aluminum housing, which is watertight, with entry ports for connection to the trunk and distribution cables. All modern trunk stations are modular, with plug-in slots for various devices. This makes service easy, since almost all maintenance can be done by simply plugging in new modules.

At a minimum, a trunk station has a power supply module, which supplies DC voltage to the other modules; a trunk amplifier module; and test points (connectors to allow attachment of signal level measurement equipment without disrupting service). The trunk station is shown in Figure 3–4.

FIGURE 3-4. Trunk station.

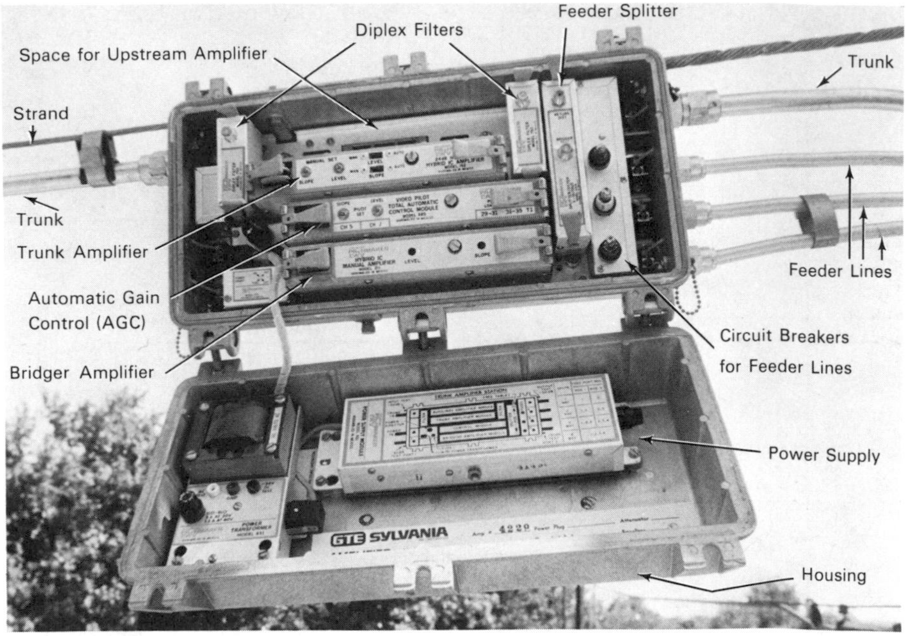

Feedforward Amplifiers

Recently, most manufacturers have introduced *feedforward* amplifiers. In these amplifiers, the input and output signals are compared to determine the distortion that the amplifier creates. A signal that is the exact opposite of the distortion, an *error correction signal*, is created, then mixed with the incoming signal. In the amplifier, the error correction signal mixes with the distortion that the amplifier creates, and the two cancel each other out. In this way, the final distortion created by the amplifier is very small.

Feedforward amplifiers are more expensive than normal trunk amplifiers but are gaining acceptance, especially in supertrunks and in long trunk cascades with 50 or more channels.

Automatic Gain Control

The attenuation of coaxial cable does not remain constant at all times. As temperature changes so does attenuation. As it gets colder, attenuation decreases; as it gets hotter, attenuation increases. At high and low temperature extremes, a 10-percent change from the normal attenuation of the cable, or about plus or minus 2 dB between each pair of trunk amplifiers is typical. Although this effect is insignificant through a single trunk span, it is additive, and at the end of 20 trunk amplifiers, the signal level will change up or down by 40 dB over the temperature range normally experienced. During the summer months, the increased attenuation decreases the signal level entering each amplifier. The signal entering the last amplifier would be 40 dB below normal, resulting in an unacceptable signal-to-noise ratio and a very snowy picture.

In the winter, attenuation decreases, causing signal levels entering amplifiers to be higher. At the last amplifier, the output level would be 40 dB higher than normal, resulting in severe distortion.

To cope with this problem, a technique called *automatic gain control* (AGC) is employed. In a typical cable system, AGC modules are installed in every second trunk station. These modules measure the strength of the television signals leaving the amplifier and compare them with a reference standard (the signal level that the amplifier is specified to put out). The amplifier's gain is then adjusted automatically to keep its output signal level at a constant point. In this way, fluctuations in temperature do not affect the signal quality at the end of the trunk line in any appreciable way. It should be noted that as channel capacity is increased on cable systems, the performance of all elements of the cable systems becomes more critical. For this reason, many state-of-the-art systems use AGC modules at every trunk station.

Passive Devices

At various points in the trunk system, the trunk must be split to serve different sections of the system. A device called a *splitter* or *directional coupler* is used. A splitter divides the signal equally into two, three, or four different legs. A coupler, on the other hand, removes a portion of the signal from the main trunk and feeds it to a single output leg. Splitters and couplers are often referred to as *passive devices* since they have no electronic amplifying or control circuitry. Amplifiers are often referred to as *active* devices.

Connectors

Anytime the coaxial cable enters an amplifier, splitter, or coupler, a *connector* must be used. Connectors are made of aluminum. A small copper tube within the connector makes contact with the copper center conductor of the cable. The outer aluminum shield of the cable slips inside a larger tube in the connector, and a nut is tightened on the connector to squeeze a small metal ring around the aluminum outer conductor to make good contact. Connectors also contain rubber seals to keep moisture from entering the cable.

The Bridger Amplifier

As mentioned, the purpose of the trunk is to carry signals with a minimum of degradation to each neighborhood in a community. To feed the signal into the neighborhood, a *bridger amplifier* module is installed in a trunk station. A small portion of the signal is taken off the trunk and is fed into the bridger amplifier, which amplifies the signal to a level substantially higher than the level carried on the trunk line. It is then split into up to four *ports* that are used to supply feeder cables leaving the bridger.[7]

THE FEEDER NETWORK

The *feeder system*, sometimes called the *distribution system*, is the network to which subscribers connect that parallels each street within a neighborhood. It is illustrated in Figure 3–5. Feeder cable is typically half an inch in diameter. Because of its smaller size, feeder cable has a higher attenuation than does trunk cable, but it is substantially lower in cost.

Unlike the trunk system, the feeder system rarely has more than two amplifiers in cascade. As a result, amplifiers can have a higher signal output than trunk amplifiers and still provide acceptable distortion levels. And again, because only two are in cascade, AGC is rarely provided. However, many feeder amplifiers have *thermal compensation*, electronic circuitry that adjusts the gain of the amplifier with temperature to provide a crude form of automatic level control.

The *feeder amplifier*—also called *line extender amplifier* and *distribution amplifier*—is much less complex than a trunk station, is substantially smaller physically than a trunk amplifier, and costs only a fraction of what the trunk station costs. Feeder amplifiers usually have 25 to 35 dB of gain.[8]

Multitaps

After the signal leaves the port on the output of the bridger amplifier, it passes through a series of *multitaps* (diagrammed in Figure 3–6). A multitap is a device that takes a small portion of the signal off the feeder cable and

FIGURE 3-5. Feeder system.

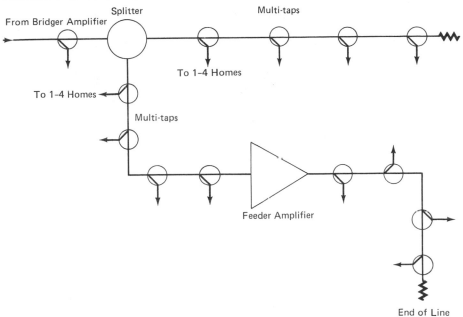

FIGURE 3-6. Multitap.

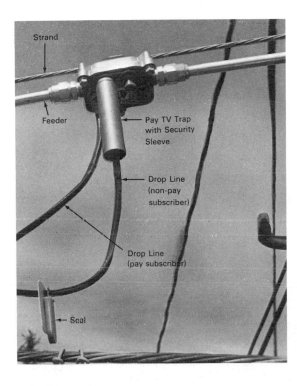

feeds it to subscribers. Multitaps are available to service two, four, and eight subscribers from a single location. The cable plant is usually designed to follow the existing telephone and electric service. Generally, if a home receives electric and telephone service from a particular location, the cable multitap will be located there.[9]

Because there will be different signal levels within the feeder at different points, multitaps are made in different tap-off values; that is, different amounts of signal are removed from the feeder line. When a multitap is located immediately following a bridger or feeder amplifier, signal levels within the feeder cable will be high, and only a small proportion of the signal needs to be taken off the feeder cable to provide the subscriber with adequate signal. Farther down the feeder cable, less signal will remain in the cable, and a greater percentage will have to be tapped off to serve subscribers. Finally, there will not be enough left in the feeder cable to service a subscriber, and a feeder amplifier must be installed or the line must end.

CABLE SYSTEM POWERING

The amplifiers used in cable systems require electric power for operation. In the early days of cable, each amplifier was connected individually to the power line. Later, methods were developed for *multiplexing* (combining) electric power with cable signals on the coaxial cable. Power supply units are now located at convenient points in the cable system. The 117-volt electric power from the power line is stepped down to 60 volts (for safety reasons) and fed into the cable using a *power inserter*. In this way nearby amplifiers are powered. Approximately one power supply is required for every two to three miles of trunk and feeder plant served.

A major source of signal outage in cable systems is the failure of electric power. If the first trunk amplifier leaving a headend is without electric power, subscribers all down the line will be without cable service. For this reason, many cable operators are installing battery standby power supplies, and most new systems being built in large cities today utilize such power supplies. A bank of batteries will take over to power the cable system for four or more hours if the electric power service is interrupted.

DUAL CABLE

Many of the franchises granted in the late 1970s and early 1980s were to companies proposing dual cable systems. *Dual cable* originated in the early 1970s, as a means of increasing channel capacity. Two 12-channel cable systems were built, each with its own cable, amplifiers, and subscriber connections. Twelve channels were carried on each cable system, and a switch was installed in the cable subscriber's home allowing access to a total of 24 channels. As franchise competition increased, cable companies, in their attempt to provide higher numbers of channels, proposed the same type of arrangement, with 50 or more channels on each cable. Dual cable systems do not differ significantly from single cable systems. In some cases a common housing is needed for amplifying equipment for both cables, and the powering equipment is common to both.

Dual cable systems, although offering double channel capacity, cost only about 50 percent more than comparable single cable systems.

INSTITUTIONAL NETWORK

Some of the newer cable systems, whether single or dual cable, include an additional cable referred to as an *institutional cable* or *institutional network*. This cable is typically another trunk cable paralleling the route of the trunk network. Its purpose is to provide interconnection of governmental and commercial institutions. (The proposed uses of the institutional networks are described in Chapter 9.) In addition, the institutional network's signals can be routed to the cable subscriber network, allowing programming to be originated at an institution and viewed by all cable subscribers.

Institutional networks are two-way (see Chapter 5); that is, they are capable of transmitting signals back to the headend from locations within the cable plant. Institutional cables are designed to carry an approximately equal number of channels in each direction. A 400 MHz institutional cable might carry 20 channels in one direction and 30 channels in the other direction.

Since institutional networks are basically trunks, they involve no feeder systems and sometimes no bridger amplifiers. Signals are fed into and tapped off the trunk as the trunk passes institutions that are connected to the network.

AMPLIFIER REDUNDANCY

Even using a hub system design, a trunk line in a cable system serves thousands of subscribers. The failure of a trunk amplifier near a hub or headend can result in loss of service to thousands and sometimes even tens of thousands of subscribers.

Cable operators are beginning to experiment with techniques of providing *redundancy* for critical amplifier locations. One method utilizes trunk stations with dual amplifiers and automatic circuitry for switching from one to the other should the main amplifier fail. Very few of this type of amplifier have been installed by the cable industry, and no data are yet available on the effectiveness of this approach.

A second approach utilizes a *bypass switch*. If an amplifier fails, a bypass jumper is automatically connected between its input and output. The signals passing through the amplifier will not be amplified, and the signal levels will be about 20 dB lower than normal. Television pictures that subscribers receive will be substantially degraded, but home security systems, which require lower signal levels, will still operate properly. This approach, as with the redundant amplifier approach, is new and is in operation in only a few systems.

It is likely that as cable systems carry increasingly important services, new methods of redundancy will be developed.

SYSTEM REBUILD

As cable technology changes and the demand for cable services increases, or as equipment deteriorates, older systems face the problem of *upgrading* facilities. Almost all the original 5-channel cable systems have rebuilt to 12-channel capacity, and very few systems still exist that utilize the braided type of coaxial cable. Most coaxial cable of the solid sheath aluminum variety can carry signals up to 300 MHz and in many cases to 500 MHz.

Cable subscribers in low channel capacity systems are now, inevitably, near large capacity systems. The customer awareness of desirable channels carried on the large capacity system that are not carried on the low capacity system puts pressure on the system management to upgrade.

Upgrading or rebuilding of cable systems can take several forms. In the simplest, a cable system carrying 12 channels can add one or more midband channels simply by adding the required headend equipment and putting the channels on the system. (For the subscriber to see these channels, a converter, which will be described in the next chapter, must be used.) Most 12-channel cable systems can be expanded by at least one channel to add a pay TV channel without removing any channels from its basic service.

To add the full nine channels to the midband in older systems often requires the replacement of amplifiers and, in some cases, relocating the amplifiers. Some cable operators faced with this problem have taken the approach of installing an HRC headend as discussed in Chapter 2. This is a headend in which the cable television channels are put on special frequencies so as to make the distortions created by cable plant amplifiers less visible to subscribers. The effect is that several midband channels can be added without replacing amplifier equipment. HRC headends are relatively expensive, but when compared with the expense of replacing amplifiers in a large cable system, the approach is generally less costly.

To go beyond midband channels requires installation of new amplifiers and, in many cases, new multitaps. Such a rebuild can convert a 12-channel system into a 30- or 35-channel system at a relatively low cost ($2,000 to $4,000 per mile of cable plant).

Finally, some cable systems must be totally rebuilt to increase channel capacity. In this case, a new duplicate cable system is built, and subscribers are switched over to the new system. The old system is removed and sold to a salvage company. The cost of this kind of a rebuild often equals or exceeds that of building a new cable system.

Frequently, other elements of the cable system, such as powering equipment or subscriber drops, must also be replaced to increase channel capacity or because they have become unreliable.

Cable manufacturers have now introduced 78-channel equipment covering the frequency spectrum from 54 to 550 MHz, and it is likely that the frequency limit will be pushed even higher. In anticipation of the need for increased channel capacity, many new cable systems today are being built with system designs to allow for upgrading to a higher channel capacity by replacing only amplifiers. For instance, many of the present 300-MHz systems are designed for eventual 400-MHz operation, and all cable and multitaps are specified to operate up to 400 MHz. At some future date, 300-MHz amplifier modules can be removed and replaced with 400-MHz modules, thus increasing the channel capacity from 35 to 54 channels.

Some systems being built today are being equipped with 450-MHz amplifiers but designed to accommodate 550-MHz amplifiers at some time in the future, permitting further expansion in channel capacity.

FCC GUIDELINES

For many years, the FCC has regulated the technical performance of cable television systems through a series of technical standards. These standards are found in Subpart K, Sections 76.601–76.617 of the 1972 rules. The FCC no longer enforces the standards. They are now referred to as *guidelines*. Cities may write the guidelines into the franchise agreement and enforce them, but they may not impose more stringent standards. Because the specifications were originally written to cover all types of systems, including those built many years before the 1972 rules were adopted, the guidelines are minimal.

The FCC's guidelines require that certain minimum and maximum signal levels be delivered to subscribers' television sets, that the signal-to-noise ratio and various types of distortions do not exceed certain limits, and that sufficient isolation at the multitap, or in the splitter for a multiple-set installation, is provided so that there is no interference between neighboring subscribers or receivers.[10]

FCC regulations (*not* optional, as the guidelines) strictly limit the operation of cable systems in the frequency bands of 108–136 and 225–400 MHz, namely, those used for aircraft communication and navigation signals. Cable operators are required to *offset*, or move, signals carried on the cable away from frequencies used at nearby airports. Cable systems must monitor signal leakage and keep accurate records of such leakage.

DISTRIBUTION PLANT CONSTRUCTION

Cable television systems are constructed on telephone and power company poles (*aerial construction*) or are buried in the ground (*underground construction*). In this section the procedures involved in each type of construction are explained.

Aerial Plant

Most utility poles are owned either by the local power company or by the telephone company. Both have contractual arrangements that allow attachments of other types of lines to their poles. For a cable system to install its lines on a power company pole, a contract is signed with the power company giving the right to attach to that pole. A rental fee is charged between $1 and $4 per year per pole. In recent years, telephone companies have been eliminating installation of new poles wherever possible. As a result, most poles now in use are owned by the electric companies.

For uniformity and safety, there are established *spaces* for attachment by each occupant of a pole and consequent *clearances* to be built and maintained by all users. These clearances usually conform to National Electric Code standards, but they can be modified by local or regional practices. Telephone lines are at the bottom of the poles, and cable lines are located above the telephone lines. Telephone lines must be located a certain minimum distance above the ground, and there must be a minimum clearance between the telephone line, cable line, and power lines. In many cases, poles were installed without cable systems in mind, and telephone lines were located too close to the power lines to allow installation of cable lines without relocation of the phone lines. The process of making space on the pole for installation of cable lines is called *make-ready*. Make-ready, which is performed by the owner of the pole, can be time-consuming and expensive. It is a major cause of delay in construction of cable systems.

The process of installing an aerial cable plant is long and complex, involving many steps. The following is a basic outline of the process.

First, a *base grid map system* is set up. Since the area served by the cable system will require many maps, a logical system for numbering and locating the maps must be set up. The base grid map system serves this purpose. It covers an area that includes not only the initial cable construction area, but all areas into which the system might be expanded in the future. Often this grid may be duplicated from a city or utility grid in existence. A utility may be most anxious for the cable company to duplicate its grid to simplify pole permit procedures.

Base maps are then made for the areas where the system is to be built, showing all streets, railroads and rivers. Then *strand mapping* is done. The entire area to be served by cable is mapped by a cable employee or contractor on the base map. In the strand mapping process, the location of all utility poles, the number of residences served from each pole, and the distances between poles are noted. Also, notations are often made as to what make-ready might be required on each pole. The strand mapping information is then placed on the base maps.

The next step is the design of the cable system. Using the specifications for cable attenuation, amplifier gain, amplifier cascade limitations, and multitap losses, the designer lays out a cable route that will efficiently provide a good picture to each subscriber or future subscriber within the franchise area. (A portion of a system design appears at the opening of this chapter.) Often the design process is aided by the use of the computer, since the calculations are long and arduous. Computer programs are also available to do the entire system design and create the needed maps.

The specification for equipment performance, cable attenuations, amplifier gains, and multitap losses come from the manufacturers of the various equipment, but many cable operators add their own safety factors to the manufacturer's specifications.

Proper cable TV system design is crucial. Many earlier cable systems were built without proper attention to the various tolerances and limitations of cable and equipment. Not only do today's requirements have to be considered, but the future must be considered as well. Some of the factors that must be taken into account in designing a future-oriented cable system include the following:

1. *Operating temperature.*   The system must be designed to meet all technical standards over the entire range of temperatures that are expected.

2. *Equipment tolerance.*   Often manufacturers' specifications are given for an average amplifier or length of cable as well as worst case specifications for the equipment. A prudent cable designer will use the worst case figure rather than the average figure in design calculations.

3. *Increased attenuation as a result of additional splices.*   A prudent designer will include a factor in the design to compensate for additional splices over time. As splices are added to the cable, losses between amplifiers will increase.

4. *New homes in existing neighborhoods.*   A cable design should include provisions for residential lots on which homes have not yet been built. A multitap should be installed for all locations, even the vacant lots.

5. *Subscriber drop length and number of TV sets.*   A designer should include an estimate of the length of the subscriber's drop and number of TV sets to be served in the design calculations. For instance, in wealthy neighborhoods with large homes on lots it can be expected that each house will have several TV sets and that a long drop line will be required to reach the house. Adequate signal strength must be present at the multitap to serve these houses.

6. *System expansion.*   As communities grow, cable operators must extend their trunk lines to cover the new areas. The cable designer should take the projected growth into account when designing the trunk system to make certain that the maximum amplifier cascade is not exceeded at some time in the future.

7. *Human and test equipment error.*   A margin of error should be built into the system design to allow for inaccuracies in test equipment and errors made by technicians in balancing and operating the system.

The difference in cost between a system designed with these factors taken into account, and one designed to meet minimal performance standards as cheaply as possible, can be substantial. In the long run, however, it is sensible for the cable operator to design the system with adequate margins. Systems designed without such safeguards, though perhaps capable of providing good-quality pictures initially, may not provide high-quality service as the franchise area grows and subscribers are added.

Once the system design is completed and put on the base maps, the cable company applies to the utility companies for permission to attach to poles that are needed. Each utility company has a different process for this application, but generally they require *proof of authority* (a franchise) to operate a cable system, *proof of liability insurance,* and, in many cases, posting of a *bond* to guarantee that the utility will be reimbursed for any damage.

Next, a joint pole inspection, commonly referred to as a *walkout* is scheduled. A representative of the cable company and of all the utilities renting or owning space on the pole go to each pole on which cable facilities are to be attached. Poles found to not have sufficient space for addition of cable lines are measured to determine what changes are necessary. Often, the pole already violates the *clearance* requirements, and the utility company must make changes at its own expense. Otherwise, the utility company prepares an engineering estimate of the cost of performing the make-ready required, and submits this estimate to the cable company. If there are no problems, the cable company then authorizes the utility company to proceed.

All make-ready costs that relate to the installation of cable TV lines are paid by the cable company. It should be remembered that all costs are estimated and that the cable company can be entitled to a refund or may have to pay more when all make-ready work is completed. The utility company will then proceed with the rearrangement of the facility on the poles. This process sometimes takes weeks and even months.

Once the make-ready is completed, the installation of strand can begin. *Strand* is a galvanized steel supporting cable approximately one-quarter inch in diameter that is strung from pole to pole. At specified intervals, the strand must be grounded and *bonded* (connected) to other utility strands for lightning and safety protection. *Guy wires* must be installed to relieve stress on the utility poles. Proper tensioning of the strand and proper grounding are important factors in the life and reliability of the system.

When the strand installation is complete, the cable can be installed. A device called a *lasher* is used to connect or *lash* the cable to the strand. The lasher wraps a very small stainless steel wire around both the strand and cable to hold the cable in place. *Expansion loops* are installed at every pole using special bending tools. These loops allow for expansion and contraction of the metal cable components. At this stage of construction, great care must be exercised by the construction personnel. Coaxial cable is susceptible to kinks and dents that will degrade the performance of the cable system.

If the cable system is a dual cable system or if there is an institutional network, all the cables are lashed to a common strand. All trunk cables or multiple cable runs should be *double lashed* for extended life. This requires a second wrap from the lasher, but the process is relatively easy since most lashers are equipped for two spools of lashing wire.

After the cable has been installed, *splicing* can begin. Splicing involves installation of the electronic components and the splitters and multitaps that will tie the cable system together. Splicing is perhaps the most critical of all cable construction steps, since an improperly installed connector can result in serious degradation of performance and moisture leaking into the cable, which can cause permanent damage. Cable companies that use contractors for installation of strand and cable sometimes use their own employees to perform the splicing, since it is regarded as so critical. Cable splicers may use a test device such as a continuity checker to verify each connection as it is spliced, thus speeding up activation time.

In certain climates, near the seacoast or areas of high industrial pollution, for instance, special precautions are taken to protect the cable from corrosion. In these locations, coaxial cable with a polyethylene jacket is used in place of one with a bare aluminum jacket. Special sealing materials are used around all connections to make certain that they are completely watertight.

Underground Plant

Underground cable systems are constructed either on publicly owned *easements* (the right to use these easements is granted in the franchise) or on private easements granted by individual property owners. In new subdivisions, easements are often granted to the cable companies by the developers of the subdivisions. In an existing subdivision, however, cable companies are sometimes required to obtain individual easements from property owners before installing cable.

As in the case of aerial construction, underground cable systems generally follow the same route as the power and telephone companies. The cable is usually buried in the same area, and cable pedestals are located near the telephone company and electric company pedestals. A *pedestal* is a junction box where cable television equipment is located.

Next, the cable route and system design are established. Most states and many countries and cities have requirements for permits before underground cable can be installed. From the system design, the cable company can determine which streets it needs permission to use and which private easements it must obtain.

Getting easements from individual homeowners where dedicated cable easements do not exist is a difficult and costly process. First, subdivision plot maps must be obtained and the ownership of lots determined. Then, a door-to-door team visits each homeowner to get the easement document signed. Often the document must be notarized and filed with the local government, and a nominal fee of $1 paid to the homeowner.

Because of the complexity and cost of this approach, many cable companies take shortcuts in the easement process. For instance, a cable operator may be willing to take the risk of not having an easement agreement notarized and filed with the county court house. In this case, the easement may not be binding on the next property owner, should the property be sold. However, the practical risk of a future homeowner's demanding removal of cable facilities is small, and when compared with the savings in the easement procurement process, the risk may be justified. When using this simplified procedure, easements may be obtained by mail.

In many cases easements are not required. For instance if a cable is to be located in the front yard, county or city right-of-way may be used for cable installation. In this situation, the cable operator is under no requirement to even notify the homeowner or resident of the intention of installing cable. A prudent cable operator will, however, mail a letter or deliver a door hanger to each home well before construction is to begin, to inform the householder of the company's construction plans.

As in the case of aerial construction, the underground construction process begins by mapping the area to be served. Utility company maps sometimes serve as a starting point, or, in many cases, subdivision plot maps are used. In this mapping process, the location of property lines, existing utilities, easement lines, and obstructions that might make underground construction difficult (such as a fence or garden) are noted.

As cable companies attempt to get permits and easements, problems may be encountered. Often homeowners may refuse to sign the cable easement, or the location of existing underground utilities may make the original cable route impossible. Underground cable systems are frequently redesigned several times before final installation, as the cable company attempts to balance the availability of easements and the problems with existing utilities against the most economical system design.

Once the final design has been completed and all necessary easements and permits have been obtained, a public relations process begins. No matter how careful cable companies are in underground construction, some disruption will occur. Residents are very sensitive about digging on or near their property, and great care must be taken by the cable company to inform residents of exactly what to expect. Although cable installations are usually quite neat, residents frequently have an exaggerated idea of the damage that will occur from experience with the more complex construction of gas, water, or sewer lines. Cable companies sometimes use slide presentations or photographs to show residents how little disruption will occur.

Just prior to the beginning of construction, the utility companies must be notified so that they may *spot* their facilities. The power, water, gas, and telephone companies mark the route of their cables by using stakes or spray paint so that cable installation crews don't disrupt the services or endanger themselves.

The coaxial cable used for underground construction has an outer jacket of polyethylene. Between this jacket and the aluminum coaxial cable is a *flooding compound*, a tarlike substance that will spread and seal any puncture that may occur in the outer jacket. Cable is usually buried at a depth of 12 to 36 inches.

Two different methods are used to bury the cable. Cable can be plowed into the ground using a *vibratory plow*. A steel blade about one inch wide is pulled and vibrated through the ground. The cable is fed into a chute on the back edge of the blade and is laid at the bottom of the slot which the blade creates. Plowing is by far the least expensive way to install cable, but maximum depth is limited to about 18 inches. In addition, there is a chance of damaging the cable, since rocks or other obstructions may be present at the bottom of the slot. Another advantage of plowing is that restoration of the ground is very simple, since little dirt is removed during cable installations.

The second method of cable installation is *trenching*. A trench 4 to 6 inches wide is dug and the cable is installed in the bottom. The dirt is then put back on top and compacted. Trenching is often used for trunks, where depths of 36 inches are desirable to reduce the possibility of the cable being cut.

Caution is necessary in laying and burying the cable. Aluminum, if left in contact with the soil, will corrode in a matter of weeks, so great care must be taken to install the cable in such a way as to avoid puncturing the plastic jacket. Rocks can cause kinks or dents in the cable, which will affect the cable system's performance.

Some cable companies use armored cable for underground construction. *Armored* cable has stainless steel bands wrapped around the outer jacket, with another jacket on top. Armored cable is an effective way of providing protection from damage, but the extra cost is significant.

In some cases, cable is installed in plastic or metal pipe, called *conduit*. In this case, a trencher digs a hole in which the conduit is laid. Later, cable is pulled through the conduit. The conduit method provides some additional protection against cable damage, but it is substantially more expensive than the direct burial method. Another advantage of a conduit system is the relative ease with which faulty cable can be replaced.

In high-density downtown areas, utilities are often in duct systems located underneath the streets and owned by the power company or phone company. Cable companies, in this case, lease space in the ducts. A similar arrangement is used for crossing rivers where conduit may have been attached to a bridge or buried under water.

The next step is the installation of *pedestals* (boxes in which amplifiers, multitaps, and other equipment are located) and the splicing of the cable. Pedestals are located above the ground, sticking up a foot or two. Sometimes, however, *vaults*, which are installed flush with the ground surface, are used. Care must be taken when installing these pedestals so that they are not subject to physical damage and do not allow water to enter the cable or its components. In underground systems, waterproofing is essential, and special sealing materials are used to protect all connectors from leakage.

After the installation of the cable and pedestals comes the most sensitive part of the installation process, the restoration of the ground. The ground must be compacted and seeded where grass may have been destroyed. Shrubs and trees that may have been damaged must be replaced. Success of the cable system's marketing effort will depend on the care taken in the restoration. Many cable systems that did a poor job in the final stage of underground construction many years ago are still living with low penetration in those areas because of the ill will created.

In some circumstances, the procedure for installation of underground cable will be slightly different from that just outlined. In new subdivisions, for instance, *joint trench* arrangements are often used. In this case, a single trench is dug by one of the utilities, and telephone, electric, and cable lines are placed in the same trench. All three companies share the expense.

Plant Costs

The costs for 400-MHz single cable plant are about $10,000 per mile in 1987 for aerial construction including labor, materials, design, and make-ready. The second cable for a dual system adds $5,000 per mile. The cost for underground cable in a residential neighborhood averages about $15,000. Underground cable in urban areas may be $50,000 per mile or more. These figures do not include the drop, which is about $30 for aerial. Dual cable and underground construction add about $10 each. Make-ready costs can vary greatly, from as little as $200 per mile to over $3,000 per mile.

Balancing and Proof of Performance

At this point, the cable system construction is completed. Electronic technicians then follow to perform the balancing and proof of performance.

Balancing is the adjustment of each of the cable television amplifiers to its specified operating level and testing to make certain that the losses between amplifiers are exactly as anticipated in the system design. As in the case of splicing, balancing is often done by cable television employees rather than by contractors.

Proof of performance testing involves the use of sophisticated test equipment on a completed cable system to make certain that the system operates as designed. Tests are usually made for signal level uniformity, various types of distortion, signal-to-noise ratio, hum modulation, and signal leakage. *Hum modulation* is distortion of the TV signals by the electrical current necessary to operate the amplifier and appears in TV pictures as dark or light horizontal bars moving through the picture. *Signal leakage*, also called *signal egress*, is the radiation of signals within the cable system into the air, which can cause interference with over-the-air broadcast services for noncable subscribers. Signal leakage can also interfere with aircraft navigation and communication channels.

If the cable system is a *turnkey* installation, in which a contractor supplies a complete operating cable plant to the cable company, the proof of performance test serves as a quality check before final acceptance by the cable operator. Whether the cable system was built by a contractor, by a combination of contractors and cable company employees, or entirely by cable company employees, the proof of performance serves as a final check before connection of subscribers.

The cable system is now ready for installation of subscribers. Many cable companies prefer to wait a period of time before connecting subscribers. Electronic components have their highest rate of failure during the first few hours of use, and allowing a cable system to *"cook"* for a few days before connecting subscribers reduces the possibility of a new subscriber being annoyed by a failure in the first days of service.

NOTES

[1]There is also a need to know the absolute level of television signals. The *decibel relative to a millivolt* (dBmv) is used as the standard measurement unit in the cable industry. The volt is a measure of electrical energy, and a millivolt represents one thou-

sandth of a volt. Zero dBmv is equal to one millivolt. 20 dBmv is ten times one millivolt, or ten millivolts, while − 20 dBmv is one-tenth of a millivolt. The table shows the relationship between millivolts and dBmv.

Relationship Between Millivolts and dBmv

dBmv	*Millivolts*
60	1,000
50	300
40	100
30	30
20	10
10	3
0	1
− 10	.3
− 20	.1
− 30	.03
− 40	.01
− 50	.003
− 60	.001

[2]Most modern cable amplifiers generate noise at the absolute level of about − 50 dBmv. It has been determined by subjective viewing tests that subscribers will rate a picture with a *signal-to-noise ratio* of 43 dB (that is, the signal is 43 dB above the noise) as excellent. To obtain a signal-to-noise ratio of 43 dB, the signal fed into an amplifier must be at least − 7 dBmv (− 50 dBmv + 43 dB = − 7 dBmv). A signal any lower than − 7 dBmv will result in a picture degraded by noise.

[3]A typical amplifier might have a cross modulation specification of − 93 dB with an output level of + 40 dBmv. This means that the cross modulation distortion will be 93 dB below the output level, or − 53 dBmv (+ 40 dBmv − 93 dB = − 53 dBmv). Again, subjective tests have shown that subscribers will not object to cross modulation levels that are − 57 dB relative to the signal level. The − 93-dB cross modulation distortion in the preceding example would, therefore, be perfectly acceptable. So, for this hypothetical amplifier, we have a *minimum* signal input level of − 7 dBmv and a *maximum* signal output level of + 40 dBmv in order to avoid objectionable degradation of the pictures.

[4]In the case of noise, to end up with a 43-dB signal-to-noise ratio (for an excellent picture) at the *end* of the system, a signal-to-noise ratio of 46 dB must be maintained at each amplifier if two amplifiers are cascaded, 49 dB for four, 52 dB for eight amplifiers, and so forth. To maintain a 52-dB signal-to-noise ratio, we now need not the − 7-dBmv signal level at the input of the amplifier calculated above, but + 2 dBmv (− 50 dBmv + 52 dB = + 2 dBmv).

The table shows the effect of cascading amplifiers and the signal level required at each amplifier to maintain a 43-dB signal-to-noise ratio at the end of the system. The table also shows the output level at which a hypothetical amplifier can be operated while maintaining cross modulation at − 57 dB at the last amplifier in the cascade.

Signal Levels Required to Maintain 43-dB Signal-to-Noise Ratio and − 57-dB
Cross Modulation for Various Amplifier Cascades.

No. of Amps.	Input Level	Output Level
1	− 7 dBmv	+ 53 dBmv
2	− 4	+ 50
4	− 1	+ 47
8	+ 2	+ 44
16	+ 5	+ 41
32	⊦ 8	+ 38
64	+ 11	+ 35

[5]Television signals enter the trunk system from the headend with a level of about + 35 dBmv on the higher channels and + 30 dBmv on the lower channels.

[6]Trunk amplifiers generally have signal input levels of around + 12 dBmv and output levels of + 34 dBmv. Gain is usually 22 dB.

[7]Bridger amplifiers usually operate with output levels around + 48 dBmv.

[8]Input levels are typically about + 20 dBmv and output levels about + 45 dBmv.

[9]A signal level of about + 6 dBmv is ideal for connection to a television set. This is high enough to avoid visible noise, and low enough to avoid overloading the television set, which would result in signal distortion. To obtain + 6 dBmv at the TV set, a level of around + 13 dBmv is needed at the output of the multitap (the 7 dB difference is the loss of the drop cable).

[10]In 1987 the federal appeals court in *New York City v. FCC* ordered the FCC to reconsider its policy on technical standards with the intent to give cities more authority to regulate signal quality. At this writing the FCC had not completed the review.

CHAPTER **4**

Home Drop

Remote controller and addressable converter.

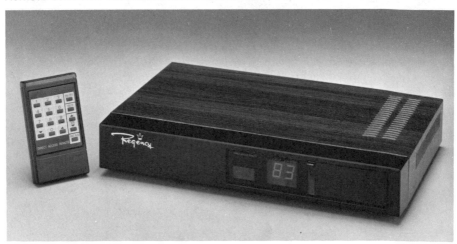

DROP LINE

In this chapter cable finally reaches the subscriber. The subscriber connection to the distribution plant is made through a *drop line*, which is a small-diameter (about one-quarter of an inch) coaxial cable leading from the multitap in the cable plant to the subscriber's television receiver.

The drop line is similar to the coaxial cable used in the distribution plant except that its outer sheath is generally braided copper or aluminum (usually with an aluminum foil wrapping) covered by a plastic jacket. These sheath materials are used instead of solid aluminum because the drop wire must be flexible.

Generally, if the cable system plant is aerial, the drop will be aerial; if the cable system plant is underground, the drop will be underground. As the cable enters the house, the outer sheath is connected to a water pipe or ground rod to keep dangerous electrical currents from reaching the TV set. The drop is then routed to the subscriber's TV receiver. If the TV set does not have a cable connection, a *matching transformer* is used to connect the coaxial cable to the antenna terminals on the back of the TV set. The purpose of this matching transformer is to mate the coaxial cable, which carries all the signals in its center conductor, with the two antenna terminals on the TV set, each of which must be provided with a portion of the signal.

If more than one TV set is to be connected, a *splitter* is used to route the signal to two or more TV receivers. If the cable operator offers FM radio service, TV/FM splitters are installed to route FM frequencies (88–108 MHz) to the FM receiver and TV frequencies (54–88 and 120 MHz and above) to the TV sets. It should be noted that some newer homes are prewired for cable.

CONVERTERS AND SIGNAL SECURITY

As noted in the discussion of cable distribution plants, a cable system with more than 12 channels carries the additional channels in areas of the electromagnetic spectrum between channel 6 and channel 7 (midband) or above channel 13 (superband). Most older TV receivers and some newer ones will not tune these frequencies, but recently most TV sets are *cable-ready*, meaning they are capable of receiving midband and superband channels.

TV sets will tune to the UHF spectrum (channels 14 through 83), the channels in the frequency range of 470 to 890 MHz. Because most signals carried on cable are below 470 MHz, it is not possible to use the UHF tuners in TV sets to tune cable channels. Therefore, to tune the mid- and superband channels carried on cable systems, a converter is required unless the subscriber has a cable-ready set. The purpose of a converter is to convert the mid- and superband cable channels to VHF television channels that can be tuned on a TV receiver.

There are two basic types of converters: block converters and tunable converters. *Block converters* are used mainly in small cable systems that carry a few midband or superband channels. Block converters have a two-position switch. In the normal switch position, the subscriber tunes 12 TV channels on the receiver's VHF selector. In the midband switch position, seven different channels appear on channels 7 through 13 of the receiver's VHF selector. Block converters are simple, very inexpensive, and offer the small cable system a low-cost way of increasing channel capacity. Another type of block converter shifts the midband or superband channels to the UHF band selector.

With the *tunable converter*, the subscriber's television receiver is tuned to a vacant VHF broadcast channel (typically channel 2, 3, or 4). Then all channel tuning is done using the converter. The converter has a knob, slide switch, series of push buttons, or a keyboard that selects the desired channel. Many of these converters allow the subscriber to turn the TV receiver on and off using the converter.

Tunable converters come in three varieties: set-top, wired remote, and wireless remote. The *set-top converter* is self-contained for location on top of the subscriber's television receiver. The *wired remote converter* consists of a unit located behind the subscriber's television receiver connected by a thin wire to a hand-held control unit. The *wireless converter* is similar to the wired remote, but uses infrared light for communication from the hand-held control unit instead of a wire.

Recently, with the introduction of inexpensive microprocessors, cable converters have become more sophisticated and complex. Some now offer the ability to control the volume or the capability of restricting channels by requiring a special authorization code to be entered before the channel can be viewed. With this feature, the subscriber is given a four-digit number at the time of installation. When one of the restricted channels is tuned, such as a channel with R-rated programming, the subscriber must enter the four-digit code or the screen will remain blank.

Some of the new converters also permit programming of up to ten of the subscriber's favorite channels in a memory and recalling them with the push of a button. The feature may, however, defeat the cable operator's objective of encouraging the subscriber to make maximum use of the service by using most of the channels available.

Addressable Converters

Many of the new converters are *addressable*; that is, they can be controlled (addressed) from the cable headend. Individual channels, groups of channels, or all channels can be turned off or on from a computer terminal. Often, addressing is controlled by the system's billing computer. As subscribers are offered a multitude of channel options, it becomes desirable for the cable operator to be able to control which services a subscriber receives without mak-

FIGURE 4-1. Addressable converter.

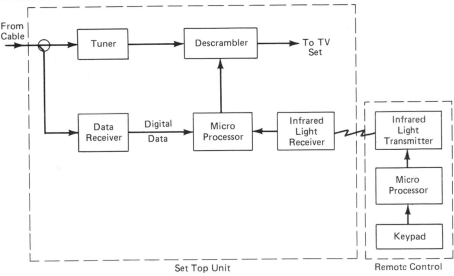

ing a service call to the home. Addressable converters provide the solution. If subscribers wish to receive a pay television service, for instance, their converter is "addressed" from the cable company's headend with instructions to allow that pay service into the home.

Addressable converters were introduced to the cable industry in the early 1980s. Early designs had serious reliability and operational problems and were substantially more expensive than nonaddressable converters. The latest generation of addressable converters, however, cost little more than the nonaddressable ones, and have proven to be quite reliable.

In the mid-1980s addressable converters accounted for a majority of new converter purchases by cable systems, but, as cable-ready television set sales increased and the number of subscriber households with several pay channels decreased, addressability lost some favor.

Traps

When pay television was introduced, cable operators were faced with the problem of how to deliver it only to the subscribers who paid for the service. Some early systems used an inexpensive converter and carried the pay TV channel in the midband. But many subscribers soon learned how to receive the pay television channel without paying for it. TV repair persons would retune the subscriber's television set to receive these channels, or converters designed for other purposes were used to tune in the pay television channel.

Today, two methods are generally used to provide security for premium TV service. One of these is the *trap*. A trap is a device that is located between the output of the multitap and the subscriber's drop. For high-penetration pay services, where more than 50 percent of all subscribers take a particular pay channel, a *negative trap* is used for each household that does not want the service. It prevents a signal from passing through.

Traps have two major advantages. First, they are very inexpensive ($5 to $10 each). If the pay channel penetration of a particular premium service is high, traps can be a very low-cost method of providing security. (Traps are put on all *nonpremium* subscribers' drops; so, as the number of premium subscribers increases, the number of traps required decreases.) If the penetration on the premium channel is low, traps are relatively expensive. Second, traps are extremely secure. To steal a premium service, the subscriber must climb a utility pole, remove the trap, and reconnect the drop line. Most cable operators use *security sleeves* and connectors that require special tools to be removed. It is very difficult, therefore, for a subscriber to steal service in a system using traps.

However, traps have disadvantages too. For one thing, they are difficult to administer. A special record must be kept of the location of the installation, and when a subscriber disconnects from a premium service, a trap must be installed in the line. Another disadvantage to traps is that, with few exceptions, a trap is required for each channel to be blocked. When only one pay TV service was typical, traps provided an ideal solution. However, with multiple tiers of pay TV being offered, traps became cumbersome and expensive. In practice, no more than two, or perhaps three, channels may be secured using traps.

Traps are also used in *lock boxes*, devices installed at the back of a television set to allow subscribers to lock out channels that may be objectionable. Lock boxes consist of a trap that can be activated by a key.

For low-penetration pay services, a *positive trap* is sometimes used to remove an interfering carrier placed on the channel for each of the minority of subscribing households that desire the service.

Descramblers

The second method of providing security for premium services is *scrambling*. The television signal is scrambled at the headend in such a way to make it difficult or impossible for a subscriber's TV set to get a good picture. In some systems, the sound is also scrambled. In the subscriber's home, a *descrambler* is provided to customers paying for a premium service, and often the descrambler is incorporated inside the converter. Descramblers are more expensive than traps. Included as a part of the converter, the additional cost of the descrambler is around $20.

Descramblers have the advantage of being capable of descrambling many TV channels with little if any additional cost above the cost of descrambling one channel. When combined with a programmable or addressable converter, they can be used to descramble an almost infinite variety of channel combinations for different subscribers purchasing different program packages.

Many techniques are used for scrambling TV pictures, with little compatibility between techniques used by different manufacturers. Most methods alter the synchronizing pulses that make TV pictures hold still on the TV screen. The descrambler then reconstructs normal synchronizing pulses.

There are two major types of descramblers: RF (radio frequency) and baseband. *RF* descramblers reconstruct the television channel directly on a cable channel. *Baseband* descramblers demodulate the scrambled channel to its original video and audio components, reconstruct an unscrambled sound and picture, and then modulate the sound and picture back to a television channel.

The latest generation of RF converter/descramblers incorporate techniques that make it much more difficult for someone to steal service. For example, one technique randomly delays the synchronizing pulses and transmits a "key" in a coded fashion to tell the descrambler where to replace the synchronizing pulses in the picture.

The baseband technique allows more sophisticated scrambling of the picture and sound. In one popular system, the polarity of the video information on each frame is randomly reversed (bright areas in the picture become dark and dark areas become bright). A key, which is itself scrambled, is transmitted to tell the descrambler which lines have had their polarity reversed. The descrambler then processes the video information on that frame to make its polarity normal. Baseband descramblers are substantially more secure than RF, but cost more too.

If the penetration of a particular premium service is low (below 30 percent of basic cable subscribers), descramblers can actually be cheaper than traps. However, descramblers have a serious liability. As more premium services have been added to cable systems, the incentive for theft of service has increased. In new major market systems, the total monthly charge for all services combined has reached $50 to $60, and a market has been created for bootleg equipment to receive these signals.

Descramblers must be relatively inexpensive for the cable operator to afford them. Yet they must be complex enough to make it difficult for a subscriber to defeat. As the incentive for stealing pay services increases, more subscribers will be tempted to purchase bootleg descramblers (which can be designed and built by an average TV repair person). In fact, designs for descramblers have appeared in *Popular Electronics* and *Radio Electronics*, two major electronic hobbyist magazines. Though the use of bootleg descramblers is illegal, it is very difficult for the cable operator to keep descramblers off the market.

It should be noted that addressable converter-descramblers do little to solve this problem. A standard converter combined with a bootleg descrambler will allow viewing of all the channels carried on the cable system. The latest scrambling systems, however, are substantially harder to defeat than earlier methods.

Off-Premise Terminals

One potential solution to the security problem with converters within the subscriber's home is to move the converter, or a portion of it, to the cable distribution plant. When located on a utility pole or in a pedestal, tampering is more difficult. Also, the portion of the converter within the subscriber's home is inexpensive, reducing the loss to the cable company if it is stolen. For years the cable industry has experimented with various types of off-premise equipment, but, because of the high costs and unproven technology, no off-premise system has gained widespread acceptance.

There are two general categories of off-premise equipment: the addressable tap and the off-premise converter.

Addressable Taps

Addressable taps have been available since the mid-1970s, but their use has been limited exclusively to apartment complexes. Multitaps were built with electronic circuitry in them to allow the cable operator to turn on or off a subscriber's drop from the headend. In addition, one, and in some cases two, channels of pay TV could be controlled. Recently, a new generation of addressable taps has been introduced that allow control of ten or more channels.

Addressable taps, theoretically, have numerous advantages over converters located within a subscriber's home. First, no scrambling is needed on the cable system, since control of what a subscriber can receive is done at the multitap. Second, if the subscriber has a cable-ready TV set or videocassette recorder, no converter is needed in the home, and, if the subscriber has a remote control TV set, that remote control can be used to change channels. If a converter is required, a simple one without descrambling capacity can be sold or rented to the subscriber. Finally, since little or no equipment is located within the subscriber's home, the potential loss due to theft is minimal.

Addressable taps are, at present, more expensive than in-home addressable converters, and they are capable of control of only ten or so channels. For these reasons, and because the technology is new and unproven, the cable industry has not installed many addressable tap systems.

Off-Premise Converters

Like addressable taps, off-premise converters are located at the multitap. Only a channel selection device, usually a keyboard, is inside the home. This device signals the off-premise converter to send the requested channel down the drop to the subscriber. Information on which channels each subscriber is authorized to receive is sent periodically down the cable from the headend to each off-premise converter.

Off-premise converters have some of the advantages of addressable taps. They eliminate the need for scrambling on the cable plant, and they require only an inexpensive control device within the home. However, one off-premise converter must be provided for each TV set or videocassette recorder connected. At present, only two off-premise converters can share one drop cable, so for the third connection within a home, a second drop must be installed. With the growing proliferation of VCRs, most homes will probably require at least two off-premise converters, and many homes will require two or more drops.

As with addressable taps, off-premise converters are more expensive than in-home addressable equipment, and the probable need for two or more drops in a typical home adds even more cost. As with addressable taps, the cable industry has been slow to accept off-premise converters.

Converter Costs

A 36-channel set-top converter without a descrambler cost the cable system about $45 to 1987. A 60-channel wireless-addressable converter was about $100. Off-premise converters and addressable taps cost about $150 per subscriber.

VIDEOCASSETTE RECORDERS

Only a few years ago, the connection of cable to a television set was a simple matter. With the rapid growth of videocassette recorder sales, cable-ready television sets, and television remote controls, cable operators are faced with significant problems in interfacing with these devices.

The owner of a VCR wants to be able to easily record a program off a cable channel while watching another channel. Unfortunately, this is difficult to do with present technology. If a cable system uses a converter and scrambling (most systems today use both), the VCR must be connected to the output of the converter in order to record cable channels, and the converter tuned to the desired channel. Without a second converter, it is impossible for the subscriber to watch another channel on the television set at the same time.

Without cable, most VCRs can be programmed to record a particular channel at a predetermined time. When connected to a cable converter, however, this feature cannot be used, since the cable converter must be tuned to the desired channel and the VCR cannot control the converter's tuning.

Finally, connection of a VCR to cable requires a multitude of signal splitters, switches, and cables. Showing the subscriber how to use the VCR, TV set, and cable converter is time-consuming and difficult. The effort may provide a hidden advantage for operators, however. Speaking of the need to help

FIGURE 4-2. Coping with the cable VCR connection.

cable customers deal with their VCRs, Steve Effros, President of the Community Antenna Television Association quipped, "Once they figure it all out they'll never unplug it. They'll never even dust it."[1]

At present, there is no solution to this problem. Some cable companies are offering switch boxes to subscribers which make the job of going from VCR to TV simpler and allow the use of many but not all of the features of VCRs to be used with cable.

CABLE-READY TV SETS

Another problem faced by the cable operator is with so-called cable-ready television sets. Most sets sold today can tune 50 or so cable channels, and are advertised as being cable-ready. If a cable system did not scramble signals, these sets could get all cable services. However, since most systems scramble at least their pay television channels, cable-ready sets still require a converter/descrambler. This makes the cable-ready feature of the set useless for pay subcribers. In addition, many sets today are sold with wireless remote controls. When a cable converter is needed, the channel selection portion of the subscriber's remote control is useless. In fact, the TV set's remote control must be used to control volume, while the cable converter's remote control is used to change channel and turn the set on and off. Again, there is presently no solution to the problems. Chapter 5 describes future developments in this area.

CABLE AUDIO

FM Radio Service

Many cable systems offer FM radio service as an option. Since cable systems carry the FM radio band, it is simple to carry 30 or so FM radio channels on a system. Usually, a splitter is installed in the subscriber's home to connect the FM radio receiver. However, some systems are now providing converters for FM radio service to increase the security of their cable audio programming. With this approach, cable audio signals are carried at 108-120 MHz, above the FM radio band and below channel 14, the first cable channel commonly in use in the midband. The converter, which is of the block converter type, converts these channels down to the standard FM radio band.

Stereo Audio

As already mentioned, stereo audio transmission is now a reality in the United States. The system which has been adopted by the television industry uses subcarriers of the television audio carrier to allow stereo and multilingual sound. Recent developments in noise reduction technology have made stereo possible, since sound quality on the subcarriers would be inadequate without noise reduction techniques. Stereo TV is still in its infancy, with manufacturers just beginning to introduce stereo receivers. It is likely, however, that stereo TV will grow at a rapid rate, since the cost of stereo receive equipment is not substantially above the present monaural equipment.

Stereo television creates problems for cable operators. Some headend equipment is not compatible with stereo and will have to be modified or replaced. More significantly, many scrambling systems are not compatible. Converters will have to be replaced or upgraded to allow stereo sound on scrambled channels.

As new services are added to cable, the home drop increases in complexity. In Chapter 5, future technological developments and their related subscriber equipment are described in detail.

NOTES

[1]Linda Hangsted, "Operators Urged to Educate Consumers," *Multichannel News,* July 21, 1986, p. 49.

The Technological Future

Full screen tabular display of financial market news.

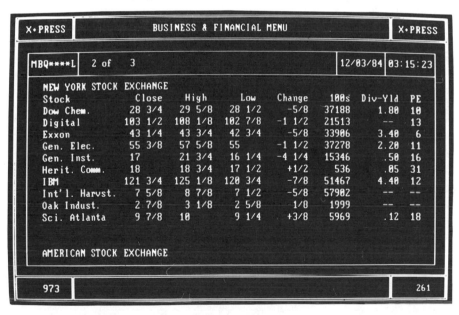

Cable systems are capable of providing many forms of nonentertainment services because of their large information transmission capacity and their bi-directional transmission capability. In this chapter the technology needed to provide some of these services is described.

TWO-WAY PLANT

Cable systems are capable of transmission of audio, video, and data signals not only from the headend to subscribers but also from subscribers back to the headend. This is possible using a technique called *frequency division multiplexing.* Cable systems use frequencies above 50 MHz for transmission of video, audio, and data information into subscribers' homes (referred to as *downstream* transmission). The coaxial cable itself can also be made to carry frequencies below 50 MHz by installing amplifiers that have bidirectional capa-bility (described later in this chapter). Signals in the frequency range of 5 to 35 MHz can be transmitted from points in the distribution network back to the cable headend (referred to as *upstream* transmission).

Two-way capability in cable systems has been in use for many years. In the late 1960s, two-way communication was used by some cable systems to transmit local origination programming from a downtown office or studio to a headend located outside of town. Some cable systems use portions of their trunk to send video signals from remote locations such as schools or auditoriums back to the cable headend, where they are then processed and retransmitted in the downstream direction on a cable television channel to subscribers. In these applications, only a portion of the cable system, usually the trunk, is equipped for two-way transmission.

In the early 1970s, cable operators began experimenting with two-way distribution plants where the entire system was equipped with bidirectional amplifiers. Three companies produced small quantities of interactive cable terminals that were tested in several cities, including El Segundo, California; Overland Park, Kansas; and Orlando, Florida. All these terminals were of the *transponder* type. A digital address command was transmitted down the cable from the headend to the home terminal. The home terminal, in turn, recognized its proper address and responded with a digital message upstream to the headend.

These systems were plagued with problems, the most serious of which was *signal ingress*—the leaking of over-the-air radio signals *into* the cable, which caused interference with the upstream cable communications.

In addition to the technical problems with these terminals, the cable in-dustry expanded too rapidly in the early 1970s and overestimated the speed with which two-way services would develop. As a result, none of these ex-periments ever proceeded beyond a small-scale test.

In the mid-1970s, two other two-way technologies emerged. One, pioneered by Tocom, involves a transponder system similar to the earlier experimental systems. Tocom installed its first system in The Woodlands, Texas, a planned residential community of upper-income homes. Because the cable system of The Woodlands was quite small, problems with signal ingress were controllable and the system has functioned reliably. Over the years, Tocom has installed several of these systems in planned communities like The Woodlands and has supplied most of the home security terminal equipment now in use in the cable industry.

The second technology that emerged is the Coaxial Scientific Corporation's *area division multiplexing* technique. In this system, which will be described in more detail later, the home terminal is not a transponder, but only a transmitter, sending out its message from the home continually. Code-operated switches (COSs) within the cable plant route these signals back to the headend. The Coaxial Scientific transmitter system was first installed in Columbus, Ohio in 1974 as part of a pay-per-view television system. At its peak, the system had over 6,000 home terminal units in operation, solving the ingress problems associated with the transponder technology and also allowed the home terminal's cost to be reduced to less than half that for the transponder-type home terminal. The transmitter system is now in use for home security and interactive TV in several U.S. and Canadian cities.

In the late 1970s, Warner Communications introduced its Qube system also, coincidentally, in Columbus, Ohio. The Qube system uses transponder-type home terminals, coupled with a code-operated switch (or bridger switch), an element borrowed from the Coaxial Scientific system. Qube's two-way terminals have proven to be reliable but quite expensive.

During 1980 and 1981, several new transponder-type technologies emerged from suppliers. All these systems utilize a transponder-type technology in one form or another, and all are similar in price.

The Two-Way Distribution System

Two-way cable distribution systems are identical to the description in Chapter 3 except that the amplifiers must operate in two directions. This is accomplished by utilizing two separate amplifier modules, one of which covers the frequency range of 50 MHz and up, and the other of which covers the frequency range of 5 to 35 MHz (called channels T7–T11). At the input and output of each of these amplifiers are diplexing filters (Figure 3–4). *Diplexing filters* split cable signals into two frequency bands, routing frequencies above 50 MHz to the downstream amplifier and those below 35 MHz to the upstream amplifier.

For a variety of reasons, diplexing filters cannot be designed that allow the use of all the spectrum. A *guard band* must be provided, and frequencies above 35 MHz in the upstream direction or below 50 MHz in the downstream direction cannot be used.

The coaxial cable itself is insensitive to which direction frequencies are transmitted, so it is possible to transmit frequencies above 50 MHz downstream and below 35 MHz upstream over the same cable. The frequencies between 50 and 54 MHz are used for control and data transmission, but not for video.

All these two-way parts are modules that plug into the trunk station or line extender amplifier housings. Most two-way cable systems incorporate some form of automatic gain control in the upstream direction. In some cases, AGC circuitry similar to that described in Chapter 3 is used. In other cases, thermal compensation systems, which regulate an amplifier's gain in accordance with the outside air temperature, are used.

The Code-Operated or Bridger Switch

As mentioned earlier, signal intrusion is a major problem with upstream communications. This is because the frequency spectrum occupied by upstream transmission is used over the air by many high-power transmitting sources, such as international shortwave broadcasts, amateur radio operators, business, and military communications. Signals can leak into the cable because of a break in the aluminum sheath or because a connector has become corroded or loose. (There is also a condition called *signal egress*, in which cable signals leak from the cable into the air. The FCC has put strict limitations on cable's use of aircraft navigation and communication frequencies because of possible interference due to leakage from cable systems.)

The code-operated switch, designed to solve the ingress problem, is located at each bridger amplifier and allows the downstream signals to pass unimpeded from the trunk into the feeder system and from there to subscribers' homes (Figure 5–1). Upstream signals from each feeder area, however, are routed through an electronic switch. A computer at the cable system headend can turn on or off each code-operated switch in the system.

In operation, the computer turns on one code-operated switch at a time, allowing only signals from one bridger area to enter the trunk and travel up the trunk to the headend. If signal ingress occurs within a feeder area, only signals from within that feeder area will be affected, since the ingress will be switched off when the code-operated switch is disabled.

The code-operated switch can also be used in the opposite configuration, with all switches normally turned on, allowing data to be gathered from all home terminals in the system. If a fault occurs, such as signal ingress, the switch serving the area in which the fault is located is turned off. In this method, the code-operated switch serves as a diagnostic tool, and to limit the effects of ingress and faults to one area of the system.

At this time, there is some disagreement in the cable industry as to whether code-operated switches are necessary. Some engineers claim that

FIGURE 5-1. Code-operated switch.

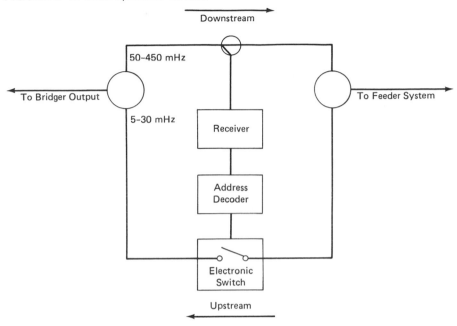

modern construction practices and good maintenance procedures can eliminate the problems of signal ingress. Code-operated switches are effective in reducing signal ingress problems, but it is difficult with code-operated switches to offer services that must be in use continuously, such as data interconnection between computers.

The Two-Way Headend

Additional equipment is required at the headend for two-way services (Figure 5–2). Although certain elements are standard in any two-way system, equipment needs are dictated by the two-way services to be provided.

The heart of a two-way cable headend is a minicomputer or microcomputer system. This computer polls code-operated switches (if used) and home terminals to retrieve the desired data. It then formats the data and presents it to the cable operator in usable form.

Such a system can be quite complex and expensive. However, if only a few subscribers are involved and if the two-way service offering is limited, systems can be relatively simple and inexpensive. For instance, a two-way system offering only home security service to a few hundred subscribers might have only a small microcomputer and floppy disk drives, a $10,000 to $15,000 investment.

If other services are offered, more sophisticated computer installations

FIGURE 5-2. Two-way headend.

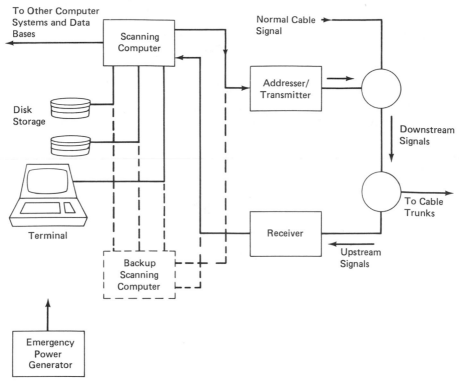

are necessary. A Syracuse, New York, system offers home security, opinion polling, channel monitoring, and system status monitoring using dual DEC PDP-11/34 computers, together with four disk drives and related equipment (over $100,000 in cost). The system automatically switches from the main computer to a backup in case of failure, thus assuring uninterrupted home security service.

Newer cable systems use even more sophisticated computers, with four or more minicomputers, costing over $500,000. It should be noted that much of the equipment and software being proposed for new systems is experimental and that the final hardware and software configurations have not evolved.

To connect the computer system to the cable system, data modulators and demodulators (or receivers) are required. A data modulator takes the digital output from the computer system and places it on an unused section of the downstream cable spectrum (typically between 108 and 121 MHz). The data demodulator takes the upstream signals from the cable system and converts them to digital signals for processing by the computer.

Two-Way Design and Construction

Although a two-way cable plant is similar to a one-way system, special care must be taken in the design process. In the past, many cable operators have designed their systems for downstream performance only, believing that upstream signals could be successfully added later.

Experience has shown, however, that calculations must be made for both upstream and downstream signals. Those cable operators with two-way experience have developed design criteria based on their operating experience, assuring that the two-way plants they build will work properly. Some cable operators have successfully upgraded existing one-way systems at a relatively low cost with good results.

During construction, care must be taken to assure the integrity of all connections, since upstream frequencies are more susceptible in ingress than downstream. Connectors must be chosen to assure good upstream performance.

Status Monitoring

Because of the special problems of upstream cable, and because of the importance of reliability in new services, especially home security, some operators are installing automatic system status monitoring devices in the cable plant.

Status monitoring involves the installation of transponder units in various places in the cable distribution system. A computer at the headend interrogates these units, instructing them to send data on signal levels and picture quality back over the upstream path to the headend. The computer tabulates the information and presents it to the maintenance personnel. In this way the cable operator knows the condition of the system at all times and can respond immediately to outages. In addition, minor problems which may not yet be visible to subscribers can be detected and cured.

Status monitoring has been available since the late 1960s but has not yet become widespread. Most status monitoring equipment is expensive ($150 to $300 per location) and will monitor only the trunk lines. Since cable operators constantly monitor the condition of their trunks using field tests by technicians, status monitoring offered little benefit. Today, however, concerns with reliability have forced cable operators to include status monitoring in new systems. Several types of equipment are available.

Many amplifier manufacturers offer status monitoring modules for their trunk stations, costing about $150. These modules provide reliable information about trunk performance, but they tell nothing about feeder performance. Also, only two channels out of the total of 35 to 50 are monitored. In most cases, this provides reliable indication of system performance, but occasionally problems occur that will not be detected by this type of monitor.

Some manufacturers offer status monitoring devices that are in separate housings, allowing installation at any point, trunk or feeder, in the plant. These units are costly, however, and feeder monitoring is rarely provided.

The Two-Way Subscriber Drop

Two-way subscriber drops do not differ significantly from one-way drops. Again, special care must be taken to assure that connectors are tight and waterproof, and drop cable must be chosen to have good shielding characteristics to prevent ingress.

SUBSCRIBER EQUIPMENT

The Two-Way Home Terminal

The two-way home terminal is a device that sends data from a subscriber's home to the cable headend. Data can be from many sources, including fire and smoke detectors, intrusion detectors, utility meters, or keyboards. Terminals are often included as part of the converter, although such designs have liabilities if home security or other monitoring services are provided (all the wiring from the detectors for these services must be routed to the back of the TV set).

There are two basic types of terminals. The transponder and the transmitter type.

The Transponder Terminal

A *transponder* terminal (Figure 5–3) consists of (1) a digital receiver, which is "talked to" by the computer at the headend, (2) a transmitter, which "talks to" the computer, and (3) a microprocessor, which controls all the functions of the terminal.

The headend computer sends a digital message to all terminals simultaneously. Part of this message is an *address*, a digital code that is different for each terminal. When a terminal recognizes its address, it responds by turning on its transmitter, which sends back upstream a digital message containing its address plus data from the various devices that are monitored by the terminal. For instance, one "bit" of the message sent might indicate the presence or absence of a fire alarm; another group of bits might contain the channel numbers to which the subscriber's converter is tuned.

After the message is sent, the transmitter is turned off, and the terminal waits for the next transmission of its address. The microprocessor in the terminal performs the address recognition, transmitter control, and data transmission functions.

FIGURE 5-3. Transponder-type terminal.

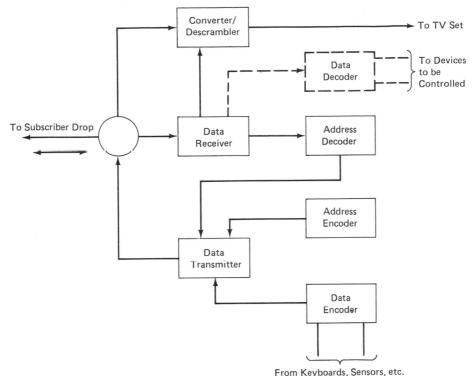

From Keyboards, Sensors, etc.

Transponder terminals are the most common form of two-way terminals in use today, being manufactured in one form or another by many suppliers. They have the advantage of allowing communications *into* the home for such services as power load management as well as *out* of the home. Their disadvantages are cost ($200 to $400) and the fact that they all share a common upstream channel, making them susceptible to signal intrusion and even intentional jamming. The use of code-operated switches (already described) in conjunction with transponder terminals solves this problem.

A variation of the transponder terminal is the *contention* terminal. When a subscriber enters data, the terminal listens to the downstream command channel to find out if the system is in use. If the system is free, the message is transmitted by the home terminal. At the headend the message from the home terminal is sent back over the command channel. The home terminal listens to its own message to make certain that it was received in an ungarbled fashion by the headend. If not, the terminal repeats its message.

The advantage of the contention-type terminal is that messages are almost instantaneously transmitted from the terminal to the headend. Another advantage is that only new messages are transmitted from the terminal so that a great deal more data can be transmitted on a given upstream signal band-width. This may be important in the future when new services are added.

The main drawback of the contention system is that failure of a terminal may not be recognized for many hours since the terminal is not continuously polled to determine its status. In a contention system, a "global" poll of all terminals to determine their status is done periodically. In this mode, each terminal is requested to respond to assure that its operation is normal. Global polling, however, takes place only periodically and in between global polls a terminal can fail.

The Transmitter Terminal

The second two-way home terminal type is the Coaxial Scientific/Rogers Cablesystems area division multiplex transmitter-type home terminal (Figure 5–4). This device consists of only a transmitter and a few digital *chips* (integrated circuits). A code-operated switch is required at each bridger amplifier.

FIGURE 5–4. Transmitter-type terminal.

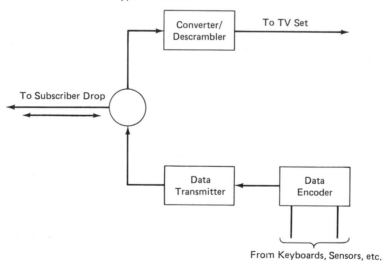

In this type of system, the home terminal sends its digital message continuously. Each terminal within a feeder section has a separate upstream frequency. At the bridger amplifier; where upstream feeder signals are routed into the trunk, the code-operated switch controls the flow of these signals into the trunk.

The computer at the headend instructs one code-operated switch (COS) to turn on. It allows the signals from all homes in that feeder section to enter the trunk and to travel to the headend. At the headend, the computer receives and stores the data from all these homes simultaneously. Then the COS is turned off. The next COS is turned on and so forth until all the plant has been scanned. The whole cable system is scanned in a few seconds.

Transmitter-type terminals are in use in several systems operated by Rogers Cablesystems. The system has two major advantages. First, and most important, is cost. With its limited capability and fewer components, a complete home security terminal costs under $50, compared with $200 for a comparable transponder-type. Second, the system is less vulnerable to signal ingress and intentional jamming than the transponder-type.

The system's disadvantages are that it occupies slightly more spectrum space in the upstream direction (4 to 10 MHz compared with 2 to 6 MHz for transponders), and that commands cannot be sent into the home, making services such as utility load management impossible. However, a receiver module can be added to the terminal to allow such services. This add-on would be provided only for those subscribers requiring the service.

Connection of Alarms, Keyboards, and Other Devices

A two-way terminal can be connected to any number of data *sources.* For home security, an alarm *panel* is installed that is an interface between the intrusion detectors (door switches, motion detectors, and the like) and the home terminal. This panel also has the necessary subscriber controls to enable and test the alarm system. (Figure 5–5.)

To connect a home terminal to a utility meter, a special meter must be installed, which electrically encodes the meter dials. A connection is made to an interface circuit, which is located inside the home terminal.

For monitoring television viewing, a connection is made to the channel selector circuitry inside the converter. Often, the keyboard part of the converter used to select channels is also used as a subscriber response keyboard. With this approach, opinion polling, interactive TV programming, or home shopping is possible.

The most sophisticated source for connection to a home terminal is the full alphanumeric keyboard. With this configuration, the terminal becomes a modem, allowing full data retrieval and transmission capability. Home computers can be connected to home terminals to provide this capability. (See the videotex section following for more on information retrieval.)

The State of the Art

Though many cable systems are capable of two-way transmission, very few are being used to provide two-way services. During the early 1980s there was a great deal of development of two-way technology. However, since two-way services have not proven to be profitable for cable operators, little new equipment is being developed or installed.

FIGURE 5-5. Connection of data sources to a home terminal.

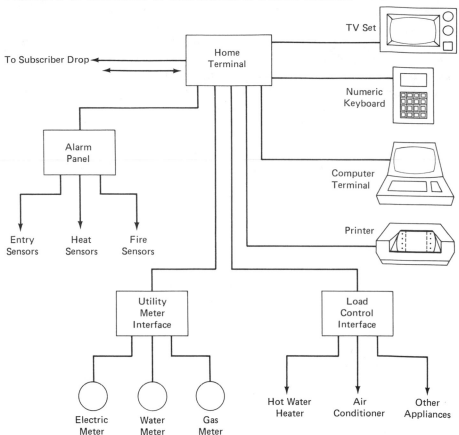

TEXT AND GRAPHICS SERVICES

Cable systems can be used to transmit various forms of alphanumeric text or graphics services, including videotex and teletext. Text services use telephone lines, over-the-air TV, or cable to distribute information to subscribers' TV screens or computer terminals. The development of text services has been most widespread in Europe, where governmental funding has subsidized their growth. In the United States, most text service experimentation has been over telephone lines, but technology has been developed for both cable and over-the-air text services.

One-Way Cable Text System

In *teletext* systems in which text information is transmitted on a broadcast or cable TV channel, digital information is sandwiched within TV pictures in an area called the *vertical blanking interval*. The subscriber to the teletext service is provided with a *decoder*, which extracts this information, stores it, and formats it as a still TV picture.

Information for 200 to 300 *pages* of display is transmitted sequentially, with the transmission of each whole cycle of 200 to 300 pages taking 5 to 10 seconds. The teletext terminal has a selector switch or keyboard that instructs the terminal to "grab" the desired page as it is transmitted. The user has to wait only a few seconds for the desired page.

When one-way teletext service is offered over a cable system, many channels can be used in this manner, so several thousand pages of information can be accessed by the subscriber.

A variation of this approach uses an entire TV channel to transmit several thousand pages. Teletext decoders are relatively expensive ($200–$500) and very few are in use in cable systems today. It is possible to build teletext decoders into TV sets at a relatively low cost, and in the future it is likely that such decoders will be available.

Two-Way Videotex

Videotex can be offered either over telephone lines or over two-way cable systems. At the cable headend (or at a convenient central location in the case of a telephone-based videotex system), a large computer stores hundreds of thousands of pages of information. Often, the computer is interconnected with other *data bases*, giving it access to millions of pages of data.

The home decoder has the ability to recognize a unique address and can store a page and format it on a TV screen. The decoder is connected to a two-way interactive terminal in the case of a cable videotex system, or to a modem with telephone videotex.

When a subscriber enters the desired page number on a keyboard, the information is transmitted back to the cable headend to the videotex computer. The computer then sends out the subscriber's digital address, followed by the desired page. The videotex decoder recognizes its address and "grabs" the page, which it then formats and displays.

Page capacity is limited only by the size of the videotex computer system, and access time is independent of the number of pages desired. As opposed to a one-way teletext terminal, terminals for these types of services are more expensive, ranging from $300 to $1,000. A home computer can be used as a videotex terminal, connected to the cable system through a two-way terminal.

Home Computers

Another method of providing videotex service involves the use of home computers. Several dial-up telephone networks allow home computers to access large data bases, such as The Source, Compu-Serve, Dow Jones. With these services, home computer owners dial a local telephone number and are connected to a large computer system. Users are billed on a per-minute basis. Several experiments in the cable industry involve simply replacing the telephone line as the communications path between the home computer and the data bases. On a very small scale, this is easy to do in a cable system. A dedicated narrow bandwidth channel is provided for each user in both the upstream and downstream direction. As the number of users increases, however, the amount of spectrum space consumed increases to an unacceptable amount.

Several experimental systems have been developed which time-share a portion of the cable spectrum between hundreds of users to allow cable systems to efficiently transmit data between home computers. One feature of these systems is the low cost of the cable *modem*, the device that interfaces the home computer to the cable.

Another way of using a home computer involves data *downloading*. Rather than having a two-way interactive communication path between the computer and the cable headend, information can be sent down the cable for storage, processing, and later display by the home computer. Data can be sent either in a cycle, like one-way teletext systems, or to a specific individual home computer using an addressing system.

In the early 1980s the Nabu network started a download service in Canada and the U.S. Video games, stock market information, news, and miscellaneous computer programs were sent by satellite to cable headends and then over a 6-MHz channel to subscribers. Customers purchased or rented a special home computer and receiver and paid a monthly fee for the service. Nabu was not successful in attracting enough subscribers and was shut down after a few months of operation.

Another downloading system, called X-Press, has been started by McGraw-Hill (the publisher), Telecrafter (a communications services company), and TCI (the largest cable operator). The technology is similar to that used by Nabu, except that IBM or Apple home computers are used. The receiver, which connects the computer to the cable, is inexpensive. Software, which is furnished by X-Press, allows the home or business computer to interact with the receiver and access the information.

System Standards

A major problem facing text services, whether on cable, over the air, or over phone lines is standardization of data formats. At the present time, two major formats exist.

The simplest system and the oldest is the *British Prestel System*—a system developed by the British Post Office and presently in use over the British telephone network and British over-the-air television. The decoders for this system are relatively simple and inexpensive, but the display is limited to simple alphanumeric characters and crude graphic symbols.

The other major format is the *North American Presentation Level Pro-*

tocol Standard (NAPLPS), developed by AT&T as an extension of the Canadian Telidon system, which allows sophisticated graphics and even simple animation. The decoding equipment for the NAPLPS format is complex and expensive, however.

Security

Many types of information retrieval do not require a high level of privacy. For instance, requests for information from a video newspaper, weather information, stock market reports, and the like could all be provided with little or no security. Services that involve financial transactions, however, are a different story. If home banking or home shopping is ever to be offered using videotex systems, privacy will become a major concern. The cable system is poorly suited for privacy since it is, in essence, a giant party line. Tapping into the system to listen to electronic conversations between the headend and the home is simple. If security is to be provided for financial transactions, some type of complex and effective data encryption will have to be installed in each home decoder.

FUTURE DISTRIBUTION SYSTEM TECHNOLOGY

This section outlines emerging technologies that may shape the future of cable television distribution plants. As mentioned in Chapter 3 on distribution systems, there has been a trend over the years toward increased bandwidth on cable systems to increase channel capacity.

Higher Bandwidth Systems

At present, amplifier technology limits cable to approximately 75 TV channels, and the upper frequency limit of cable amplifiers is 550 MHz. Equipment is becoming available for use up to 600 MHz.

There is little difficulty in increasing the frequency coverage of amplifiers to 500 MHz and above. In Europe, many cable systems cover the entire UHF spectrum (up to about 800 MHz). The major limiting factor to upward expansion of channel capacity in cable systems is the distortion characteristics of amplifiers. Because amplifiers with higher upper frequency limits carry more channels, distortion is greater and the amplifiers must be operated at lower output levels, resulting in system designs with shorter trunk cascades. As hybrid amplifier module design progresses, distortion will be reduced, and a larger number of channels will be possible. It should be noted that increasing cable system bandwidth is a very economical way of adding additional channels. Going from 300 to 400 MHz, for instance, increases the cost of the cable system by only about 10 percent, whereas channel capacity increases from 35 to 54 channels, an increase of almost 60 percent. An increase from 400 to 500 MHz would add 17 more channels at a nominal increase in system cost.

There is no theoretical limit to the number of channels in future cable systems, and it is likely that, over the next decade, systems will be built with 100 or more channels carried on a single cable. Of course, converter technology has to keep pace.

Bandwidth Compression

As already mentioned, a standard television channel occupies 6 MHz of spectrum space. It is possible to compress a television picture into a smaller amount of spectrum space. This is done by eliminating *redundancy* in the television picture.

At the present time, there is only one commercially available *bandwidth compression* system, the General Electric Comband system. Comband uses a complex method of separating video, audio and color information, transmitting these elements from two separate program sources sequentially. In this way, two channels can be sent over cable in the space normally occupied by one. Comband is still experimental and expensive. However, if an operator is faced with rebuilding a plant to increase channel capacity, Comband could be used with the existing plant to provide the extra channel capacity at a lower cost.

Another approach uses digital techniques. Rather than transmitting 30 complete individual television pictures per second, as the present TV system does, each picture is compared with the previous picture and only the differences are transmitted. The difference between any two pictures, or *frames*, will be transmitted except when there is a scene change, in which case an entire new picture must be transmitted. The total amount of information transmitted, however, is less, and, as a result, the bandwidth required for transmission is reduced a great deal.

Although this bandwidth compression technique has been in existence for some time, the decoding equipment is still extremely expensive. In the next ten years, however, integrated circuits will likely be developed that will bring the cost of the decoder down to a reasonable figure.

If a television picture could be compressed to 2 MHz instead of the present 6 MHz, a cable system could carry substantially more channels than at present. For instance, a 400-MHz system could carry 150 channels.

A second advantage of bandwidth compression is that it is in itself a form of scrambling, since a normal TV receiver would not get a picture on a cable system using bandwidth compression.

Fiber Optics

Over the last ten years, a whole new communication transmission medium has emerged. Rather than transmitting electrical impulses through a wire as the telephone and cable networks do, optical fiber systems utilize a thin strand of flexible glass. A light source at the transmitting end, such as a small laser, and a photodiode at the receiving end allow the transmission of impulses of light through fiber optic cable.

Fiber optic systems have many advantages over coaxial cable. First, very great bandwidths are possible—several times that of coaxial cable. Second, ingress and egress problems are nonexistent. Third, illegally tapping a fiber optic cable is much more difficult than is tapping into a coaxial cable or telephone wire. Fourth, signal loss through optical fiber is substantially less than through coaxial cable, requiring fewer *repeaters* (the equivalent of amplifiers in coaxial systems). Finally, some time in the future, fiber optic systems are likely to be cheaper than coaxial cable systems. Fibers are made of glass, a relatively inexpensive material. Fiber cost has been falling, whereas coaxial cable costs have been rising over the years. Fiber is also lightweight and small.

Most fiber optic development in the past has been in the telephone industry. The telephone company presently utilizes copper wire or coaxial cable to tie together switching centers within cities. In the major cities, these cables are in ducts located beneath the streets. As ducts fill up, and new ducts must be installed, tremendous capital investment is necessary. Since a single optical fiber can carry as many telephone conversations as a massive cable with several thousand individual wires, telephone companies have seen the advantages of replacing present wire circuits with fibers to increase the communication capacity in existing ducts. In addition, fiber optic lines are now cost-effective with microwave and coaxial cable for long-distance telephone lines. As a result, a high percentage of new intercity phone interconnection is now done with fiber optic links.

The advantages of fiber optics for individual telephone circuits into residences and business are not as great. It will be some time before optical fibers are installed in place of wires for individual telephone circuits. This is because of the cost of the equipment to be located in the subscriber's home. In addition, phone lines typically have useful life of 30 to 40 years. Even if optical fibers become economically feasible for individual lines by 1990, it would be another 20 years before a majority of lines are optical fiber. In certain specialized applications, such as high-speed digital communications, optical fibers may be in use relatively soon because of their high information carrying capacity. If new services are introduced by the phone company, such as picturephone or even some form of entertainment TV service, the incentive to add fibers would increase, since twisted wire circuits cannot handle such services.

There has been some use of optical fibers in cable systems. Most have been "supertrunk" situations where two headends are to be tied together or headends are to be interconnected with hubs. At the present time, the cost of these systems is approaching the cost of a comparable coaxial cable system, and reliability and performance are better. It is likely that there will be a significant movement to fiber optics for supertrunking in the near future.

FIGURE 5-6. Tree, star, and hybrid architecture.

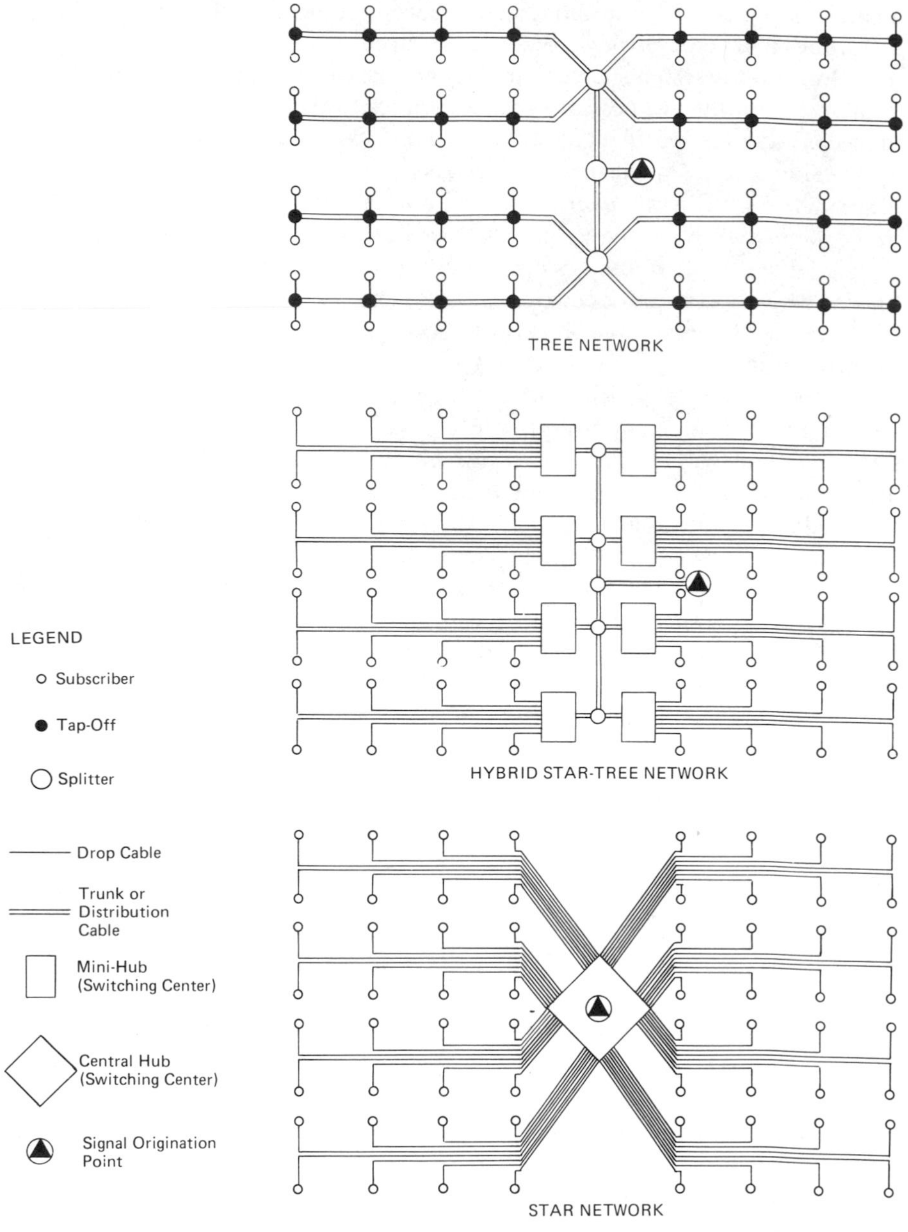

TREE NETWORK

HYBRID STAR-TREE NETWORK

LEGEND

○ Subscriber

● Tap-Off

◯ Splitter

────── Drop Cable

══════ Trunk or Distribution Cable

▢ Mini-Hub (Switching Center)

◇ Central Hub (Switching Center)

🔺 Signal Origination Point

STAR NETWORK

Fiber systems for trunk and feeder use are far less developed. The telephone network is called a *star network* (Figure 5–6). In a star network, individual lines run from a central point (the switching center) to each home.

Every subscriber has a private line (except for party lines, where several homes share one line). Party lines have become much less common over the years.

A cable system is a *tree network*, with common signals all leaving a central point (the headend) on a single trunk. The signals are branched out into individual feeders and then to homes. In essence, all subscribers share one large party line.

A *hybrid star-tree network* uses conventional coaxial cable trunks to deliver all channels to switching centers located in neighborhoods and in apartment complexes. Each switching center serves from 20 to 200 homes.

Hybrid star-tree networks are not new to cable TV. For many years Rediffusion, a British cable operator, ran hybrid star-tree cable systems in Hong Kong and England. These networks use small switching centers serving a few hundred subscribers each, with individual drop lines from the centers to homes.

In the early 1970s, a Rediffusion system was installed in Hyannis, Massachusetts. Another system was tried in a suburb of San Francisco. Although both systems worked relatively well technically, the cost per subscriber was unacceptable.

Recently, Times Fiber has installed a hybrid star-tree system in Alameda, California, serving about 20,000 homes. Times had previously developed this technology for apartment complexes and modified it for use with individual residences. The Alameda system, like previous star-tree systems, proved to be too expensive, and it is unlikely that more of this type will be built in the near future.

Fiber optic equipment manufacturers have concentrated on technology for the star network, and little has been done to develop equipment for tree networks. When optical systems are manufactured in large quantities, only star hardware components will be available at reasonable costs. Therefore, if fiber optic networks are used by the cable industry, it is likely that a star network system design, not the traditional cable tree network, will be employed. However, large-scale fiber systems will probably not be in use for cable-type services until the mid-1990s. Video frequency, large-capacity switching equipment remains to be developed. Substantial capital will be necessary, and the capital investment already made in cable is also a deterrent.

With a star network, channel capacity is unlimited. Rather than sending all channels down a single cable and allowing each subscriber to choose one, as is done in a tree network, the star network sends only one channel at a time down each individual line, as requested by the subscriber.

A star network need only have a two- or three-channel capacity (one for each TV set in the house) on each line. Channel offerings can be as great as is desired, limited only by the amount of headend equipment rather than by subscriber equipment. Also, star network designs solve security and privacy problems, since only authorized channels are sent down a subscriber's line.

In the distant future, cable systems and phone networks may be almost indistinguishable in capability. Both could be star networks utilizing optical fibers. Both could have sufficient bandwidth to handle video, audio, and data services, and both could be switched allowing point-to-point communications. In fact, by the year 2000, either network could handle both telephone and cable services. Whether both these networks would survive side by side and which services would be offered on what network are questions to be answered by regulatory and economic conditions in the future.

SUBSCRIBER EQUIPMENT OF THE FUTURE

Scrambling and Encryption

Newer scrambling systems rely on *encryption* of video and audio using techniques pioneered by the military to provide secure data and voice communication. This technology, which is digital rather than analog (as is television audio and video), is coming down in price rapidly. The scrambling technique developed by M/A-COM (now General Instrument Videocipher II) and used by all the major program suppliers for their satellite transmission uses digitally encrypted audio. It is likely that the same technology will become available at a low cost for cable security in the near future.

Recently, the EIA (Electronic Industries Association), a trade group representing television manufacturers, developed a *universal decoder interface* standard. The UDI is a plug that will be installed on the back of all television receivers made in the future. This plug will allow a cable descrambler to be connected to the receiver, thus eliminating the need for a cable converter. When this plug is available (probably not before 1990 on most TV sets), television remote controls will work for cable, and the need for duplicate controls will be eliminated. (See Cable-Ready TV Sets in Chapter 4). This plug will also be installed on videocassette recorders, solving the problems of interfacing VCRs and cable.

High-Definition Television (HDTV)

The present U.S. TV system was developed in the 1930s and 1940s. In the 1950s, the system was adapted for color. In other parts of the world other systems were developed, some of which are far superior to that in the United States.

The U.S. system, called the *NTSC (National Television System Committee)*, has several disadvantages. First, the *resolution*, or sharpness, of the TV image is limited. With small-screen TV sets, the fuzziness is not noticeable; on large-screen sets, however, and especially on projection-type sets, pictures are truly inadequate.

Second, the color technique used by the NTSC system is not very stable. It is difficult to maintain natural looking color with the system. In addition, *color resolution* is even poorer than black and white, resulting in smeared areas of color, especially in lettering and small detail.

Finally, when three-dimensional technology is perfected, it is unlikely that the NTSC system can be modified to work.

Over the years, many improvements have been made to transmitting and receiving equipment. However, the limits of improvement with the present system are close to being reached now.

Changing TV standards creates two main problems. First, a new standard would probably not be compatible with the NTSC system. For many years, as TV sets of the present style would be phased out, programming would have to be transmitted on two different channels, one for each standard. Both Britain and France did this in the early 1950s when they switched to newer, higher-resolution systems.

Second, a new standard will almost certainly require more bandwidth than the NTSC system. There is already a scarcity of spectrum space at TV broadcast frequencies, so it is not likely that this new system will be broadcast over the air. The two best candidates for transmission of high-definition TV (HDTV) are DBS (direct broadcast satellite, discussed in Chapter 18) and cable.

Several organizations, including the Society of Motion Picture and Television Engineers (SMPTE) and the International Radio Consultative Committee (CCIR), have been studying high-definition television. Field tests have been conducted in Japan and the United States.

The overall goal of the SMPTE, for example, is to provide a television picture equal in quality to 35-millimeter theatrical film. To do this, they have concluded that the new standard should:

1. consist of a TV picture with about 1,100 lines compared to the present 525 lines,
2. have an *aspect ratio* (the ratio of horizontal to vertical screen dimensions) of 5:3 or 2:1 as opposed to the present 4:3 ratio, thus making the screen dimensions resemble the movie screen,
3. improve the method of transmitting the color information to provide greater color sharpness and detail, and
4. increase the sharpness of detail by increasing the bandwidth utilized for video information.

These changes will require a bandwidth of up to 30 MHz per channel as compared to the present 6-MHz channel width. Bandwidth compression techniques (described earlier) would reduce the spectrum required, but it is unlikely that a high-definition TV picture could be compressed to fit in the present TV channel.

Before a new standard can be established, international agreement will be necessary to avoid repeating the problems created in the past by each country adopting its own television standard. The frontrunner is the NHK system, developed in Japan and adopted by U.S. broadcasters and manufacturers. However, objections from European interests may prevent this system from becoming a world standard. It is likely that no agreement will be reached on HDTV standards until 1990 or even later.

Three-Dimensional Television

Three-dimensional movies, of course, have been around for many years. Three-dimensional television has not emerged because of the cost and complexity of the television receiver required to display 3D images. Recent developments, however, may speed the introduction of 3D.

A sheet of plastic containing liquid crystal elements, arranged as a polarizing filter, is placed over a standard color picture tube. By changing the electrical signal sent to the crystal elements, the crystal passes either horizontal or vertical light rays. The viewer wears glasses with a vertical polarizing filter over one eye and a horizontal filter over the other.

Liquid crystal elements are now inexpensive, making this system practical. It will first be used for industrial and scientific applications, where three-dimensional images are desirable, and it may ultimately be used for consumer television.

Component Television Receivers and Integrated Systems

Today, a typical cable home contains one or more TV sets, a converter, and a telephone. As two-way and information retrieval services develop, more hardware will be added, such as data terminals, videotex decoders, keyboards, and home computers.

Many of these devices could be consolidated into one unit. For instance, a TV set could incorporate a converter at a minimal extra cost. The same is true of some forms of videotex decoders. Typically, a significant proportion of the cost of a device is in the power supplies and cabinets, so that any design that consolidates two or more devices that share these costs will be cheaper than the combined cost of the separate devices.

The problem with integration of hardware lies in the way in which services develop. Each service must be tested, implemented, and proven, one at a time. Only after a service is well established, and standardization accomplished, can its hardware be consolidated into another device. One solution lies in individual TV components (as in stereo components), with the consumer building up the desired TV system using separate components such as monitors, converters or tuners, videotex decoders, stereo sound systems, and VCRs.

It is possible that more than one communication center will evolve for the home. The telephone of the 1990s will most likely contain a computer terminal and video display. A "Home Information Center," consisting of a telephone, microcomputer, display screen, and modems (modulator-demodulators) for connection to both cable and telephone networks, may evolve, which will serve the information retrieval, storage, and display functions within the home. All the services forecast for cable (and the telephone network) will be handled by the center.

Future Audio Systems

The success of the compact audio laser disk has increased interest in high-quality audio systems for cable. Present FM radio broadcast technology cannot deliver the full signal quality of compact disks, and carriage of FM over cable further degrades performance. As a result, several new systems for transmitting high-quality audio have emerged.

These systems fall into two categories; analog and digital. Analog systems use conventional frequency modulation techniques, but use a much higher bandwidth (.5 to 1 MHz per channel, compared to .2 MHz per channel for standard FM radio). Digital techniques convert the analog audio information to a digital data stream, transmit it on a relatively wide channel (around 1 MHz), and convert the digital data back to analog at the receiver. Both techniques are capable of extremely high-quality audio.

INSTITUTIONAL NETWORKS AND TELEPHONE BYPASS

Many of the services carried on institutional networks use the same equipment as is required for the subscriber network. Video programming, for instance, is put on the institutional network via a modulator, and a standard television receiver is used on the receiving end. Some services, however, require specialized equipment.

Interconnection of computers over institutional networks requires high-speed modems, capable of transmitting data at speeds of 1.5 megabits (million bits) per second or higher with very low error rates. Such equipment has been developed for use on *local area networks* (cable networks for interconnection of computer equipment within a single building), and adaptation for cable institutional networks is not difficult. Some cable operators are installing these modems on their institutional networks, but regulatory and economic problems have kept data transmission services experimental.

A second service requiring specialized equipment is telephone bypass. With the introduction of competition in the long-distance telephone business, alternate carriers, such as MCI and Sprint, are looking for ways to tie businesses into their microwave and satellite networks. Cable institutional networks are ideal for this purpose. For data circuits, the modems described previously are used. Voice circuits can be handled in two different ways. One method is to digitize the voice and transmit it as digital data using modems. A second way is to transmit it in analog form, using modulators and demodulators. At present, equipment for analog voice transmission over institutional networks is experimental.

CHAPTER **6**

Over-the-Air and Community Channels; Text and Audio

Public acess studio and control room, East Lansing, Michigan.

Cable operators are gradually building and rebuilding so that most areas of the United States will eventually have high-capacity systems. For now, the nature of the cable service in any community depends on the technology available when the system was built or last rebuilt.

The large-capacity cable systems carry broadcast television signals (sometimes called *over-the-air* or *OTA* signals), several satellite networks, local origination community channels, access channels, a number of text and graphic channels, and audio services. Some of the services are marketed separately as add-on *pay* or *premium* channels. All of the others are referred to as *basic* channels.

Basic channels may be offered together as a single package, or split into two or more levels or *tiers*. In the latter case, some of the more popular satellite networks are put into a second tier called *expanded basic, extended basic, satellite tier*, or another name created to signify a separate and higher level of service. In some systems the technology creates the tiers. The first tier of basic may include only 12 channels for which the subscriber uses the detent tuner on the television set. The next levels of service require a converter.

Premium, or pay, channels require subscription to basic cable and an additional monthly fee for each channel individually, *a la carte*, or in *packages* with other pay channels.

Cable systems also offer *ancillary services* (additional outlets for second, third, and more TV sets and VCRs, hand-held wireless remotes, FM outlets, and printed program guides).

It is important to keep the new television programming vocabulary straight. Now, the term "television network" can be used without qualification only if it is meant to refer to *all* types of television networks. More specifically, there are *broadcast networks* and *cable networks*. Among cable networks there are the *basic networks, pay networks* and *pay-per-view networks*. The cable networks may be further categorized into *advertiser-supported, religious, sports, regional, ethnic, entertainment,* and *information networks, superstations* and *interconnects*.

Chapters 6 and 7 of this part of the book cover the tremendous diversity of basic channels: Chapter 6, over-the-air television, data channels, local programming, automated channels, and cable audio; Chapter 7, the basic satellite networks; Chapter 8, the premium channels; Chapter 9, the more advanced services possible with the cable technology; and Chapter 10, the consumer and audiences for cable services.

OVER-THE-AIR TELEVISION

The foundation of the basic cable television service in some communities is still the delivery of off-air television signals. This is not true, of course, in the metropolitan areas that have an abundance of clear television station signals available to any television receiver with rabbit ears. But in towns that have strong signals from *fewer* than the three networks, a principal appeal of cable is the supply of the missing broadcast television network. As a matter of fact, commercial network television, with its news, sports, and popular entertainment series and specials, is the most watched category of television service, even in cable households with 30 or more basic and premium channels. Surveys of cable audiences with a full range of cable services available consistently find the predominance of viewing time to the commercial broadcast networks.[1] Independent (non-network-affiliated) broadcast stations are also brought in off-air by cable systems. These stations offer an additional variety of off-network syndicated programs, sports, and movies.

FCC Signal Carriage Rules

In 1965 the FCC began a series of rules limiting the number of distant broadcast television stations a cable system could carry.[2] In 1980, the FCC finally ended the regulation of distant signal importation.[3] The 1965 rules also mandated cable carriage of *local* broadcast television signals. This was referred to as the *must-carry* rule. In 1985, the U.S. Court of Appeals declared these rules unconstitutional, a violation of the cable system's First Amendment Rights.[4] The court suggested that the FCC might develop new must-carry rules that would conform to the constitution and, after some false starts, in 1987 the FCC did adopt new rules. These rules are summarized in the Chapter Notes.[5]

According to the FCC, the object of must-carry is, "to maximize the availability of program choices by competing providers of video services, both off-the-air and on cable . . ." There has been a concern that if cable bypassed local stations, those stations might not survive, having lost the cable audience. The long-standing FCC policy of providing *local* broadcast voices would suffer. Although these fears are largely hypothetical, they have been a determining factor in FCC policy toward cable since the beginning (see "Regulatory History," Chapter 11).

The 1987 must carry rules again failed to meet the First Amendment test and were struck down by the U.S. Court of Appeals in December 1987. Although an appeal of this decision or still another try by the FCC at constitutional must carry rules is possible, it is now most likely that the concept of "must carry" will be dropped until there is some concrete evidence that cable signal carriage policies violate the public interest.

Other FCC signal carriage rules are the network nonduplication and sports blackout rules. The *network nonduplication* (or network exclusivity) rule prohibits duplication of local network signals by a distant station on the same network. The idea is to prevent loss of audience for local advertisers buying local availabilities in network time. Presumably it would not matter to the cable subscribers whether they saw the network programs on one channel or the other.

The FCC has *proposed* to reimpose its *syndicated exclusivity* rules. These rules would recognize exclusive contracts between syndicators of programming and broadcasters so that cable systems might have to blackout imported syndicated programs. For example, if a broadcaster in San Diego obtained exclusive rights to the "Cosby Show" in off-network syndication, then a cable system in the San Diego market could not import the "Cosby Show" from a Los Angeles station it carried. The cable system would have to black out the program on the Los Angeles station. This, of course, would be upsetting to cable subscribers and the cable industry is strenuously objecting to syndicated exclusivity.[6]

Also still in effect is the *sports blackout* rule.[7] Cable systems serving 1,000 or more subscribers may not carry live, home team at-home sports events carried by distant stations if the same sports events are not broadcast by local stations. For example, suppose that the Detroit Tigers baseball team were playing the Chicago White Sox in Detroit and Chicago station WGN were broadcasting the game, but none of the Detroit television stations were broadcasting the game. A cable system within 35 miles of Detroit could not carry it even if WGN were a part of its regular cable service. The cable operator cannot violate the rights of the sports franchise owner. The franchise owner has the right to deny television rights if considered good business in protecting the gate or maintaining the overall television coverage plan.

DATA CHANNELS

To fill excess channel capacity and supply information for ready reference, cable systems have always offered certain automatically functioning alphanumeric and graphics channels. These may be called *data, text, alphanumeric,* or *automated* channels. The input to data channels is from a character generator which is programmed locally by keyboard or fed by telephone lines or satellite.

Program Guides

Cable systems provide so many program options that a separate channel to index all the others is useful for several reasons. Many cable subscribers do not use printed guides, or the guides may be lost or out of reach. The printed guides may be too confusing and too large. Some viewers prefer quick access to program listings displayed succinctly on the screen.

The program guides can be created locally with a character generator or delivered to the cable system by satellite from a syndicated service. The syndicated services customize the listings so that the individual system channel numbers appear. Only local origination programs, if they are to be listed, need to be supplied by the cable operator. The service puts in the cable network programming and even the local broadcast stations. These local stations

FIGURE 6-1. Television directory.

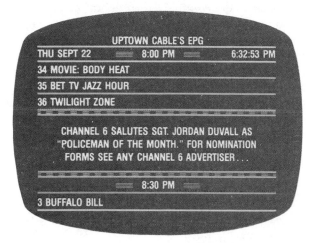

are usually willing to mail the listings to the electronic program guide supplier. Different color blocks within a program listing page can be reserved for advertising and program promotion. Cable operators can change advertising copy or program listings 24 hours in advance by telephone.

The two large syndicated services are EPG (for Electronic Program Guide) and TV Decisions. EPG is exclusively a program guide supplying cable subscribers with that service continuously on demand. TV Decisions has the program listings twice hourly and adds trivia games, soap opera summaries, and movie descriptions at other times.

Weather

The earliest form of automated channel that positioned a rotating camera in front of weather indication gauges has given way to character-generated weather (temperature, forecast, wind velocity) and time. Upper and lower portions of the weather screen may also be used for other information such as the program highlights, news, and advertising.

The Weather Channel, a satellite network, transmits localized weather from more than 700 different zones that are inserted into the channel's programming by the affiliated system's character generator on cue from the channel.

News Channels

AP News Cable, AP News Plus (with graphics), Reuters (pronounced Roy'ters) News View, UPI Data Cable, and others offer news services to cable systems. The services are usually segmented, with world and national news, business, sports, and, in some cases, state news from regional bureaus. The news services may be customized (such as an all-sports channel) by recycling the segments. Graphics may be employed to enhance the presentation. The services feed character generators through telephone lines or via satellite. A line or two is available for advertising messages, if desired.

Charges for the news services are five cents per subscriber per month, or less, depending on the number of subscribers served and the graphics.

Users of these channels get a brief budget of news, in summary, that is as current as any source, including all-news radio. It is accessible 24 hours and can be used by subscribers to fill time before programs start or between programs.

Cable systems may also program local news in text form, usually in cooperation or through a leasing arrangement with a local newspaper. This may be a separate channel or integrated with the syndicated services just described.

Classified Advertising

The future of electronic classified advertising may be in videotex where the user has an index to find desired categories of ads. Nevertheless, cable can serve a general interest in classifieds such as "garage sales" and the kinds of items that come up on "tradeo" programs on radio, as well as the traditional interest in automobile and real estate classifieds. Because the viewer may want specific items and not have the patience to read through a cycle of ads to get to a particular category, some systems program categories by time period; for example, real estate on the hour, help wanted at :15, used cars at :30, household at :45. The scheduling scheme can be inserted at a break between each category and occasionally within a category. The schedule can also appear on the program channel and in printed listings.

To enliven the text, graphics and photos can be used. Personal computer-based video imaging systems, capable of transforming printed material into video ads, are now in use in some systems. The Fayetteville, North Carolina system has a drive-in window, in a shopping center where a color picture will be taken of the item or visual material can be delivered along with the copy. The Fayetteville channel is on 24 hours per day offering classified ad buyers at least one exposure per hour per day. In one two-week period, the system claims to have sold 26 used cars for its classified customers. Cable advertising sales people go to the commercial customers rather than wait for them to come in. Realtors are major classified advertisers in Fayetteville, and, indeed, some cable companies have channels that are exclusively real estate.

Shopping

Shopping channels may take a variety of forms. The channel may present comparison marketbaskets. Commonly purchased grocery items are shopped at major local supermarkets each week. The prices are presented by category of item—meats, produce, frozen, canned goods, and so on. A total price for each category adjusted to a quantity suitable for an average family size may conclude the listings. Another plan is to list all the local shopping specials for a week, by category and store.

Bulletins

Announcements of community events, meeting agendas, schedules, and other information may be programmed on separate channels for that purpose or as a part of other channels such as government and education access channels. These channels may be relied upon by substantial numbers of subscribers for up-to-date information about the community, schools, and government services. The cable system itself may program a community bulletin board, with announcements fed to the cable system via phone lines from the source of the information—a school district headquarters, city hall, or the like.

People don't buy cable subscriptions to get the data channels, but such channels are a bonus to be offered in the sales presentation. Once subscribers are aware of their existence, some become regular users, which may become a factor in the subscribers' general satisfaction with the cable service and their retention as cable customers.

ACCESS CHANNELS

In its 1972 rules, the FCC required larger cable systems to make available separate channels, free, for the public, education, and government (PEG). These channels are known collectively as the *access* channels or the *PEG access* channels. The requirement was imposed in response to frustrations with the broadcast system, which did not accommodate much local government and education programming and which could not offer adequate citizen access.

In a case known as *Midwest Video II*, the federal court struck down the FCC access rules.[8] But the concept had already been established. Many cable franchise agreements and city ordinances required public, education, and government channels separately or in combinations. The 1984 Cable Act certifies these requirements. "A franchising authority may establish requirements in a franchise with respect to the designation or use of channel capacity for public, education, or government use. . ." (Appendix E). Under the Act, all existing requirements for PEG services, facilities, equipment, and financial support may continue to be enforced. For a new franchise PEG access requirements may be spelled out in the request for proposal. Offers, beyond the requirements, made by the successful applicant may be built into the franchise agreement. The franchising authority must adopt a plan for allowing the cable operator to make use of the access channels if they are not used by the public, educational institutions, or government agencies for whom they are intended.

The cable operator is *not* permitted to exercise any editorial control over the channel use. Since this would leave the cable operator in a vulnerable position in case of civil or criminal proceedings under obscenity and libel laws, a separate provision of the Cable Act releases cable operators from any liability for PEG programs. The *users* of the channels, of course, *are* responsible.

Public Access

The public access channel is one of the most interesting developments in cable television. In modern times the concept of free expression, in the mode of Tom Paine, is impractical. The population is too vast, printing is too expensive, and broadcast channels are scarce. Public access channels on cable were conceived as an opportunity to provide free expression to any individual. In the abundance of cable channels, setting one or more aside for this purpose, on a trial basis, seemed to have merit. Many of the public access channels went unused. But in a few places, the public access channel took hold to become a vital and diversified communication channel for the community.

It is appropriate here to elaborate on the public access concept, although its implementation may vary from community to community. Generally, the public access channel is available to any resident of the community on a first-come/first-served basis, to communicate whatever is desired. To many people some of the content may be thought to be totally without merit, indeed even harmful. Nonetheless, in the concept of public access, this is irrelevant. No one is to judge another's message. Minority ideas and tastes, as well as eccentric communication, have as much status in public access television as do communications from the mainstream of society. In fact, the public access channel may be one of the few places in which novel ideas may be introduced to the "marketplace of ideas."

To maintain a public access channel, where no one imposes a judgment on the merit of another's communication, requires great tolerance on the part of the cable company, citizens, and the franchising authority. The public access channel in most communities will contain what some people in the community think is indecent, tasteless, trivial, stupid, weird, or unpatriotic. Seldom is any of it illegal, but nonetheless intolerance of unconventional television presents a major public relations problem for both the cable company and the city that provides the franchise. The city and the cable company are put in the position of explaining constitutional law and the idea of public access while they may *appear* to be defending the offending content. Perhaps, over time, the existence of public access channels will develop a more mature understanding of public access and free expression.

The public access channel in most towns is the only available outlet for expression. In times of commmunity stress it may be a safety valve. Furthermore, the public access channel has instrumental value to individuals and groups. It is an inexpensive and effective means of communication available to the entire community. In various places it has been used by politicians of all stripes, the Boy Scouts, the Visiting Nurses Association, the entire spectrum of religious groups, amateur entertainers, public affairs commentators, PTAs, minority groups, handicappers, 4-H Clubs, arts groups, artists, human services agencies, fire chiefs, and so on.

In speaking to the cable industry at the 1980 National Cable Television Association Meeting, FCC Chairman Charles Ferris said,

> Your industry's experience with local access channels has shown that there are in many communities local video producers willing to provide exciting and innovative program ideas, and that there is local interest in cable programs on community issues and events . . . while you look to the sky for new programming ideas [a reference to satellite networks] I hope you are not ignoring them in your own backyard.[9]

In large communities, it is now relatively certain that the public access channel will be active and filled with original and taped material produced locally or elsewhere. More and more people are capable of producing television programs. Periodic workshops sponsored by public access channel coordinators increase the numbers with that competency.

A number of large-capacity cable systems have multiple access channels, sometimes called *theme* channels. A channel may be devoted to the elderly, one or more ethnic groups, fine arts, and the like. A theme channel has the advantage of an unambiguous identity and can be promoted and recognized for its specialty. However, it may be difficult to fill each channel. Some public access coordinators advocate a *single*, horizontally programmed public access channel to represent a composite of the community. Public access may help various subcultures within the community to maintain their distinguishing characteristics, yet at the same time expose others in the community to their uniqueness.

Administration of a public access channel may be accomplished in a number of ways: through the cable company, the city, or a nonprofit corporation. Often the franchise agreement requires the cable operator to build and equip public access facilities, and in some cases provide staffing for instruction and coordination. In other cases, some unit of government, such as the public library, administers the channel and may house the studio. Increasingly, nonprofit corporations are designated to manage the public access channel for the community. In channels administered by government and nonprofit corporations, the cable company may make a capital contribution to the facility and grants for operational expenses.

There are strengths and weaknesses in any of these administrative schemes, but from the variety of experience with all of them, it is clear that any system can be made to work.

Whatever the administrative structure, two elements are crucial in public access: a coordinating person or staff and an operating budget. The coordinator may be an employee of the city or the cable company. The principal functions are to *promote* and *facilitate* use of the channel. The first task is to get people to use the channel. After a few years, as people become aware of the channel and facilities through the promotion and public relations work of the coordinator, the job becomes more administrative and instructional. The coordinator teaches people to use the tools, and programs the channel so that it is of maximum value to users and audiences.

Operating budgets come from several sources. Most common are the cable company, as mentioned, private and government grants, on-channel fund raising, dedicated portions of franchise fees, memberships, contract productions and *underwriting* (where businesses contribute and receive brief credit mention at the beginning or end of the program). A relatively secure operating budget is important to the success of the channel.

It is necessary that certain rules of procedure be established to assure an orderly function of the public access channel and also to prevent any individual or group from monopolizing production facilities and channel time. These procedures must not be overly complex and thereby forbidding. Further, there should be no clause that permits disguised censorship. For example, a rule could state that materials presented for play on the public access channel must meet certain technical standards. Ostensibly, this rule would maintain a basic level of quality on the public access channel, but could also be used as a device for rejecting materials not suitable to someone on grounds of *content.*

A sample set of rules that has evolved over several years of application and amendment is included as Appendix A.

Ordinarily, public access channels do not charge for initial production facilities and channel time. Some charges may be made for studio and equipment use beyond a maximum free time. Since most public access channel users are people of modest means or members of nonprofit groups that cannot afford to buy time or space for their messages in other media, free or low-cost access is important to the purpose of the channel.

Equipping a public access channel is a problem. Durability is highly desirable since some users may be inexperienced. Maintenance is not only costly but puts the equipment out of service. Simplicity is also important so that public access does not become a private domain of the technical elite. Fortunately, there is now midline color equipment that suits most production needs and holds up well under public access channel use.

An inevitable function of the public access channel personnel is training. Even an established public access channel will have a continual flow of new users needing basic training in television production. Most cable public access channels have a routine training sequence as part of the checkout procedure for first-time portable equipment users. Workshops are presented periodically for prospective users of the studio equipment and those who want to refine skills. Many public access channels publish "how-to" primers for channel users.

Public access channels use different categories in accounting for channel activity when reporting to franchising authorities and others. Hours of *original* programming produced weekly or monthly is one category. It indicates the amount of production activity and distinguishes this programming from repeats and externally produced materials. Another figure is *total hours* of programming for a time period, which indicates the amount of programming scheduled. A subset of total program time is *unduplicated* program hours. This is a figure that has meaning only in a broader context. Repetition of programs is important in public access as in other cable programming. It provides the opportunity to suit the convenience of viewers and to accumulate audience for a program over time. Therefore, several repeats may be desirable. A further figure, the number of *individuals and groups* involved in public access programming, indicates the breadth of the channel use across the community. Access channels also tabulate the number of people trained in each reporting period.

While this kind of quantitative approach to account for channel time may not seem very substantive, it is not possible, without violating the concept of public access, to make qualitative evaluations. The vitality, the uniqueness, the informative and entertainment value, and other benefits of the programming may be important criteria for assessing some television programs, but not public access. Only the access user must be satisfied that the effort has merit.

Educational Access

Educational access channels are used mainly for *external communication*—community relations, the essential communication between public school systems and citizens and taxpayers. School millage issues, new curriculum plans, and other general interest information may be on the channel. This could include school board meetings (live and/or taped for replay) and school board informational programs and community debate on issues (such as redrawing of school boundaries). More narrowly, the channel may serve the homes of students with instrumental information such as school menus and the calendars of events. Meetings of parents' associations are sometimes conducted by cable.

Further use of education channels may be in continuing education for adults. This is often a responsibility of the secondary schools and almost always a responsibility of community colleges. Where the nature of a course is not principally social (a chance to get out of the house and meet new people), and the instruction can be adapted to television, cable may be used to lessen the travel burden of students and limit the physical plant needs of the educational institution. Community colleges are developing the use of telecourses as a means of nontraditional education suited to the needs and convenience of the community. These courses are ideal for cable.

A special category of use for educational channels is in the training of the homebound and the handicapped. Several experiments have established that people can learn well and respond favorably to one-way and interactive instruction to the home via cable.[10]

Television production may be a course in the schools. A good argument can be made for bringing elementary and secondary school students into the modern communication age by giving them experience with television. The education access channel is an outlet for television production students.

Students in the performing arts courses or extracurricular activities may use the channel. A wider audience through television and the excitement of being on television give their work additional meaning.

Education access channels may be used for school reports. Although cable systems sometimes do sports on other channels, educational institutions may elect to put competitive athletics on their own channels, promoting the use of the channel and perhaps covering television costs by inviting underwriting by local businesses.

For a bulletin board announcement of the school calendars, menus, board agenda, and so on, it is only necessary for the school system to have a character generator. In a few minutes, a staff member can type in the messages each day. For use of cable as a distribution system for audiovisual materials, the schools must have a videocassette playback machine for materials prepared for television and a film chain for instructional films and slides.

Operation of studio and remote equipment requires at least one professional. Students can be trained to serve as crew, and schools that offer television production as a curricular or extracurricular activity quickly develop a cadre of conscientious and competent "video freaks." It becomes an engrossing experience for a number of students and can lead to careers in communication.

In planning for use of cable, it is essential for educators to budget for maintenance and equipment replacement. A conservative sum (perhaps 1 percent per month of the original purchase price) should be set aside for maintenance and replacement or the system will soon be useless.

It is most convenient if the school system feeds the cable system from its own location. This can be accomplished by use of a special cable between the school system studio or audiovisual center and the cable headend. In most newer cable systems, the link between the schools and the cable system is the institutional network, described in Chapter 3. For each channel in use, the educational institutions must provide a modulator. Cable systems may want to maintain the modulators themselves so that the signal quality meets their standards. Almost all cable franchises specify a free cable drop for each school building in the cable service area. If the schools wish to connect more than one room, the schools' maintenance staff can do the wiring to cable system specifications or the cable company will do it, usually at cost.

In a small community, administration of education channels is not difficult. The channel is turned over to the public school system, which generally serves an area at least as broad as the cable franchise. If there is a private school in the community, its needs are accommodated through the public schools. In larger communities, several school districts, community colleges, and universities may all desire a channel or channel time. Depending on the cable system design, and the franchise, there may be one channel each for the community college, the universities, and the public school districts. Where there are many users for a single channel, some coordinating body is necessary. Establishing such a body is the responsibility of the franchising authority or users, not the cable operator.

Government Access

Government access channels can serve purposes quite similar to those of the educational channels—communication between government and taxpayers, interconnection of geographically separated units (such as fire stations, police precincts, district courts), and training programs.

Many government channels are used as bulletin boards—with such mundane but essential information as trash pickup schedules for holiday weeks, tax notices, city council agendas, street closings, city job listings, and availability of recreational facilities.

These bulletins are produced on character generators from information usually at hand in various printed reports and from information supplied especially for cable by the city departments. The character generator should have enough page capacity so that, on heavy information days, old but current items do not have to be bumped to accommodate new items. On the other hand, the time for a complete cycle of all information cannot be too great or users will have a long wait for the information they desire. Some cable bulletins contain one index page to inform users of the contents at any time.

Usually the bulletin is updated daily by a clerical person designated to collect the materials, but it can be updated more often if desired. The cable bulletin is extremely useful in government public information programs, especially for cities and suburbs swallowed up in metropolitan areas where there is no other local media source for detailed information.

The automated bulletin is available 24 hours per day on most cable government channels except when preempted by full audiovideo information programming. These nonbulletin programs are principally live and taped replays of city council and other government meetings, a valuable part of the government cable service. In East Lansing, Michigan, for example, the average viewership for each council meeting is 4.9 percent of subscribing households (about 250).[11] The capacity of the council meeting chamber is only 100.

Other informational programs are created especially for cable on topics of interest. These include explanations of new ordinances, budget presentations, employment opportunities, inservice training of government personnel, regular phone-in programs for access to elected and city department personnel, public safety programs (such as fire and crime prevention), and annual reports of various government units. According to a survey of 45 cities by the Cable Television Information Center, the majority of municipal channels cablecast an average of 10 hours of original programming per week. (As a general rule, the more subscribers, the greater the budget and the more programming.)[12]

The nonautomated programs are usually scheduled at an established time so that the community can come to expect these programs. They are repeated often to maximize the audience. Since the city council meetings are viewed most heavily, brief spot announcements may run at breaks promoting other government programs.

It takes many *years* to establish a government channel. Both the government and the citizens must learn to use the channel. Through time, the city, with the help of experience and user feedback, makes better use of the channel. Very gradually cable subscribers become aware of the service, discovering the channel accidentally while sampling cable offerings or hearing about it by word of mouth and publicity. A council meeting dealing with a controversial issue for a neighborhood, or the need to know the schedule of leaf pickup in the fall, may be the first exposure to the channel. Eventually (after several years), most people in the community are aware of the channel and make good use of it on an irregular basis.

It is possible for the government channel to become a platform for the promotion of individuals and points of view. The City of East Lansing has established a policy that the channel not be used for advocacy:

> Programming shall be information rather than advocative. This does not preclude the cablecasting of public hearings, city council meetings, and other governmental meetings where advocacy may take place.[13]

In East Lansing, the city council members appear on the government channel only as a group, not individually. If a city council member wants to advocate a particular point of view, the public access channel is used.

The government channel is especially effective for the training of city employees. Firefighter training is a good example. Firefighters train continuously to stay familiar with infrequently used equipment and information and to learn new equipment and techniques. Cable, usually providing a free drop to each fire station, can be used to train firefighters in their stationhouses where they are free to respond to an emergency in their districts. The instruction may be live, using telephone contact with the instructor, or via videocassettes, which can be used conveniently for each shift of firefighters and as often as necessary for new recruits and refresher training.

Using the government access channel for training has some disadvantages since the channel is also available to the general public. For example, certain elements of an emergency medical training program might be upsetting to some adults and children. In this case the channel may be scrambled or, if available, institutional network channels, which are received only by the institutions on the network, may be used.

The cable system may have an emergency notification system. If so, the public safety director or police chief takes responsibility for its implementation. This service provides a one-line crawl message and/or an audio message on every single channel in case of an emergency (flood, hurricane, and tornado warning, gas leak, and the like). A visual message can be prepared on the same character generator that is used for the automated bulletin. An audio message can be taken over the phone from the emergency dispatch office and may be somewhat more convenient for public safety personnel.

The full use of the government channel is costly. An original outlay for equipment is necessary. In the previously mentioned study of 45 government access operations, the average budget was $150,000 and the average number of full-time employees was $3\frac{1}{3}$. The full-time staff members are usually supplemented by part-time employees and interns.[14]

Leased Access

Federal law requires cable systems with more than 36 activated channels to make channels available for "commercial use by persons unaffiliated with the operator" (see Chapter 11). Local franchises, in effect prior to the 1984 Cable Act, may have provisions requiring similar channel designation for cable systems with fewer than 36 activated channels.

This kind of arrangement, where the cable company makes the distribution system available on a common carrier basis to anyone who can pay a fair price, has traditionally been referred to as *leased access.*

Actually it would be quite awkward for another business to operate, say, a premium channel, maintaining a separate, and costly, billing, marketing, and promotional structure. But one could imagine a sports franchise, or group of sports franchises, leasing a channel to market telecasts of their games. In this case the sports franchises would be retaining their own telecasting rights. They would be able to maintain control of the marketing and not have to be dependent on the cable operator to set the price, develop the marketing plan, and carry it out effectively. Any other company with rights to program material or a service idea they think the cable operator is overlooking, could lease a channel and try to make it work. Newspapers do lease cable channels to extend their news and classified advertising services. Others have leased channels for shopping services. A few entrepreneurs lease cable channels, produce programming, and then sell advertising time intending to cover the costs and make a profit. In Athens, Georgia, the *Athens Observer*, a weekly newspaper, is attempting to create a commercial cable television station in a town that has no broadcast TV station. The newsgathering and advertising sales experience and contacts at the newspaper can be applied to the television business.[15] Nonprofit organizations have also leased channels for a variety of purposes, usually at a very low rate.

Sampling Access Programming

The listing below samples the 62 winners of the 1985 National Federation of Local Cable Programmers "Hometown USA" awards. The flavor and diversity of access programming can be sensed from the program descriptions; the words of Robert DiMatteo, programming critic for *CableVision* magazine, who made the presentation at the annual NFLCP convention.[16]

Portland, Oregon, "Good Listeners Make Better Mayors." This feisty interactive show turned out to spark a real controversy, figuring in an historic local political upset during the Spring primaries of 1984 in Portland, Oregon.

Austin, Texas, "All Pro Video Rasslin." A funny take-off on pro wrestling—as if pro wrestling isn't already a take-off.

San Rafael, California, "Beyond Technology: Students Question the Humanities." A special project of 13 Marin County High School students, this show questions the need to study the humanities in school and their relevance to current society. Those who turn up include Francis Copolla and the late Dr. Frank Oppenheimer (in his last interview).

Kalamazoo, Michigan, "Every Child a Wanted Child." A down-to-earth, sensitively handled series that covers various aspects of reproductive health and sexuality—all the way from birth control to herpes.

Pittsburgh, Pennsylvania, "'84 Three Rivers Regatta." Highlights from activities of Pittsburgh's annual 3-day celebration of its waterways. Everything from aerobatics planes and hot air balloons to fireworks and the Pittsburgh Symphony figure in this enlivening documentary.

Hamtramck, Michigan, "Lynn Kinsman." A telethon as homespun as Jerry Lewis' is slick. Made in Hamtramck to raise money for a liver transplant for a young girl named Lynn Kinsman. It lasted 3 days, most of which was covered by this special cablecast.

Burnaby, Canada, "The NDP Convention." On May 18-20, 1984 the New Democratic Party of British Columbia met in Vancouver to select a new party leader. This program provides a comprehensive look at that political process of selection.

Irving, Texas, "A Life of Roses." A study of responses to death, this original dramatic production focused on a little girl whose mother had died, studying how the girl and her father try to cope with their sense of loss. Then the father himself dies. The program is now being shown in local hospitals as a way of helping staff and patients deal with death.

New Orleans, Louisiana, "Music City." An extraordinary series—42 hours at last count that documents that wide range of indigenous New Orleans music—especially the music of the older jazz and blues musicians.

Darland, Texas, "The Music and Art of Stiv McGregor." A striking collection of video art pieces that blends both sound and image. Mr. McGregor wrote and recorded all of the music himself for this program.

Downers Grove, Illinois, "Inside Business." Inside Business is a magazine format program that provides a rare inside view of Downers Grove local businesses. Industry people are interviewed and we are taken on informative tours of the various local plants.

Lowell, Massachusetts, "Local Cable News." Over a dozen stories a night, all about the Lowell, Massachusetts community. A fine example of narrowcasting in the news format.

LOCAL ORIGINATION

Technically, any signals that begin at the cable system headend are *local origination*, distinguishing such signals from off-air, telephone line and satellite-delivered signals. But, the term has come to mean specifically those programs that are produced by the cable operator. Local origination, or *LO*, channels are programmed by the cable system, or under the auspices of the cable company, and where the company has *editorial control.* This is opposed to the access channels, which are programmed independently by educational institutions, government, or the public, and the cable company has *no editorial control.* However, since the cable company may employ volunteers in local origination and do many of the same types of programs, actually, the difference between LO channels and access channels may not be apparent.

Local origination serves a number of purposes for the cable operator. It may be fulfilling a requirement written into the franchise agreement. LO programming is good public relations. It gives the local cable operation, which usually has an absentee owner, some community roots. The various individuals and groups whose interests are represented in the programming become a constituency of supporters for the cable company. In periodic reports to the franchising authority, the cable company can point to its community involvement.

Under most franchises, access channels are noncommercial, but no such constraints are placed on the LO channels. In a few cases advertising on LO channels contributes to the profits of the business, or at least it helps to offset the LO costs. In almost all cable systems that sell advertising time, the local origination staffs also produce the local commercials that are inserted in the LO and cable network programming in the system's line-up thus helping to support themselves.

A few cable systems have *super LO* channels. These are essentially the same as an independent broadcast television station programmed with older movies and syndicated programming. A large multiple system operator (MSO) or a regional cable interconnect may program a super LO channel for several systems. The kind of programming on a super LO channel is a relatively good vehicle for most LO advertising because it attracts larger audiences.

There are also regional LO networks. Several cable companies in the same area get together to share programming, or a single system serves as the production hub for the network. Spreading the cost of production over a wider subscriber base permits a greater investment of production resources probably enhancing program quality. A larger subscriber base for LO makes local advertising sales a better possibility. Statewide networks have been formed in New Jersey and Texas. They are fed by LO programs contributed by the affiliated systems.

Two types of programming constitute the bread and butter of LO: local sports and information.

In earlier days of cable origination, *sports* events, such as regional high school tournaments, were the principal programs. There were dependable audiences, local sponsors to cover the out-of-pocket costs, and usually volunteers to handle the production. Most cable systems that do local origination programming cover high school, youth, and adult leagues, community college, and some college sports. Because of a fear of damaging gate receipts, and sometimes because of league rules, most games are cablecast on a tape-delayed basis.

Information and news is an important product for LO channels. Many communities served by cable are engulfed by metropolitan media. Only the most heinous crimes and unusual human-interest stories for the community would ever come up on broadcast stations and in the daily newspapers. For example, metropolitan Pittsburgh encompasses 698 local governments. The City of Pittsburgh, itself, has only one-fifth of the population of the Pittsburgh television market. But, on average, Pittsburgh broadcast television stations give Pittsburgh a 40-percent share of the news coverage of the market. The other 697 localities with four-fifths of the population divide up the rest of the time, so that it is a rare day that any one is even mentioned.[17]

To solve this kind of problem, Cablevision Systems Corporation serving Long Island operates a $6-million-per-year, 24-hour news service reaching over a half-million subscribers. Broadcast stations covering Long Island are preoccupied by New York City news.[18]

There are also many cable markets that are out of reach of television journalism, that is, in cities on the edge of, or wholly outside of, broadcast station coverage. If these places are large enough to justify the cost of gathering and producing news programming, cable can fill the gap.

Broadcasters have learned through repeated trial and error that the news format is the most attractive way to present public affairs programming. Attempts to win audience for locally produced documentaries, panel shows, and other information formats have repeatedly failed. Cable is probably the same—the best way to build an audience for information on LO channels may be in the news format.

Whether or not a cable system carries regularly scheduled newscasts, it may offer special coverages of community news—disasters and other crises, civic affairs, political campaigns and elections. Cable may afford a forum for political debate at a grassroots level, giving people and issues valuable television exposure.

In programming LO and other channels where the cable company has editorial control, the cable company must follow the same rules that the FCC and the U.S. Criminal Code prescribe for broadcasting by candidates for public office, lotteries, wagering, obscenity, fraud, and sponsorship identification.[19] The full statement of these rules is in Appendix B.

The FCC applies its rules to all *origination cablecasting*—that is, not only LO, but also anything over which the cable company has control, even satellite-delivered program services for which the company contracts. The only television programming *not* covered by the FCC rules is PEG access and over-the-air television.

Briefly, the rules on political candidates require that, if the cable system permits any legally qualified candidate for any public office to use the LO channels, then all other candidates for the same office must be afforded equal opportunities. This rule excludes newscasts, news interviews, news documentaries, and on-the-spot coverage of news.

Cablecasts of advertisements or information concerning lotteries, gift enterprises, or similar schemes offering prizes, dependent in whole or in part upon chance, are prohibited.[20] Advertisements, lists of prizes, and other information concerning a lottery conducted by a state are exempted from this prohibition.[21]

For any commercial cablecast for which money, service, or other consideration is received, at the time of cablecast, an announcement must be made that the matter is sponsored, either in whole or part, and by whom.[22] This rule applies to paid political cablecasts and, if the political cablecast is more than five minutes long, the announcement must be made at the beginning and at the end of the cablecast.

According to the U.S. Criminal Code, whoever utters obscene, indecent, or profane language is subject to fine and imprisonment.[23] Schemes using information transmitted by wire to defraud are also prohibited by the U.S. Criminal Code.[24] Using wire communication for wagering or betting, except where legal under state law, is prohibited in the U.S. Code.[25]

The FCC rules on candidates for public office, lotteries, and sponsorship identification *do not* apply to access channels. Criminal code violations *do* apply to access channels, but are the liability of the access user, not the cable system.

A clear picture of the diversity and creativity of the best in local origination is represented by a sampling of the National Cable Television Association's Awards for Cable Excellence (ACE).[26] These are 1985 winners:

Overall commitment to local programming. Continental Cablevision Cook County, Elmhurst, Illinois serves 19 communities in the suburban Chicago area, producing more than 800 hours of local programming each month. The local origination channel provides more than 45 hours of locally originated programming weekly, and, through its four theme channels, it cablecasts more than 150 hours of access programming each week.

Documentary series. Cablevision of Boston, Massachusetts. The ''Neighborhoods Series'' celebrates the strengths of Boston's distinct neighborhoods and shares with the entire city the struggles and victories of each neighborhood. In their own words residents present issues and activities that affect their neighborhood.

Public affairs program. Cox Cable, San Diego. ''Speak Out! With Stephanie Donovan: Victims' Rights.'' Part of an ongoing public affairs series, this program was promoted by local reaction to the San Ysidro massacre and focused on victims of various crimes rather than criminals and their acts.

Community events coverage. Mile Hi Cablevision, Denver. ''Center Stage Breakin'' was a 28-minute slow-motion stereo sound production of a Denver break dance contest. The show emphasized the versatility and fluidity of dance and featured Rock Steady, New York City professional breakers.

Community events series. Cablevision Systems of Long Island, Woodbury, New York. A nightly news update service, ''Cablevisionews'' consists of five-minute reports each half-hour between 4:30 P.M. and midnight. Inserted into CNN Headline News, at least 5 of the 16 nightly reports are live, and updated as events develop.

About sports, series. Suburban Cablevision, East Orange, NJ. In its ninth season in 1985, ''Time In'' is a live, weekly, one-hour, call-in sports talk show featuring timely reports from the high school, college, and professional sports ranks.

Comedy/variety series. Group W Cable, Dearborn, Michigan. A live, weekly, four-hour music video cable show for kids, ''Backporch Video'' not only stars kids but is written, edited, and shot by kids who decide program content. The format includes PJ's (Porch Jockeys) who play nationally produced music videos, locally produced music videos and "wacky footage" played to records.

Educational program. Part of an on-going series about San Francisco landmarks, "Landmarks of the City: Fisherman's Wharf" tells, through photographs and reminiscences, the story of immigrants who began California's fishing industry.

Children's programming. Times Mirror Cable, Orange County, Laguna Niguel, California. "Finding Home" was an original musical, theatrical parable designed to teach children to be more understanding of recent immigrants to southern California. The theme is presented through a story about a California surfer, Burger, who surfs his way to Aboland where both the customs and language are strange.

ASSESSMENT OF COMMUNITY PROGRAMMING

The content of PEG access channels and local origination channels is collectively referred to as *community programming.* Community programming has avid supporters and serious detractors. Its supporters assert the importance of television in modern communication environments. The development of *local* television has long been a goal of public policy. Because of its more limited service area, cable fits the ideal of local television better than broadcasting. Wherever communities are buried in a large metropolitan media market, or outside the broadcast television's journalistic range, there is a need for cable television news. In any market, cable has the channel capacity to reflect the problems, issues, and achievements of its community in much greater depth than broadcasting. And because time is less dear in cable, experimentation is possible. Community programming does not have to fit the mold of conventional television. There is an opportunity for a community to look at itself through television, for people of all walks of life in a community to get to know each other.

Without the time constraint, scheduling is flexible. A 30-minute local newscast can be repeated for 24 hours until its next update. Public meetings, court trials, special community events, speeches, and debates on community issues can all be run in their entirety and rerun.

People can interact with television by telephone, or come to a public access studio and produce a program. Small businesses serving a limited market are given their only chance at television sponsorship by LO programming.[27]

Young people, and adults too, are excited by the presence of persons they know on television. Until community programming on cable became well established, it was incredible to see a person known in the flesh, or a neighborhood place, featured on television. Conventional television dwells on national and international events and personalities to an excess. The television viewer lives in a world, not his or her own, experienced vicariously. Community programming may restore the balance. It may help people to identify with their own community and their own immediate problems.

This, of course, is idealistic. Realists have difficulty in looking at television this way. They have been trained in the scarcity of spectrum. Television spectrum is not to be squandered on small audiences. The highest production values are expected of television. And if the realist is not offended by violations of these standards, there is the cost. Public access equipment must be housed and maintained. LO news gathering and program production are demanding professional activities and LO programming is seldom self-sustaining or profitable through advertising sales. The costs of community programming, which may be used by only a few people, must be subsidized by all subscribers.

Those who support community programming in the cable industry and outside, must hope that it survives the formative years and is recognized for its purpose and achievements. It may be viable only under various combinations of circumstances.

For public access:
- An effective coordinator/promoter.
- An energetic body of volunteer production workers, with reinforcements available as burnout occurs.
- Dependable funding.

For local origination:
- Advertisers and underwriters who recognize the value of community and program identification.
- Cable subscriber awareness and appreciation of the LO programming as an important value in the service.
- Creative and competent producers.
- An advertising/sales department need for production facilities and staff (helping to justify the production resources).
- A fairly large subscriber base.

For all community programming:
- A commitment to the community programming concept by the operator.
- A franchise requirement for community programming.
- Programs vigorously promoted and routinely listed in program guides so that they are accessible to appropriate audiences.
- Community interest and support (an active constituency).

Not all of these conditions must be present for community programming to succeed—each success has its own formula—but, most are prominent in the best examples.[28]

CABLE AUDIO

FM Connections

For many years cable companies have been offering FM connections. Distant signals are brought in by antenna and added to local signals. Short-wave signals are also possible. Theoretically, the system could offer the BBC and Radio Moscow. Some daytime AM stations have found a way to extend their service past sunset by leasing an FM channel on cable. The FCC has no signal carriage rules for cable audio.

Most of the cable systems charge about $2.50 per month for the audio service, although some give it away as an incentive to subscribe. The National Association of Broadcasters estimates that 64 percent of the cable operators offer an FM service averaging four local off-air radio stations, 11 distant stations, and two satellite-delivered signals.[29]

Satellite Audio

Since each cable satellite transponder is capable of carrying at least six subcarrier channels, hundreds of audio signals could be provided to cable systems by satellite. Several are, including radio superstations WFMT, Chicago (a classical station) and KKGO, Los Angeles (a jazz station).

A number of companies have attempted to develop audio packages distributed by satellite to cable systems. Each signal in the package is a distinct commercial-free format, such as jazz, comedy, progressive rock, big bands, rhythm and blues, country, easy listening, and classical. Cable systems may offer these services as a *package* or *a la carte* and as *pay* or *basic* services. Some of these services are taking advantage of the interest in high-quality audio inspired by the *compact disc* technology. CD-quality satellite transmissions, that cannot be duplicated by conventional FM broadcasting, can be received if the subscriber has an audio decoder supplied by the cable operator and good speakers.

Audio Background for Data Channels

Most cable systems back some or all data channels with audio. The background could be a local station, a popular and locally unavailable FM station from a distant city, one of the audio superstations, an audio origination service (discussed later in this chapter), or a special network such as Lifestyle, an upbeat form of Muzak.

Audio Origination

A few cable systems originate audio services. These may be college or high school stations, public access audio or commercial stations.

Some college stations have taken to cable because the FCC has deleted 10-watt noncommercial FM. Cable radio has similarly substituted for, or transmitted, college carrier-current stations. Cable eliminates the need for transmitters, is capable of stereo, and is not controlled by the FCC.

Public access audio services have arisen in a few places because they are faster, cheaper, and easier than video.

For similar reasons, a few commercial stations have been established in cable. A minority audience may be served by cable either when making the capital investment in broadcasting is impractical or when a license is not available. A cable system may give an FM channel to a commercial audio service in return for spots promoting the pay cable channels.

Arbitron has discovered 30 "cable-only" radio stations in its radio audience measurement diaries.[30]

Simulated TV Stereo

Many of the premium and basic cable networks also offer a separate satellite feed of their audio, in stereo, for FM receivers. The signal is real stereo, simulating stereo TV. This practice was begun by MTV to enhance the value of their music service and also to give operators a hook for selling FM connections.

Assessing the Market for Audio Services

FM connection and other audio services are of principal value in places where few broadcast radio signals are available. In these locations only some of the common radio formats are accessible. Cable can bring in a diversity of radio comparable to large metropolitan areas. Still, cable operators have difficulty selling the FM service because so many cable subscribers connect themselves with a splitter from an electronics supply store. Premium audio services require a descrambler in the home and additional equipment at the headend.

Commercial-free audio may have some appeal in the bigger cities, even in competition with broadcast radio, but theft remains a problem.

NOTES

[1]A.C. Nielsen Company, "Source of Household Viewing—Prime Time (Monday–Sunday 8–11 P.M.)," *85 Nielsen Report on Television*, p. 12, CableTelevision Advertising Bureau, 1985 Cable TV Facts, p. 6.

[2]First Report and Order in Dockets 14985 and 15233, 38 FCC 683 (1965).

[3]Report and Order in Dockets 20988 and 21284, 79 FCC 2d 663 (1980).

[4]*Quincy Cable TV, Inc.* v. FCC (D.C. Cir. No. 83–1283).

[5]Amendment of Part 76 of the Commission's Rules Concerning Carriage of Television Broadcast Signals by Cable Television Systems, Docket 85–349 (1987).

The rules have two major components, must-carry and the A/B switch. The must-carry rules create a pool of local stations that *"qualify"* for carriage. The cable system must carry a *quota* of stations prescribed by the FCC from among those qualified. To qualify for mandatory carriage, a broadcast station must be within 50 miles of the cable headend *and* also have a 2-percent share of audience *and* a 5-percent net weekly circulation in noncable homes in the county in which the cable system is located. (These numbers are based on commercial audience figures for the county. A "2-percent share" means that the station gets 2 percent of all the viewing of television in the county, not including cable households. The "5-percent net weekly circulation" means that 5 percent of the noncable households in the county watch the station at least once during the week.) *New broadcast stations*, because they have not had a chance to establish an audience, are automatically qualified for one year whether or not they meet the share and circulation criteria. All noncommercial educational stations within the 50-mile radius qualify without regard to share and circulation.

Stations do not qualify unless they deliver a high-quality picture to the cable headend. However, a station may pay a cable operator for the equipment necessary to provide the high-quality picture.

The quota is based on the number of active channels on the cable system. A system must carry at least one qualified noncommercial educational channel and, if it is 54 or more channels, two. A system with up to 20 active channels need not carry any more than one broadcast station (probably a noncommercial station). Systems with over 20 active channels must carry about 25 percent of their capacity in broadcast channels. For example, a system with 34–37 channels would have a quota of 9 broadcast channels. When there are more qualified stations in the 50-mile radius than the quota, then the system may choose among the stations to reach the quota. It should be noted, however, that the number of stations in the quota is not dependent on the number of stations in the cable system's area, but on the system's channel capacity. Therefore, high-capacity cable systems in broadcast markets with only a few stations would have to carry them all.

Cable systems may not charge stations for carriage when they are part of the must-carry quota, except for costs to provide a high-quality signal, as noted above, but may charge the broadcast station for carriage if the station qualifies but exceeds the quota.

All must-carries, carried to meet the quota, must be on the lowest-priced tier of service. In other words, a must-carry station may not be placed on an expanded basic tier for which the subscriber pays a premium.

The must-carry portion of these rules has a "sunset" provision; the rules expire on June 10, 1992.

The A/B switch rules are part of the FCC plan to maximize the availability of program choices. The intent of this rule is to educate subscribers to the fact that not all local broadcast stations may be available via cable and to make some provision for cable subscribers to also receive these stations (the A/B switch). The *A/B switch (input selector)* allows the cable subscriber to change easily between cable service and broadcast signals. The FCC requires an annual notice to subscribers to inform them of the switch option and its purpose.

The switch may be sold or leased to the subscriber for a monthly fee. The cable operator's employees may not suggest that the subscriber remove the television antenna. Subscribers ordering an A/B switch must be informed that it has to be connected to an antenna.

The requirement to offer the A/B switch expires in 1992 along with the must-carry rules, but the rules on consumer information about cable and broadcast carriage will continue indefinitely.

[6]First Report and Order in Dockets 14895 and 15233, 38 F.C.C. 683 (1965).

"In the Matter of Amendment of Parts 73 and 76 of the Commission's Rules relating to Program Exclusivity in the Cable and Broadcast Industries," Notice of Inquiry and Notice of Proposed Rule Making, Federal Communications Commission, April 23, 1987.

[7]47 C.F.R. 76.67

[8]*FCC v. Midwest Video Corp.*, 440 U.S. 689 (1979).

[9]Charles Ferris, address to the National Cable Television Association Convention, Dallas, Texas, May 1980.

[10]Erling S. Jorgensen, Thomas F. Baldwin, Stephen L. Yelon, and John B. Eulenberg, "Final Report, Project 'CACTUS'—Computers and Cable Television in a University Setting," Michigan State University, East Lansing, September 15, 1976; Peg Kay, "Social Services and Cable TV," NSF/RA–760161 (Washington, D.C.: G.P.O., 1976).

[11]"Cable Television Public Channel Viewership Survey," City of East Lansing, Michigan, 1979.

[12]_____, "CTIC Releases Municipal Programming Study," *Community Television Review*, Vol. 8, No. 3, 1985, pp. 14–17.

[13]"Programming Policy for Channel 22," City of East Lansing, Michigan, East Lansing Cable Commission, January 18, 1978.

[14]"CTIC Releases Municipal Programming Study," op. cit.

[15]Searey, Chuck, "Leased Access Flourishes in Athens," *Community Television Review*, Winter, 1984, p. 20.

[16]National Convention of the National Federation of Local Cable Programmers, Boston, Mass., July 11, 1985.

[17]William C. Adams, "Local Television News Coverage and the Central City," *Journal of Broadcasting*, Spring 1980, pp. 253–265.

[18]For a discussion of these issues see David A. Patten, *Newspapers and New Media*, Chapter 6, "Community Television," Knowledge Industry Publications, 1986, pp. 81–88.

[19]The FCC's Fairness Doctrine once also applied to LO channels, but was rescinded for cable because of the wide diversity of channels and the availability of a public access channel in many systems.

[20]18 U.S.C.A. 1304.

[21]18 U.S.C.A. 1307.

[22]*Cable Television Report and Order*, 36 FCC 2d 143, 24 RR 2d 1501 (1972).

[23]18 U.S.C.A. 1464.

[24]18 U.S.C.A. 1343.

[25]18 U.S.C.A. 1084.

[26]ACE Awards presentation, NCTA, 1985.

[27]Randy Welch, "Slouching Toward Utopia," *Channels*, July/August 1985, p. 8.

[28]See also: Susan Tyler Eastman, Sydney W. Head and Lewis Klein, *Broadcast/Cable Programming: Strategies and Practices*, Wadsworth, 1985, pp. 529 and Susan Tyler Eastman and Robert A. Klein, *Strategies in Broadcast and Cable Promotion*, Wadsworth, 1982, pp. 355.

[29]Richard V. Ducey and Mark R. Fratrick, "The New Audio Marketplace: Challenges and Opportunities for Broadcasters," National Association of Broadcasters, September, 1985, p. 15.

[30]_____, "Radio Today: 1984 Edition," Arbitron Ratings Co., New York, New York, March 1984.

CHAPTER **7**

Basic Satellite Networks

Communications satellite in orbit (drawing), RCA Satcom I. Photography courtesy of RCA.

More than 40 program services are available to cable systems and other users, by C-Band satellite. Most are 24-hour networks. *On Sat* and *Orbit* magazines print monthly programming guides and listings of all cable satellites, transponders, and networks.

Large-capacity cable systems take nearly all of the basic satellite networks including one or more superstations. Sixteen of the basic networks are available to more than half of the U.S. cable households. The big cable networks have thousands of affiliated cable systems.

The almost instant jump from 3 television networks to 40 has produced some failures and successes. For the most part, it has been a hard struggle in developing a program strategy, program product, and adequate economic base but now most of the major networks are in the black. We will describe the extant basic cable networks and their economic structure in this chapter. The networks are grouped into superstation, news and information, religion, ethnic, childrens, sports, educational, music, and general categories although few fall neatly in a single category. Appendix C briefly describes the programming of each of the nationally distributed basic satellite networks, its affiliate and subscriber counts, audience size, economic structure, and hours of programming.

SUPERSTATIONS

Superstations are independent broadcast television stations that have been retransmitted via satellite. The first of these, beginning in 1976, was WTBS (then WTCG), of Atlanta, Georgia. The others are WOR and WPIX in New York and WGN in Chicago. KTVT in Dallas/Fort Worth, Texas is a regional superstation. These stations are licensed broadcasters serving their city of license and are not viewed differently, under FCC rules, than any other broadcast station.

Ted Turner's WTBS deliberately set about to become a superstation—to acquire a national distribution of the signal through cable systems. It is programmed for a national audience and sells advertising to national advertisers on the basis of its extended audience. The others are not intentionally superstations. They are sometimes called *passive* superstations because their signals are picked up and distributed by an independent company, which in turn markets the programming to cable systems for a monthly per-subscriber fee. Of course, these stations are not unaware of their status as superstations as they pursue advertising sales and programming, but they do have huge local advertising markets that provide their economic base and they do not realize direct income from cable affiliate fees.

The programming for all the superstations is quite similar. They rely on the staples of independent broadcast stations, huge movie libraries, off-network series, and sports. Because of its large national audience, measured by A.C.

Nielsen, WTBS also carries some originally produced programs and competes aggressively with other cable networks and broadcast networks for television rights to sports events. Sports are also a major attraction to the other superstations since each broadcasts its own area's professional and college teams. These games, however, are most attractive to nearby cable audiences and the cable audiences in the regions of opposing teams.

KTVT, Dallas/Fort Worth offers similar programming to the other superstations and regional programs such as the Texas Horseman's report, National Finals Rodeo, Texas Rangers baseball, and Southwest Conference college football and basketball. KTV: The Rocky Mountain Superchannel has a similar function for its area. More regional superstations could be developed. Strong independent broadcast television stations are already widely distributed to cable systems within their area on the strength of local sports and other programming that may be tailored to the particular region.

Much of the successful format of the superstations is now being programmed by the other cable networks, blurring program product differentiation. The USA Network, CBN Cable Network, Nickelodeon and others carry off-network syndicated series. Two networks carry "classic" (old) movies exclusively. Sports channels are now in the bidding for the television rights to local athletic events. All of these networks are competing for a limited pool of programming and then for a limited amount of viewers' time.

Another problem for superstations is the copyright fees operators must pay to carry the programming. The fees are charged for distant, non-network broadcast signals which include the superstations. In some circumstances these fees are very high (see Chapter 11). The copyright charge must be weighed in connection with the per-subscriber affiliation fee in deciding whether or not to include a superstation in a system's channel line-up.

NEWS AND INFORMATION

In 1980, Turner made another dramatic new entry into cable programming—the Cable News Network (CNN). It provided news and news features around the clock. All-news *radio* was well established in the bigger markets, but skeptics could not believe that there was a demand for all-news television or that its costs were feasible. Four years later CNN was in the black and gaining recognition as a major factor in world news coverage and supply. Not too long after CNN began service, the Satellite News Channel, a joint venture of ABC and Group W, initiated a television headline service. Before this service was actually up on the satellite, Turner Broadcasting had put together its own headline service (CNN-2) in competition. After heavy losses, SNC was sold to Turner Broadcasting. Most of the affiliates of SNC were folded into the Turner channel, renamed CNN Headline News.

The two all-news television channels are different in concept. CNN sup-

FIGURE 7-1. CNN news coverage, Turner Broadcasting Systems, Inc.

plies regularly scheduled newscasts of some depth and adds special reports on sports, business, entertainment, weather, and other topics. It may suspend much of this programming to cover major news stories. For example, the 1985 TWA hijacking and hostage crisis in Beirut filled most of the day for several days and earned record audiences for the network. CNN has now established news bureaus throughout the U.S. and the major news centers of the world.

CNN Headline News cycles through the international, national, business, sports, and weather news, as well as some brief features, every 30 minutes. The weather and features segment may be dumped by the cable affiliates to insert a few local headlines. The headline service is a constant source of news summaries that may be tapped by cable subscribers at any time of the day or night and at any point in the cycle.

Compared to each of the news divisions of ABC, CBS, and NBC television, CNN has a few more employees (about 1,500 vs. about 1,200), about six times as much news programming and about one-third the budget.[1]

Turner Broadcasting encourages cable systems to offer both CNN and CNN Headline News to serve the two distinct functions. The channels are sold together as a package to national advertisers.

C-Span (Cable-Satellite Public Affairs Network) is a nonprofit cooperative of the cable industry with one channel each for live presentation of the U.S. House of Representatives (C-SPAN I) and the U.S. Senate (C-SPAN II). The networks also include committee hearings, major public addresses, interviews, and call-in shows with public figures. The networks make a major effort to inform educational institutions about the programming. C-SPAN I and

II are excellent public relations for the cable industry with strong supporters in the cable industry, government, and education as well as a growing body of fans variously called "cult-like," "political junkies," and "news nuts." Cable operators, who first welcomed the original C-SPAN as a public service and a good talking point during franchise proceedings, are now recognizing the channel as a unique component of the basic service that can be used to attract and help retain subscribers.

More specialized 24-hour information channels are the Weather Channel and the Financial News Network. As with CNN Headline News, these are *reference* channels. It is not expected that people will spend a great deal of time with them, but rather check in periodically for specific information needs. A common mode of cable viewing, where channels are checked periodically or continuously, enhances the value of these services.

RELIGION

Almost as soon as satellites became available for cable system interconnection, religious networks emerged. They are free to cable systems or actually pay a persubscriber fee for carriage, sustained by viewer contributions. A very interested group of viewers uses the services, as cable operators discover if they attempt to drop one of the channels. These channels were added to cable systems as they became available and before other programming options in basic and pay were anticipated. Some operators now have second thoughts about the channels. Many carry all or more than one of the channels, but worry about redundancy. There is a community concern that the religious cable networks are draining funds from local churches. Religious network operators deny this, arguing that the networks stimulate interest in religion and result in more money to both places.

A second concern is the evangelical fundamentalism that characterizes most present channels (such as Eternal Word Television Network, Trinity, and The Inspirational Network). Although these religious networks do have programming representing all faiths, mainline religions have hesitated to participate because of a reluctance to be associated with the evangelical programming and their own greater interest in community-based religion. Therefore, the cable company carrying one or more of the channels is presenting a somewhat unbalanced perspective on religion.

In response to these problems some regional networks representing local churches have been established (such as Bay Area Religious Channel— BARC—in San Francisco, Atlanta Interfaith).

A final issue for cable is the political activism of many of the cablecasts. Some of the religious programs are slanted decidedly to the political right, giving cable systems carrying the networks another dimension to the balance problem.

Whatever the difficulties, cable is a useful medium for religion, reaching people with many more opportunities for worship, teaching, and discussion of theological issues than broadcasting, which limits religion to undesirable fringe times or brief "thoughts for the day."

Two "inspirational" networks (The Inspirational Network, formerly PTL, and TBN) are cablecast 24 hours per day, and a third (EWTN) is on during prime time daily. At the start, all three networks were anchored by talk shows hosted by the network founders: The "Jim Bakker" show on PTL, "Praise the Lord" on TBN with Paul and Jan Crouch, and "Mother Angelica Live" on ETWN. Jim Bakker resigned from PTL after a sex and money management scandal which prompted some of the cable affiliates to drop the network. TBN and The Inspirational Network sell time to other program producers who are usually evangelical preachers or "teachers." Some of the same teachers (Jimmy Swaggart, Robert Schuller, Oral Roberts) tend to be on both PTL and TBN and also available on broadcast stations. The networks support themselves by asking for donations and selling program time to other "ministries" that are supported by donations solicited on their programs.

One of the originals in cable religion was the Christian Broadcast Network. Now CBN Cable Network, it too featured its founder, Pat Robertson, on a talk program, "The 700 Club." But the Monday-through-Saturday programming is mainly reruns of broadcast network programming with commercials. American Christian Television Service (ACTS) is similarly programmed.

National Jewish Television, a part-time network, is programmed with religious, educational, and cultural material. It is accessible to several million cable subscribers.

ETHNIC

Minority audiences, aggregated over hundreds of cable systems, may be served by cable-satellite networks. This is true for any kind of minority audiences with special interests, but particularly important for ethnic minorities who are under-served by broadcasting. Two national basic cable networks have emerged. Black Entertainment Television (BET) and Spanish International Network (SIN). Both are 24-hour services. Each of the two networks will substantiallly increase the subscriber base as cable develops in the major urban markets.

BET features black college sports, news, lifestyles of rich and famous blacks, children's programming, off-network series, and music videos. The music video programs highlight an urban contemporary style in rhythm and blues, pop, soul, gospel, jazz, reggae, and country that is not prominent on the cable music networks. BET claims to do very well in audience in its own universe of black households.

SIN is a hybrid cable/broadcast/low-power TV network. It is particular-

FIGURE 7–2. BET "Video Soul" host Donnie Simpson with Vanity.

ly proud of its World Cup Soccer coverage, live from Mexico City in 1986. The Spanish language programs are heavy in "novellas" the popular programming form in that language. The network can draw on syndicated programming from the highly developed television production centers of Central and South America and Europe. Galavision, originally a pay cable network featuring Spanish language movies, is now a basic service.

CHILDREN

Nickelodeon was once a part-time, commercial-free children's network. It is now a 24-hour, commercial family network. The move to a 24-hour service necessitated a broader concept for the network, from exclusively children to the family. This was achieved by introducing old movies and off-network series (such as the film "Mr. Smith Goes to Washington") and series programs ("The Donna Reed Show" and "Dennis the Menace"). The evening programming is referred to as "Nick at Nite."

Nickelodeon, along with the other MTVN services (MTV Network: MTV, VH-1, and Nickelodeon), prides itself in a coherent "look" or "personality" manifest in a visual continuity of distinctive animated logos. This is intended to promote a clear identity.

The daytime children's programs on Nickelodeon are scheduled according to the *block programming* plan. Each block of programming fits the appropriate daypart for an *age-specific* audience (such as preschoolers in the morning). The intended audience for the block is a narrow, developmentally homogeneous age group. Television programming for children has often been criticized for attempting to attract the whole 2-through-15 age span at once and not satisfying the needs of any specific age group within that range.

Other basic satellite networks program entertainment for children, particularly the USA Network and CBN Cable. In 1984, all the basic satellite networks provided 226 hours of chilidren's entertainment programming in a sample week. This excludes family programming aimed at both children and adults.[2] Most of the religious networks supply instructional and enrichment programming for children.

SPORTS

There is only one *national* basic satellite sports network—ESPN. It is nearly a universal service in cable with almost as many affiliates as there are cable systems. ESPN must find its place in competition for programming rights and audience amidst local and network broadcasting, pay and basic regional sports channels, superstations, and general cable networks. The network can program events of national interest where successful in bidding for the rights against others. There are still far too many sports events even for all of the television outlets. And ESPN caters to sports interests that have seldom been featured by broadcast networks in other than the sports anthology series—volleyball, swimming, full contact karate, soccer, Australian rules football, women's softball, and PFB arm wrestling. In addition, ESPN may take over the first rounds of golf, tennis, and other tournaments that are carried in the later rounds by broadcast and pay cable outlets. ESPN is also a major source for live coverage of sports news events (such as the NFL player draft). In 1987 ESPN became one of the television outlets for NFL football, making a special surcharge to cable affiliates who desired to carry the games.

ESPN may broaden the American knowledge of sports, extend the breadth of what *qualifies* as television sports, and provide an around-the-clock opportunity for the eclectic sports fan.

The Madison Square Garden Network (MSG Network) is an example of a *regional* basic satellite network. It serves New York State, New Jersey, and Connecticut. Programming includes New York Rangers hockey, New York Knickerbockers basketball, college basketball and football, pro and amateur boxing, plus tennis, horse racing, track and field events, the roller derby, and pro wrestling.

While most regional sports channels are premium channels, in regions such as the New York area with its tremendous population, the broad appeal of sports may lift basic cable subscriptions to a point where the additional subscription and advertising revenues may exceed the potential of pay revenues from a much narrower base.

EDUCATION

Almost all of the basic networks provide some educational programming. The Learning Channel and The Discovery Channel draw on a tremendous reservoir of existing programming material that has been designed for formal instructional purposes or educational enrichment. At the same time, the channels may find a television audience that seeks knowledge or perhaps, put another way, an escape from television entertainment.

The Learning Channel is oriented toward mature audiences. Its programs are designed to teach skills, how to cope with life (the income and parenting programming segments), and develop pastimes. College credit may be earned for some of the courses. The Learning Channel makes most of its revenue from per-subscriber fees paid by affiliates.

The Discovery Channel is intended to provide all age groups with entertaining nonfiction films on science and technology, nature, history, travel, and adventure. The channel is commercial. The network occasionally takes advantage of special events to feature unusual programming. During the ABC broadcast network miniseries, *Amerika*, a fictional version of a Soviet takeover of the United States, The Discovery Channel programmed 66 hours of Soviet originated television.

MUSIC

MTV was almost an instant success. As a result it has spawned imitations in its own field of rock music as well as other types of music, in cable, broadcasting, and—if one is willing to stretch the point a bit—in theatrical films, some of which have been called extended music videos.

MTV began by putting available promotional videos together in a program format. The videos were free to MTV as "visual support" for record sales. Not all recorded music was accompanied by videos, often only groups that were not well established or not getting significant play on radio stations made videos. MTV quickly proved that it could create hits for groups that produced interesting videos. Now much of the successful rock music has a video counterpart. MTV receives about 30 videos for consideration each week. There is even a concern within the music industry that the videos are driving the music, instead of the other way around.

The music videos are of two kinds. The *concert video* is a visual record of the group in a performance of the music. The *concept video* is a visual accompaniment directly, or loosely, related to the theme of the music and lyrics—in rock, usually an adolescent sexual fantasy. The concept video directors have used all of the new wave television commercial techniques and added some of their own. They are characterized by multiple images, surrealistic special effects, extreme close-ups, slow motion, hard light, saturated colors, low-key lighting, computer enhancement, unusual camera movement, and quick cuts.

Richard Caplan summarizes the criticisms of the "exotic," "post punk," *"avant garde"* concept videos. He suggests that "violence, victimization, gender portrayal and subservience are all played out in a bizarre, seemingly unrelated fashion."[3] The videos have also been called racist and sexually angry. The complaint is that women are presented as sex objects, prostitutes, and "whipping posts." Caplan's research, a content analysis of 139 MTV videos, found that women are no more likely than men to be the victims of violence. He found .841 violent acts[4] per video and 10.18 acts of violence per hour of music videos, almost twice the level of acts of violence on commercial television. But women were no more likely to be the victims of violence, nor were men more likely to be the perpetrators. Research by Baxter et al. finds 60 percent of a sample of MTV videos portrayed sexual feelings or impulses, but these were "understated, relying on innuendo through clothing, suggestiveness, and light physical contact rather than more overt behaviors." This study also found violence in the majority of videos, again understated: "Violent action in music videos often stopped short of the fruition of the violent act."[5] Some parents forbid MTV, or avoid cable altogether because of MTV, but the majority accept it, albeit perplexed, and may even be persuaded to buy or retain cable on the strong attraction of the channel to their children.

Advertisers love MTV. One financial analyst says, "They [MTV] are the closest to delivering on the promise of cable—they actually attract an identifiable audience you can't get anywhere else. There's a great deal of underlap on MTV."[6]

As competition for attention in music videos escalates, they get *very* expensive to produce. MTV is now paying record producers for an *exclusive window* for videos. The network gets the first rights to show the video.

MTV Networks has a newer music video network, this one for the 25–54 age group, with contemporary, soft rock, rhythm and blues, and cross-over country. The channel is Video Hits One (VH-1). It has far less material from which to select clips than MTV. The producers of this music are not yet ready for VH-1. Some of the skeptics expressed concern that adults would not spend their time with music videos. Nonetheless, VH-1 built up a large subscriber count quickly after launch and the charter advertisers have expressed satisfaction with the medium.

Hit Video USA competes with MTV, claiming somewhat milder content. It does not have the *"brand image"* (name recognition and reputation) of MTV, but the per-subscriber monthly fee to operators is lower.[7]

Country music has its music video channel in Country Music Television. It has had an identity problem, in attempting to establish itself as an exclusive music channel, different from The Nashville Network which is music and other country entertainment.

Music video channels for every type of music have been contemplated.

Some have already come and gone. (The Turner Broadcasting music video network lasted only a few weeks.) The USA Network, WTBS, and BET have all had music video programs. Broadcasters have picked up the format. By now it is a fair certainty that music video television format, a cable phenomenon, is a cable mainstay.

GENERAL

Some of the basic satellite networks, such as USA and CBN, are quite general in their programming, behaving much like independent broadcasters or superstations without the broadcast station base. John Silvestri, USA Network's Vice President for Advertising Sales, says, "We're a broad-based entertainment package, programming to a vast variety of audiences. Advertisers want to be on a widely viewed network. So we don't intend to narrowcast."[8] USA programs off-network series from years ago ("Dragnet") and even off-network soap operas ("Edge of Night"). The programming appeals to women and children in the appropriate dayparts. The network has been a major factor in the resurrections of television wrestling ("Wrestling TNT," "All American Wrestling," "Prime Time Wrestling").

CBN Network is similar to program strategy to USA except that it attempts to project a wholesome, family entertainment image rooted in its Christian broadcasting heritage.[9] CBN features off-network series, especially Westerns, which are actually morality plays that closely fit the network's world view. Some original programming is produced by the network.

Through off-network syndication, both USA and CBN have a nostalgia appeal for older audiences and at the same time are presenting fresh material to new generations of television viewers. In a multichannel television environment, it is much easier to promote an audience for familiar programs and stars than to establish a new program, although both USA and CBN program some original series.

More narrowly focused is the Lifetime network, which bases its appeal on lifestyle issues aimed at women, usually presented in a talk television format. Strong personalities (Dr. Ruth Westheimer, "Good Sex") anchor the shows. The Nashville Network offers a wide variety of program types, all intended to have a country flavor.

The Arts and Entertainment Network is the surviving "highbrow" network in cable. Two other efforts, CBS Cable and The Entertainment Channel (NBC/RCA), failed despite their substantial backing. A&E includes a broad range of the performing and visual arts, movies, BBC programs, and some U.S. off-broadcast-network programs that were attractive to an intellectual audience but were unsuccessful in winning a sufficient share of the mass audience. It is mainly an *acquisitions* network, obtaining rights to existing programming, but it has co-produced some original material with the BBC, retaining the U.S. rights.

Providing a broad range of program types, targeted to the special audience of deaf or hearing-impaired is the Silent Network. Its program line-up has included talk shows ("Off-Hand"), musical variety with a deaf dance troupe ("Musign"), news ("Beyond Sound"), aerobics, and drama. In addition to entertainment and information, the network attempts to present role models for deaf children and adults—interesting and courageous people. The network is supported by advertising. The programming is syndicated to broadcast stations as well as playing on the cable network.

SHOPPING NETWORKS

Shopping channels became the new wave of satellite network programming in 1985 and 1986 after several years in which the concept was incubating at local cable systems. They achieved almost instant success as the shopping network sales volume, and popularity excited the cable industry and Wall Street as well.

Shopping networks offer close-out merchandise at cut-rate prices in live, glitzy, dramatic formats. Products are described and demonstrated by hosts. Buyers may call toll-free 800 numbers to order. Incoming calls are highlighted by visuals of the telephone operators on the set and occasional dialog between show hosts and callers. Sound effects register the sales for the audience. The productions are designed to put a shopping opportunity in a fast-moving, audience-participation, entertainment package. The articles are delivered to the home of the buyer in as few as three days.

The Home Shopping Network (HSN1) offers general merchandise of all types. The Cable Value Network has programs that categorize the merchandise, such as cooking and sports. Other shopping networks may be more specialized. For example, the Innovations in Living (HSN2) network, sometimes called the "yuppie network," presents high-tech, higher-priced products that might be found in a Sharper Image catalog. Sears, Roebuck is associated with the QVC shopping network. The J.C. Penney Company has made a major commitment to the development of a shopping channel. The presence of these prominent retailers lends credibility to the shopping chanels.

Specialty shopping channels have also emerged. The Travel Channel, owned by Trans World Airlines, and the Fashion Channel quickly signed affiliates aggregating millions of subscribers.

Cable system affiliates receive a percentage of the sales within the franchise, averaging about 5 percent depending on the affiliate and the network.

CHALLENGES AND PROMISE

Reemphasis on Basic

After a few years of infatuation with the dramatically developing pay networks, as described in Chapter 8, the cable industry rediscovered *basic* cable, for a number of reasons. Pay channel revenues have not met somewhat exaggerated expectations. After the initial marketing and each remarketing of pay channels, there is a regression of subscriber penetration rates to lower numbers. Pay subscribers tend to add, drop, and change channels frequently and for many cable operators this means costly service calls. Household viewing time devoted to pay cable channels has stabilized or fallen off a little from the peaks, while the basic cable networks have steadily increased their share of viewing time. Finally, the subscriber may perceive a greater value in a package of 20 or more basic channels at a single price than one pay channel at about the same price.

Over the years, the relationship between the pay and basic subscriber rates have gotten somewhat out of whack. Until 1987, franchising authorities were permitted to regulate basic cable rates, but *not* pay rates. As a result, the cable operators used pay channel rate increases, which they could control themselves, to meet the revenue demands of the business. Now that rates have been deregulated for most cable markets (see Chapter 11), basic rates are expected to rise to a more natural level. This increases the interest of the cable operator in the development and promotion of the basic services.

Cable operators receive the greatest proportion of their operating income from basic cable. Whereas pay cable program suppliers take about half of the gross revenues from pay subscriptions, the basic channel suppliers take much less, in some cases nothing at all. Rights charges paid by cable operators for *all* of the basic channels may add up to less than $1 per subscriber, while the charge for *one* pay channel may be $4 per subscriber. This imbalance is likely to change.

Programming/Audiences

Basic cable presents fascinating programming challenges. The plethora of channels creates both opportunities and problems. Some basic networks try—with mass appeal programming, as commercial broadcasters always have—to take a piece of the very broad middle of the U.S. television audiences. This audience may be large enough to be cut up several ways between broadcast and cable networks. Other cable networks seize their opportunity in finding a narrow program niche appealing to much smaller audiences. Gene Jankowski, President of the CBS/Broadcast Group, describes the two approaches to audience:

There are essentially two dynamics at work in communications. Let's call them audience aggregation and audience disaggregation. *Aggregation* brings forth mass audiences for those relatively few movies, television programs, books, records and so on whose great appeal cuts across boundaries of age, sex, income, education and taste. Its opposite, *disaggregation,* once confined largely to print, has now become an electronic process, thanks to the new technologies. It aims for smaller audiences of like-minded people who share specialized interests. These two dynamics serve two different human needs: the need to belong and the need to be an individual. They are not competitive; they are complementary. [Italics supplied.]

The disaggregation strategy of programming is well established and accepted in cable. Large channel capacity has made specialty (niche) programming possible, and, as we have seen, much of it has been successful. But the attempt of cable to compete head-to-head with broadcasting in aggregation of mass audiences has been more controversial within the industry. This requires large investments in programming. The pennies per subscriber paid by cable operators in 1987 for each mass appeal channel would turn to dollars. Some cable operators believe that this is necessary. The urban operators, particularly, feel the competition from broadcasting and videocassettes most acutely. Urban areas experience the highest levels of churn. Without cable, there are many available broadcast stations—and several video stores within a short radius of each subscriber's home. Some of these operators see direct competition with broadcasting as a means of survival, and some see it as a golden opportunity.

To see this opportunity clearly, one must review the comparative status of cable and broadcasting. Television broadcasting has peaked in revenue potential. Advertising agencies have finally balked at the annual increases, exceeding inflation, in broadcast television time rates. The commercial broadcast networks, all under relatively new management, have sought to reduce expenditures. Although they deny the cuts will affect the product, programming people insist that they will. It can be argued that advertising revenues alone cannot sustain quality mass entertainment in an era of audiences fragmented by cable and videocassettes.

Cable is in a different position. The industry is still far short of its maximum revenue potential. Cable is not entirely dependent on advertising dollars. The consumer is willing to pay *more* for programming than the advertiser is willing to pay for that consumer's presence in the audience. (This is the fundamental problem in the commercial broadcast television business where advertising is the *only* source of revenue.)

When the U.S. cable industry was free to raise subscriber rates for basic services, those rates were increased to what the industry thought were more natural levels, expecting to reach monthly rates of about $16 or $17. To achieve this rate level most operators acknowledged that new or better-quality programming would have to accompany the increase. Most of the major MSOs were willing to put some of this new revenue back into programming.

If the FCC were to adopt the proposed syndicated exclusivity rule and the Congress were to repeal the compulsory license both of which permit importation of distant broadcast signals, cable carriage of off-air television broadcasting would be substantially disrupted (see Chapters 6 and 11). Under these circumstances cable would become even more reliant on cable-only programming.

Furthermore, in 1987, cable passed only 75 percent of the U.S. television homes. In a few years cable will pass 90 percent of the homes. At current levels of penetration this would add more than 7 million homes and billions in revenues. These additional homes would produce more advertising revenue. Thus, unlike broadcasting, cable is able to support programming by *both subscription and advertising revenue.*

As some of the new revenues from increased basic rates, new subscribers, and additional advertising time sales are put into programming, cable will be able to attract the best writers, performers, and producers of mass entertainment. The cable industry could establish a *flagship network.* That is, a single network with super programming that would carry the mass entertainment programming flag for cable. Or the programming money could be distributed over several networks in *breakthrough* (or "punchthrough") programming such as NFL football on ESPN. At this point cable subscription becomes more compelling. To get all of the best of television programming, cable subscription would be necessary. In such a case, cable penetration moves up from 60 percent in 1987 to the 70s, generating still more subscription and advertising revenue.

In the meantime, broadcasting is suffering. The basic advertiser-supported cable-only network share of viewing time among subscribers rises sharply. Broadcasting would still get its 100-percent share of the viewing among nonsubscribers, but with a 70-percent penetration of 90 percent of all U.S. television homes passed, there are only 32 million broadcast-only homes. Broadcast audiences would be substantially reduced. The broadcast television networks could lose so much advertising revenues that they would no longer be in the bidding for top programming against the cable advertising-subscription system.

The logical role for television broadcasting under these circumstances is to become a *second-run* medium, with cable the *first-run* medium. Cable would play the best mass appeal programming first and then syndicate it to broadcast television, to help recover some of the costs. The people who are paying the high basic cable prices are getting the first chance at viewing the best of the programming. This seems more consistent with the U.S. economic system than the present situation where people paying the most for television—

the cable subscribers—have no better access to the best mass entertainment programming than the people who are receiving it "free."

The consequence of this scenario is a very powerful cable industry, buying all of the best in programming for its subscribers, limited only by the subscribers' willingness to pay—a willingness that may expand with a newly created isomorphism between program quality and subscription dollars.

At this writing, the U.S. cable industry was not thinking, at least publicly, in such bold terms. But the search for breakthrough programming was already underway. Cable operators were becoming more willing to be venturesome in the programming area, and they were starting to develop cooperative mechanisms for programming investment.

Nonetheless, the cable industry may not have the patience to wait for the rewards from gradual programming improvement—accepting the lag between the initial investment of programming dollars and the increase in subscribers and revenues.

Other Program Funding

In addition to operator per-subscriber fees and advertising, the basic networks are developing other markets, sometimes called the *aftermarket* to help finance programming. Original programming for some of the networks is placed in syndication for sale to U.S. and foreign broadcasters and foreign cable operators. CNN makes its news services available to domestic and foreign radio and television outlets. Joint ventures (or *coproductions*), involving a cable network as one party, are also common. Foreign production resources have been used by CBN to minimize costs.

These efforts are building viable basic service for cable. In relation to the 40-year history of television, the development of all these channels has taken place in a very short span.

NOTES

[1]N. R. Kleinfield, "Making 'News on the Cheap' Pay Off," *The New York Times*, April 19, 1987, Section 3, p. 8.

[2]David Atkin, Michelle Siemicki, Bradley Greenberg, and Thomas Baldwin, "Nationally Distributed Children's Television: The Cable Contribution," *Journalism Quarterly*, Winter 1986.

[3]Richard E. Caplan, "Violent Program Content in Music Video," *Journalism Quarterly*, Spring 1985, p. 144–147.

[4]Incident Classification and Analysis Form System. "Force or compelling threat of force that may result in harm to life or to valued objects. Violence involves behavior which violates, damages or abuses another person, animal or valued object." Includes threats and natural or human accidents.

[5]Richard L. Baxter, Cynthia DeRiemer, Ann Landini, Larry Leslie, and Michael W. Singletary, "A Content Analysis of Music Videos," *Journal of Broadcasting and Electronic Media*, Summer 1985, p. 336.

[6]Eric Schumuckler, "Basic Cable: Five Years Later," *View*, June 1985, p. 61.

[7]At press time for this volume, the survival of Hit Video USA was in doubt.

[8]Eric Schmuckler, op. cit., P. 64.

[9]At press time the CBN Network was considering a name change representative of its family programming focus.

CHAPTER 8

Pay Cable

Program supervisors and technical director monitor east and west coast transmission feeds at the Cinemax master control in New York City. Photograph by Lou Manna, courtesy of HBO, Incorporated.

Pay cable, or pay TV, is the term applied to services for which the subscriber pays an additional amount, after the basic subscription fee, to receive programming, usually by the channel and sometimes by the program. By the program, pay TV is referred to as *pay-per-view (PPV)*, which is just beginning as a major cable service. There are a great many pay channels, some national and some regional. Cable systems with 20 or more channels will usually carry several pay channels.

Pay TV is very important to the cable industry and to cable consumers. Pay cable was responsible for expansion of cable into the urban markets and the mushrooming of earth stations, which, in turn, have opened up cable to all the other satellite services described earlier. Pay cable is an interesting topic because of its incredible growth, the battles between Hollywood and pay cable distributors, the programming strategies, competition from VCRs and rental cassettes, and the sometimes surprising consumer appetites.

BACKGROUND

Seeking a cable system for its new pay TV service in 1972, the Home Box Office (HBO) Company signed cable pioneer John Walson who owned the Service Electric Cable Company in Pennsylvania.[1] Walson was the first to agree to try the pay service that took on other affiliates in the Northeast. Tapes of movies supplied to the cable system were played and repeated several times and then returned to HBO. Walson charged $6 per month and supplied a mimeographed program guide.

As noted in Chapter 5, a pay-per-view service called Telecinema started in 1974 in Columbus, Ohio on the Coaxial Communications system. Subscribers were permitted a few minutes of viewing of each film and then were billed for the film at about $2.50 if they viewed further. A two-way system monitored the set tuning so that viewing in each household could be ascertained. The headend computer automatically added each viewed movie, at the correct price, to the subscriber's monthly bill. The Telecinema system was abandoned when it became apparent that much greater subscriber penetration and revenues would result from per-channel pay television. Four years later, in a different part of the same city, Warner initiated the Qube system with the pay TV offerings on a per-program basis. Few other systems attempted PPV until the mid-1980s.

In 1975, HBO began transmitting by satellite to earth receiving stations owned by cable systems. Since the receiving stations cost about $100,000, only the big systems with 10,000 subscribers or more could make the capital investment. Others, large and small, continued to operate pay services by videotape.[2]

Another event in the development of pay cable was the 1977 U.S. Court of Appeals' decision overturning the FCC's pay cable rules, which had, in essence, prohibited pay cable use of older movies and sports events that had been shown on conventional television during the previous five years.[3]

Early pay cable entrepreneurs were plagued by problems with *churn*. The basic cable subscribers would take the service for a month or two and then quit; others would be persuaded to sign up and then *they* would drop, so that there was a constant churning or turnover of pay subscribers. The best that pay cable could do would be to hold about 25 percent of the basic subscribers, but these retained subscribers came at the high cost of installation, usually provided free as an incentive to subscribe, and disconnection for those who tried the service and did not stay with it.

Several problems accounted for the churn and low subscriber penetration. Pay cable was almost always oversold. The blockbuster film titles would be featured in advertising and promotion. The subscriber would realize only after the service started that there were not many blockbusters and that "B" pictures were used as fillers. The movie industry does not produce enough quality films to fill a daily schedule. One unsatisfied subscriber wrote to HBO's president: "I hope that King Kong ascends the Time & Life building and passes gas into your office. Then maybe you will get the drift of what I think of your schedule."[4] Furthermore, Americans were not accustomed to *paying* for television. It took several years to establish the idea that television was something of value and that unedited, uninterrupted movies, even just a couple a month, were worth $7 or more. And, of course, promoting the unfortunate name "pay cable" played into the hands of TV network competitors who labeled their own service "free TV." Moreover, the system of paying for a whole channel, when only a few programs are viewed, violated a consumer code for some people; there was too much waste.

Many cable operators just as vehemently rejected pay TV, making it difficult for pay suppliers (networks) to build a cable subscriber base. These operators felt that their business was delivering over-the-air television—a business that was already doing nicely. Some others were afraid to get mixed up with Hollywood and some feared a subscriber backlash from the uncut movies.

A major boost to pay cable came with the FCC's reduction of the required size of the C-band earth station receiving dish from 10 meters to 4.5 meters. The price dropped to as low as $3,000 for antenna and preamps, with an additional cost of $1,000 in receiving equipment for each channel. At $4,000 for a single channel, this put almost every operator in the pay TV business. Each dish was capable of receiving not only pay channels but several other network services from the same satellite as well, further justifying the capital investment.

In 1987, almost two-thirds of the basic cable subscribers were pay subscribers. Some are *multipay* subscribers with more than one pay channel.

Pay Suppliers

At first, some large cable systems—and some too small to construct a receive antenna—ran *standalone* pay channels. That is, the system would negotiate its own movie rights and secure videotapes for play on the system. This required a major administrative effort, staffing and equipment for playback, and the production of a program guide. These systems soon joined one of the pay networks.

For a time, it appeared that several pay channels could exist, each serving a large multiple system operator, with an ownership interest in the pay service, and licensing the service to other independent or smaller MSOs. This plan was not feasible. The surviving national networks, programming current movies and big event specials, are HBO, Showtime, The Movie Channel (TMC), and Cinemax. These were called *foundation services* and usually a cable system would feature only one in its marketing even though it carried the others. The other pay channels are specialized. The affiliate and subscriber counts for national and regional channels are listed in Appendix D.

An important discovery for the cable industry was made in 1978 when a Wometco cable system in Thibodaux, Louisiana offered *both* HBO and Showtime, presumably to give the subscribers a choice. Instead, many households took both.[5] Somewhat incredulous, but afraid to miss the boat, other systems tried dual-pay offerings with the same result. Soon many systems took the supermarket approach, offering every available pay cable service and most of the other satellite-delivered networks as well. Showtime began to promote itself as a second service to those systems already receiving HBO. In response, HBO created a second service of its own, Cinemax, deliberately *counterprogrammed* to HBO; that is, offerings opposite each other were designed to appeal to different tastes. Since all the foundation pay services draw on the same pool of movies, there is a clear subscriber perception of the duplication across services, but as long as the duplication is not in the same time period, the *multipay* subscriber may be satisfied.

PROGRAMMING

Product

Programming of the pay networks has been a problem. The most fundamental of the difficulties is the lack of *product*. Product means programming, particularly the output of Hollywood. In 1986 Hollywood released only 138 new films that earned their distributors more than $1,000,000 in rentals in the U.S. and Canada; less than a dozen per month.[6] Of these, only 20 earned more than $20,000,000 in rentals accruing to the distributor.[7] These are the *cover films*, featured on the covers of the pay TV guides and most heavily promoted, such as *Top Gun, The Karate Kid Part II, Crocodile Dundee*. At the bottom of the list are films that are relatively unknown and can only be used to fill out the schedule such as *Turtle Diary, Jake Speed, My Chauffeur*. Since this time, Hollywood production is up slightly to meet the demand of expanding distribution markets, a trend that could help pay cable considerably.

The pay suppliers use theater box office success as an index of which films to feature (and also as an index of how much to pay for cable rights). But sometimes a film that has failed at the theater box office will play well to pay cable homes. This may not be too surprising, since the principal theater audience is under 24 years of age and the pay audience spans the whole spectrum (still weighted at the young end). In some cases, success on cable has caused the movie producer to re-release a film to theaters.

Both HBO and Showtime have experimented with *exclusive windows* for theatrical films. The practice comes in and out of favor depending on costs and competitive circumstances. For a very big price, the network can secure exclusive access to a major film for an early period (window) while there is still consumer interest generated by publicity and word-of-mouth during the theatrical release. HBO used its economic power in an attempt to cement its first place position by securing exclusives. Struggling in its number two position, Showtime negotiated exclusives to several of Paramount's big films (such as *Terms of Endearment*) during a period when Paramount was enjoying an unprecedented string of successes. It was rumored that Paramount favored Showtime with these exclusives in an effort to stimulate greater competition against the dominant HBO. In 1987 Showtime announced its intention to strive toward 100 percent exclusivity by 1988.

Repeats

Faced with limited product, pay suppliers must repeat what they have available to fill time. Repeats, of course, are a convenience to subscribers who can look at a month's schedule and pick the most convenient time to view. Or, for those who do not plan ahead, if a film is missed on one occasion, another time will come. Tony Cox, while at HBO, said a big challenge of cable marketing is "in learning how to sell the asset of repeat programming and to overcome the kind of historical ingrained television viewing habits in which the consumer presumes the repeats are a bad buy." He speaks of "positive" and "negative" repeat experience suggesting that repeats are positive up to a point and negative thereafter.[8]

The utility of a film or program, aggregated over all subscribers to a service, increases in small increments after the first couple of showings. After about five or six plays, all the people who have wanted to see a program have had their chance. Repeats, thereafter, become negative in value. Few new viewers are added, and subscribers get irritated that the title continues to fill time that might be given to new programs. Too many repeats may lead to a disconnect.

Because movies are of variable length and are not cut by the pay cable networks, they seldom end on the half hour or hour convenient for programming the next show. Therefore, the time between the end of one movie and the next program is filled with *interstitial* programming. These are various short subjects produced for other purposes that are purchased at very low rates by the pay channels. Interstitial programming does give an outlet, if not a substantial income, to independent producers. Some of the video techniques and subjects are quite innovative.

Beyond Movies

Another problem for the pay movie services, arising from the limited product in theatrical films, is *duplication* of programming. Since all of the networks dip into the same small pool, they tend to have the same programming except for exclusives and minor films, where each service may have somewhat different bookings.

Two of the major pay services, HBO and Showtime, have gone well beyond theatrical films for programming. The Movie Channel is a 24-hour, exclusively movie channel although it occasionally has some nonmovie programs *about* movies. The "other" programming on HBO and Showtime serves two major purposes: to supplement insufficient movie product and to differentiate the services. *Differentiation* is important if HBO, Showtime, and other networks are to be considered by consumers in multipay packages. It is also valuable to have some unique features as selling points in marketing the pay channels.

One of the nontheatrical movie forms is the *made-for-cable* film. This is a film produced to have its first run on cable, although it may be released subsequently to theaters, broadcast television stations, and videocassette stores. HBO and Showtime produce several per year. These films are not big budget efforts (that is, $4 to $5 million) in comparison to the major theatrical films, but they often have name stars (such as HBO's *Finnegan Begin Again* with Robert Preston and Mary Tylor Moore). The made-for-cable films can also cater to a cable audience, where most theatrical films must be made to attract the youthful theater audience. They are heavily promoted in the program guides, on the channel, and through advertising and publicity so that it is hard for a subscriber not to be aware of the film.

Something like a made-for-cable film is the *miniseries*, a film that is spread over several episodes. The miniseries is a limited-run series, a program form used extensively in Britain. According to HBO Chairman and Chief Executive Officer Michael Fuchs, better talent is available for the short run than a long series, and the miniseries are more distinguishable from the broadcast network programming.[9] Examples are HBO's *First and Ten*, a 17-program

FIGURE 8-1. Publicity photo for the original cable program ''Hansel and Gretel.''

comedy about a football team taken over by a woman in a divorce settlement, and Showtime's *Tender is the Night,* six hours from the F. Scott Fitzgerald novel.

The conventional television *series* also appears on HBO and Showtime, although Showtime seems to be more favorably disposed to this form. Showtime pioneered the series on pay, picking up *Paper Chase* and continuing it with first-run new episodes after it had been dropped by a broadcast network. The pay network now has several original series and has rerun early installments of the "Honeymooner" series with Jackie Gleason.

Pay channels also use the *anthology series.* HBO's *The Hitchhiker,* a continuing adult thriller, was successful. Showtime's *Faerie Tale Theater,* produced by Shelley Duvall using distinguished actors and directors, has received critical acclaim and many awards.

Music, comedy, and sports *specials* are critical elements in the nontheatrical film line-ups of the pay channels. The music and comedy specials feature some of the big names in entertainment in a one-hour concert format— Tina Turner, Phil Collins, Olivia Newton-John, Hall and Oates, Kenny Rogers, Dolly Parton, George Carlin, Rodney Dangerfield, Whoopi Goldberg, Gallagher, "Wierd Al" Yankovic, Joan Rivers, Marvin Hagler. They are obviously expensive for the pay networks, but they can be promoted as "big events" and attract different audience segments than theatrical movies. Pay cable has also made a place for broad satire, such as "Not Necessarily the News," nearly absent from broadcast television or basic cable.

SPECIALTY PAY NETWORKS

Specialized pay networks have been developed to serve more narrow audience needs. While the major pay movie channels serve 50 percent or more of the average system's basic subscribers, these channels usually penetrate about 10 percent or less. The specialty channels can be used handily to create distinguishable packages for marketing purposes. Because they are not duplicative, they can be added to a foundation pay service to form a multipay package that is relatively stable.

Family

The Disney Channel is perceived as a children's channel, but is intended to appeal to families. Of course, it has the famous Disney library to draw upon, although it reserves some of the major titles for periodic, exclusive release to theaters or home video. The network also produces original material for the channel in much the same form as the other pay channels: made-for-cable (such as *The Blue Yonder* with Art Carney and Peter Cayote), miniseries (like *Return to Treasure Island*), series (*The Enchanted Musical Playhouse* with Marie Osmond). The network appeals to adults in the evening hours with off-network series such as *Ozzie and Harriet* and *The Dr. Joyce Brothers Show* discussing family problems.

Home Theater Network seeks an audience among families that reject R-rated theatrical films. It runs only PG, PG-13, and G-rated films.

In 1987, HBO introduced its Festival service using alternative *studio* releases of films with less violence, fewer four-letter words and fewer sexual situations. The service is designed for families who reject the HBO or Cinemax content.

Fine Arts

The Bravo network features English-language films (U.S., British, Australian) similar to those that play in the art theaters. The network also features fine-arts specials. It is sold to operators at a low price (around $2), as are many of the other specialty pay networks, and then retailed at a low price (around $5) or put in a package with other pay services.

Adult

Send the kids to bed. The kind of R-rated movies your customers want to see . . . when they want to see them. This stuff is explosive! Look at the trends in videocassette sales. Check the popularity of the "jiggle" and action shows on network programming. Think about the failure of pay TV channels without adult-oriented fare. America wants . . . and will pay for . . . action-packed, revealing movie entertainment in prime time [trade ad for Escapade which became Playboy].[10]

Despite the promises of this trade ad, the adult specialty pay channels have not been a particularly good business for the cable industry. There is a lack of material with any kind of story line. Cable systems are reluctant to show X-rated material or its equivalent for fear of community or franchise authority reprisals; therefore the programming is soft in comparison with the hard-core films on videocassette and in adult theaters. Some general pay cable services play similar adult material late at night. These may be enough to satisfy some of the demand short of actually buying a full adult channel. As a result, many systems do not affiliate with adult channels. For those who do, there is a very high churn rate.

On the other hand, adult content has always been a major appeal of the general pay networks. Uncut, R-rated movies earn the biggest shares of audience among the pay networks. In the old Telecinema pay-per-view system, adult movies were the biggest draw.[11]

Sports

Regional sports pay channels are emerging in several areas. To understand this form of programming, it is necessary to look at all sports television.

The cable potential of sports is intriguing to sports entrepreneurs and cable programmers, but the risks are great for both. The approach-avoidance conflict is most intense among franchise owners (professional and college). They are dependent on the gate receipts and the *appearance* of a packed house that asserts excitement and fan interest. This is particularly important for outdoor sports with huge seating capacities. Because it is so difficult to test, owners are never quite certain how TV affects attendance.

Sports promoters agree that broadcast TV is critical to creating interest in a team. Because of limited accessibility, cable does not serve the purpose of *exposure* very well. Basic penetration of 40 percent, or so, in urban areas and pay sports penetrations of 5 to 15 percent of basic do not compete, in mass exposure, with the universal broadcasting penetration.

Some fear, however, that *overexposure* is a danger. The court decision that broke the NCAA monopoly on football rights, resulting in round-the-clock live college football on broadcast TV and cable, heightens this concern. National cable networks, superstations, and the regional sports networks put out so much sports that satiation is a possibility and fragmentation of audience is a certainty. Big advertisers, forced to absorb big rights fees resulting from competitive bidding for sports, are appalled at falling commercial network ratings. To get the necessary audience reach, they now need to make some cable advertising buys, but they are still paying the same prices for broadcasting.

If advertisers can get down to prices they consider efficient for sports programming, the regionalization and fragmentation should be to their benefit. If all of the sports audiences, from all of the television outlets are added up, the numbers should exceed the old days of commercial broadcast network programming of major sports of national interest and a few regional interest contests. Regionalization of sports programming is probably more satisfying to sports fans, who can identify with the nearby teams, making those games a better environment for the advertiser. Fragmentation of audience across national and regional games and across major and minor sports has the additional benefit to advertisers of providing more direct advertising targets.

Sports franchise owners and promoters, of course, seek to maximize rights fees, consistent with the goals just mentioned of gate protection and optimum exposure. Cable makes some contribution to revenues and promises more. But how cable fits in this delicate balance has been subject to a number of trials and some costly errors.

Cable is obvious for minor sports that were once relegated to occasional coverage on the broadcast TV sports anthology shows. Basic cable networks seem to be able to find interested advertisers and need these events anyway to fill schedules. The place for cable in major U.S. sports, football, basketball, baseball, and hockey is most controversial. When basic networks enter into bidding for game packages and win, some cable operators object that the rights price is too high and brace themselves for higher per-subscriber fees.

Most pay cable sports networks have had a struggle. Milwaukee, Chicago, and Pittsburgh area regional networks have folded from a failure to attract enough sports events, enough cable affiliates, or enough cable subscribers. The Sports Time network, of well-heeled, knowledgeable sports/cable parentage (Anheuser-Busch, Tele-Communications, Inc., and Multimedia, Inc.) did not succeed with a super-regional channel covering Kansas City, St. Louis, and Cincinnati. It was a pay channel in the three city areas and basic cable on the fringes. Sports Time was apparently too broad in its coverage.

Cable operators have been hesitant to push fairly expensive regional pay sports channels because their subscribers may trade off pay movie channels for sports. Subscribership has also proved to be seasonal with homes churning out at the end of the season for the key sports franchise in the package. Perception of the values of premium sports channels is also closely tied to the quality of the team. A couple of bad years for a team can certainly prejudice the future of a sports channel. Furthermore, sports channel operators have had a tough time putting all the major sports teams into one package; thus the product is weaker than it might be. In Boston, one pay channel has the hockey Bruins and another the basketball Celtics.

The regional cable sports networks are now getting together to explore mutual interests and jointly bid for product. The first effort was a six-network consortium to buy and telecast a scholastic gymnastic competition in the summer of 1985.

Another cable sports option is pay-per-view. Some in the cable industry see it as a convenient way for subscribers to put together their own sports menu. Others look at it as a way to hook the subscriber to a channel subscription that in the long run is more economical for the consumer and promises more stable revenue for the operator.

Some programmers are looking at *basic* cable as an appropriate placement for the regional sports channel. Operators would not have to get much more revenue per basic subscriber to equal the revenues from a 10-percent or less penetration of a regional sports pay channel. Meanwhile, some of the pay sports networks are behaving like basic networks and selling advertising time.

It seems inevitable that cable will gradually increase its share of the sports programming market. We can look to a long period of negotiation and experimentation, which should develop an important place in the sports entertainment market for cable in basic, pay, and pay-per-view.

PAY CABLE AND VCRs

Does the videocassette recorder boom in the U.S. spell doom for the pay cable business? There is some reason for concern in the cable industry. The proportion of U.S. households with VCRs is approaching 50 percent. The price of the machines has come down and may drop further.

The major use of the home VCR is in recording programs for later viewing, *timeshifting*. But this is changing to an increased use of the machines for playback of purchased or rented tapes. Rental charges have dropped dramatically for a one-day rental of a theatrical film. Rental convenience has improved to the point where there is a nearby video store in almost every neighborhood. Major titles may also be found in grocery and convenience stores.

The principal offerings of the home video stores are the theatrical films. They are available in an earlier window than pay cable. Pornography, also a staple, is more explicit than that available on cable. Music videos are now only a small percentage of the home video market but are predicted to increase. Before the VCR, all of these things used to be cable exclusives in the home.

TABLE 8-1. Movie Market Order of Distribution and Windows

Order of Distribution	Window
1. Theatrical	3 to 12 months*
2. Home video and PPV	6 months
3. Pay cable	12 to 18 months
4. Broadcast networks	3 to 4 years
5. Syndication (to individual broadcast TV stations, cable superstations, "classic" pay channels, and the like)	Unlimited

*In all of the distribution channels, the length of the window depends on the particular film. As a rule the greater the popularity of a film, the longer the window at each level of distribution.

Cable operators and pay suppliers fear that the VCR has accounted for some of the recent leveling in pay cable growth. A second and third pay channel, often desired for scheduling flexibility, is not necessary if recording for timeshifting purposes is done from the first channel. Rental and purchase of tape costs may also substitute for the monthly cost of the pay channel.

VCRs have an advantage over cable in the control of content and viewing time. The whole store inventory of home video titles is available with new titles added each month. Any of these can be seen at the convenience of the VCR owners as long as the title is in stock at the neighborhood video outlet.

But the VCR need not be competitive with cable. According to A.C. Nielsen's Home Video Index, pay cable subscribers are more likely than any other group to own VCRs.[12] Although this could be an ominous sign for pay cable, if VCR use were eventually substituted for pay cable, it could also be a sign that the two are compatible.

People buy cable and VCRs because they like video entertainment and information. The VCR can enhance the value of the cable subscription. In one sense, the more television that is available to a home, the greater the need for timeshifting. The value of a movie channel is much greater if the household can tape the most desirable films for showing at the *ideal* time. The blockbuster film that may play the cable channel only once or twice in prime time can be held until the family tires of it or put into the home library.

Furthermore, basic and pay cable offer a good deal of programming not available in the home video store—sports, news, series, and the like. This programming becomes more important as the pool of home video material gets "used up" by the VCR household and the VCR novelty wears off.

Nonetheless, the cable industry is not discounting the negative potential of VCRs. Some systems are planning a restructuring of pricing, partly in response to the VCR. The idea is to increase the price of basic cable and hold the line or even reduce the prices on pay cable channels. Faring best, under VCR competition and new pricing schemes, are expected to be the differentiated specialty pay services such as the Disney Channel and regional sports. These services are not easily duplicated by the video stores. Disney, incidentally, releases many of its titles to its home video subsidiary *after* the cable run.

The general pay movie services may shift some emphasis to exclusive products, cable-only movies, specials, and series. If this programming is attractive enough, it may hold the multipay households in line and may itself be desirable for recording by the VCR.

Another hope for cable against the march of VCRs is pay-per-view. The release of movies to PPV is usually simultaneous with the home video release date. The VCR owner with PPV can tape these films, which are current, and hold them in the family library as desired. The cable subscriber without a VCR has almost the same access to major movies as the VCR owner/video store customer and at a comparable price. The PPV electronic distribution of video product seems to be more compatible with some lifestyles than the home video store distribution, which requires travel to the store, often waiting for service, and the travel to the store again for return.

The final answer of the cable industry to the competition from video stores is to join it.[13] Some MSOs are experimenting with videocassette sales and rentals as a logical extension of the cable business. One is even financing the purchase of the VCR. It is important for cable operators to help subscribers use VCRs with cable, offering how-to connection literature and, in some cases, interface equipment. If, indeed, VCRs can enhance the value of cable service, the latter step seems logical. If the compatibility of cable and VCR functions can be successfully marketed, a cable industry that offers the whole package may inherit the video future.

PAY-PER-VIEW

Although pay-per-view as a cable business is still a way over the horizon, interest is now very high. A major attack on the operating problems of PPV is mounting on several fronts—from the film industry, event promoters, cable operators, hardware manufacturers, and an assortment of middlemen.

The problems are substantial, but they may be resolved by the changing attitudes of the industry and consumers and by the latest technology. We'll look at the problems here, one by one, and the most current approaches to resolution.

Program/Product

Until recently PPV did not get much respect from film producers and promoters. Movie release dates favored the theaters and the video stores. Promoters and producers demanded large guarantees and the lion's share of receipts. Now the movie industry is beginning to encourage PPV with earlier release dates. The cable PPV release can be coincidental with videocassettes or earlier. By experimenting with more favorable release dates for PPV, Hollywood is trying to maximize revenues. There is recognition that PPV could eventually generate more revenues than have cassette sales and rental. As PPV grows, it may push back the per-channel pay TV release to make PPV even more attractive.

Pay-per-view is a formidable factor in some sports events. The Sugar Ray Leonard-Marvelous Marvin Hagler fight achieved a 13 percent buy rate at prices as high as $40.

At first most sports promoters and film distributors were looking for 60 percent or more of the pay revenues before expenses, along with a minimum guarantee. This put the operator at significant risk for very little return after expenses. Now the deal is more reasonable from the operator's perspective: usually a 50/50 split and no guarantee.

Consumer Attitudes

Consumers have to learn to accept pay-per-view as another means of receiving value from cable. PPV may reach some subscribers who might not otherwise want premium channels and add a few dollars to the conventional pay TV bills for those who are pay subscribers. Some cable people fear that PPV will cannibalize pay channel revenues (such as subscribers switching from pay channels to PPV) and not add enough of its own to make up the difference. HBO, with a great deal to lose, has been a major proponent of a "go slow" attitude toward PPV.

In the most mature PPV systems about half of the households use pay-per-view at one time or another. The average offering draws from 2 to 4 percent of the households. Warner Amex's Qube, with the longest experience in PPV, claims that viewers shun historical themes, concerts, and children's films. Adult-oriented material, where it has been used, is very popular.[14] Some operators are insistent that movie PPV systems must be set up to match consumer viewing (or taping) convenience with a film running all day long.

A few sports programmers find markets for the full package of games on a per-channel basis among some subscribers and for individual games among other subscribers. For example, a sports channel may make a package of 30 games available through the season at $10 per month, but will also offer individual games at $4.50.

Technology

There are two principal approaches to handling PPV. One is to have customer service personnel take orders by telephone and *address* the program ordered to an addressable converter. A problem arises in this method because most orders are *impulse* buys, that is, the subscriber decides at the last minute to view the program. Cable system operators are not able to take the calls fast enough, the phone lines jam, and many orders do not get through. Devices are being developed to automate the process, taking a large volume of calls and addressing the proper households in just a few seconds. An external supplier of such a system may charge from 25 to 50 cents per transaction.

Another approach is to implement the two-way capability of the cable system. The subscriber does not have to order the program, just begin viewing. The viewing of that program is noted and the program price is added to the monthly bill.

Some of the PPV impulse buying technologies involve high capital costs and operating expenses that are not yet fully understood.

Distribution

One of the reasons film studios have lacked interest in pay-per-view in the past has been the transactional costs of doing penny ante business with the various experimental standalone PPV efforts in cable. Now that addressability is reaching a critical mass, some distributors or PPV networks are ready to take advantage of economies of scale in distributing PPV by satellite. Both HBO and Showtime have been acquiring PPV rights to films. Other organizations are attempting to serve the classic middle-man function in the PPV business. It is anticipated that these networks may be able to deliver a PPV service to the operators at a better profit than a self-booked PPV channel by realizing economies in booking and distribution and obtaining access to better products. Another of the presumed advantages of a large PPV supplier would be in the ability to provide marketing materials and assistance to operators. A PPV network might also sell advertising between programs. On the other hand, standalone systems are free to experiment with programming and marketing techniques to test their own markets.

Marketing

The selling of PPV is a big task for operators with many uncertainties given the present level of marketing knowledge. Timing is important. The promotion of ordinary movies, which are principally an impulse purchase, must coincide with the date of showing. Here, program guides, program channels *(barker channels)*, and properly timed promotional spots in the unsold local advertising availabilities of cable networks are most effective. A blockbuster movie, however, may be treated somewhat differently, with more *advanced* promotion. Promotion of a major, high-ticket event, such as a boxing match or concert, must begin about six weeks before the event. This marketing involves newspaper and broadcast advertisements, bill stuffers, and publicity. It may cost upwards of a dollar per subscriber. The movie studios that are taking the greatest interest in pay-per-view are helpful in supplying operators

FIGURE 8-2. Pay-per-view fact sheet describes the deal between programmer and cable operator on a special event.

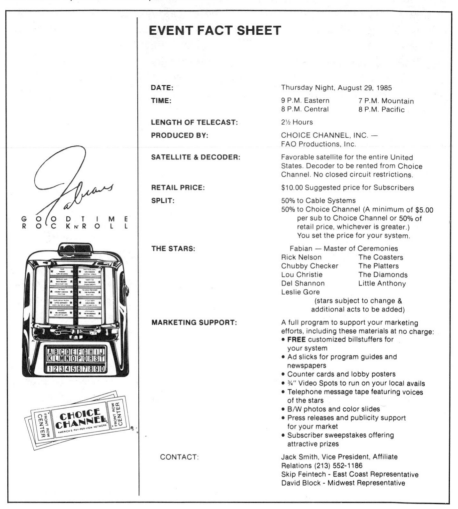

EVENT FACT SHEET

DATE:	Thursday Night, August 29, 1985
TIME:	9 P.M. Eastern 7 P.M. Mountain 8 P.M. Central 8 P.M. Pacific
LENGTH OF TELECAST:	2½ Hours
PRODUCED BY:	CHOICE CHANNEL, INC. — FAO Productions, Inc.
SATELLITE & DECODER:	Favorable satellite for the entire United States. Decoder to be rented from Choice Channel. No closed circuit restrictions.
RETAIL PRICE:	$10.00 Suggested price for Subscribers
SPLIT:	50% to Cable Systems 50% to Choice Channel (A minimum of $5.00 per sub to Choice Channel or 50% of retail price, whichever is greater.) You set the price for your system.
THE STARS:	Fabian — Master of Ceremonies Rick Nelson The Coasters Chubby Checker The Platters Lou Christie The Diamonds Del Shannon Little Anthony Leslie Gore (stars subject to change & additional acts to be added)
MARKETING SUPPORT:	A full program to support your marketing efforts, including these materials at no charge: • **FREE** customized billstuffers for your system • Ad slicks for program guides and newspapers • Counter cards and lobby posters • ¾" Video Spots to run on your local avails • Telephone message tape featuring voices of the stars • B/W photos and color slides • Press releases and publicity support for your market • Subscriber sweepstakes offering attractive prizes
CONTACT:	Jack Smith, Vice President, Affiliate Relations (213) 552-1186 Skip Feintech - East Coast Representative David Block - Midwest Representative

promotional packages of press kits, music tracks, ad slicks, promo videos, and advice.

Pricing of programs must be keyed to the presumed intensity of interest in the product, of course. Depending on their box office success, films may range from $1 to $5. A film that had only modest success in the theater run may do fairly well in PPV at a $2 price. The box office hits command much more. Sports events start at about $4 and go to $25 and are somewhat related to how often the events occur. Lowest are baseball or basketball games that are on the schedule two or three times a week. Football, which may appear weekly or every other week, may be higher priced. The once-a-year prize-fight, matching superstars, may bring $20 to $25.

In the past, both film distributors and fight promoters have wanted a guarantee from pay-per-view cable operators mainly to give assurance that the operator will aggressively market the program. The fact that these guarantees are now less frequently asked suggests that operators in PPV are becoming more skilled in marketing.

Profit

At this writing PPV has not been a significant contributor to cable profits. Revenues average less than $3 per month per subscriber in most systems. The distributor or producer gets at least half of this, leaving very little to cover marketing and administrative expenses. The companies most enthusiastic about PPV are still talking in terms of the future. There is a long chain of contingencies. If a product becomes available early and on favorable terms, if consumers see value in early access to movies and want more sports options, if addressability grows rapidly, if the technology for impulse buying comes on the market at a cost compatible with PPV earnings expectations, if new PPV networks improve the efficiency of distribution and marketing, and if marketing of PPV can bring in the critical numbers of households, *then* pay-per-view will be profitable.

NOTES

[1]For a discussion of the earlier history of pay television and pay cable, see Barbara Ruger, "Will Baltin and Pay TV: The Thirty Years' War," *CableVision*, September 22, 1980, p. 49.

[2]"The Emergence of Pay Cable Television." Volume II. *Background Data and Final Report,* prepared by Technology and Economics, Inc., for National Telecommunications and Information Administration, Washington, D.C., July 1980, p. 19.

[3]*Home Box Office v. FCC*, 567–2d 9, cert. denied 434 U.S. 829 (1977).

[4]Howard Polskin, "Inside Pay-Cable's Most Savage War," *Panorama*, March 1981, p. 57.

[5]"The Emergence of Pay Cable Television," op. cit., p. 96.

[6]"Big Rental Films of 1986," *Variety*, January 14, 1987, p. 25.

[7]*Top Gun, The Karate Kid Part II, Crocodile Dundee, Star Trek IV, Aliens, The Color Purple, Back to School, The Golden Child, Ruthless People, Out of Africa, Ferris Bueller's Day Off, Down and Out in Beverly Hills, Cobra, Legal Eagles, An American Tail, Heartbreak Ridge, Stand By Me, The Color of Money, Police Academy 3, Poltergeist II.*

[8]Barbara Ruger, "Pay Executives Weigh Value of Danger of Duplication," *CableVision*, September 22, 1980, p. 28.

[9]"What to Look Forward to on Cable," *Broadcasting*, June 10, 1985, p. 92.

[10]Ad for Escapade in *CableVision* magazine, January 26, 1981.

[11]Michael O. Wirth, Thomas F. Baldwin, and Jayne Zenaty, "Demand for Sex-Oriented Cable TV in the USA: Community Acceptance and Obscenity Law,"op. cit., *Telecommunications Policy*, December 1984, pp. 314–320.

[12]John Motavalli, "Study Says Pay Subs Most Likely to Own VCRs," *CableVision*, October 28, 1985, p. 46.

[13]Jones Intercable conducted a test in four communities offering basic and one of three packages, HBO, Cinemax, and American Movies Classics; or HBO, Showtime, Disney, and American Movie Classics; or Showtime, The Movie Channel, and American Movie Classics. Also included are remote control, printed cable guide, and rental of the VCR (the subscriber owns it after two years). The monthly cost was $44.95. Some local TV-VCR dealers were upset by the encroachment saying that cable had an unfair advantage in the VCR market. "Jones' VCR-Cable Package Test Reinforces Concern," *CableVision*, July 22/29, 1985, p. 35.

[14]Michael O. Wirth, Thomas F. Baldwin, and Jayne Zenaty, "Demand for Sex-Oriented Cable TV in the USA: Community Acceptance and Obscenity Law," op. cit., p. 318.

CHAPTER 9

Two-Way Services

Two-way impulse pay-per-view system in San Antonio, Texas.

The cable industry is experimenting with the two-way technology and its applications. Skeptics inside and out of the industry can point to some challenging technical problems and economic issues yet to be addressed fully. Few of the two-way services, mainly field trials, can be considered in themselves a *business*, the term used in the industry to label a service that has crossed the line from research and development to profit contributor.

The cable industry has been forced into experimenting with two-way services by the demands of franchising authorities and competition for franchises. Only the most committed of these companies are likely to sustain the efforts in the absence of early realization of a demand that can be met economically. But all of the promises of the earlier, more reckless days of cable development are still there. Two-way services are emerging, here and there, on a small scale and in several distinct areas.

One of the great problems in two-way cable is the need to aggregate services to justify the capital investment and operating costs in the home terminal, distribution plant, and headend hardware and software. Development of cost-effective two-way systems may also require a large population of subscribers within the service area. The demand for any one service among a relatively small number of subscribers served may be insufficient, but the demand for *several* two-way services, however infrequent or limited, over a *large body* of subscribers may support the cost. Many different services can use the same hardware on a time-sharing basis.

PRACTICAL DEFINITIONS OF TWO-WAY CABLE

Most cable systems built since 1972 have *two-way capability*. At that time the FCC required a two-way capability for all but the smallest systems. This FCC rule is no longer in effect, but equipment manufacturers maintain the two-way capability design and most cable system franchises require it. Two-way capability does not mean that the system can provide two-way services. It means *only* that duplex filters that separate the downstream (from headend into the distribution network) and the upstream (from the distribution network to the headend) signals are in place. And, since separate amplifiers are necessary for downstream and upstream signals, a space is reserved to plug in the return, upstream, amplifier in each amplifier housing.

An *operational two-way* system has the return amplifier modules in place, plus terminals in the household. Located at the headend is transmitting, receiving, and computer control equipment. The nature of the home terminal and headend equipment is dependent on the two-way services offered.

There are several types of two-way operational systems. Most common is the system in which only the *trunk line*, or a portion of the trunk line, is two-way. The trunk line is capable of video, audio, and data communication. The two-way trunk is most commonly used for televising community events, sports, school board and city council meetings from their origin, upstream to the cable system headend, and then downstream to the subscriber network on education access, government access, or LO channels.

Very few cable systems are two-way operational from the *subscriber network* (to the headend) through the home drop, feeder lines, and trunk lines. Only *data* comes upstream from the subscriber network (not video or voice as in a two-way operational trunk).

The third type of two-way cable service uses the *institutional network*, one or two entirely separate trunk lines paralleling the subscriber trunk. The extra trunk, sometimes called the *B trunk*, serves businesses and institutions. About half of these channels are upstream and the other half downstream. They are capable of carrying video, audio, and data.

To use the trunk line and institutional network, each location must have a source (data terminal, microphone, television camera), a modulator to introduce the signal to the cable, and a receiver (FM radio, television set, or a modem to interface with a computer). Most institutions have all this equipment except the modulator, which costs about $1,000 and may be portable to serve more than one location.

Programming that originates on the trunk or the institutional network may be channeled into the *subscriber network* at the headend. Thus, the institutional network and trunk are a means of feeding the entire cable system remotely, from any location on the network. Since most of the institutions are served by their own channels on the subscriber network (such as government or education access), the communication system is complete.

Within the industry the terms *two-way* and *interactive* have come to mean essentially the same thing and are used interchangeably. More narrowly, the term "interactive" is used to describe a communication system in which a signal from one direction affects the return signal from the other direction; in other words, the signals interact with each other. Not all of the two-way services and technologies described here are of this latter category.

INSTITUTIONAL NETWORK SERVICES AND MARKETS

Institutional networks were built in many of the systems franchised in the late 1970s and early 1980s. One or two institutional networks of 50 or more channels were promised in almost all of the new systems serving large urban populations. In some cases these institutional networks (*I-nets*) are installed in response to a franchise requirement for whatever public service use may develop. The cable operator may not have any immediate plans for developing business or private use applications.

Where institutional channels exist under a cable franchise, usually several of the channels in each direction have been reserved for public use at no public cost. The franchise provision will ordinarily permit commercial use of these channels until such time as public institutions exercise their option to use them.

In the past, I-nets have been largely unregulated, a situation not cheerfully accepted by the *regulated* common carriers. The 1984 Cable Act makes it clear that the FCC and the state *may* regulate "any communication service other than cable service whether offered on a common carrier or private contract basis." In response to some states that tried to eliminate cable common carrier services, the FCC subsequently ruled that state regulation of noncable services cannot be used to *prevent* their implementation. *Cable service* is defined as limited to: "the one-way transmission to subscribers of video programming, or other programming service, and subscriber interaction, if any, which is required for the selection of such video programming or other programming service." (See Appendix E.)

A number of business and private uses of institutional networks have been attempted. In some cases the primary impetus for development has come not from the cable operators, but from companies in need of communication facilities. MCI, the long distance phone company, has experimented with cable as the local connection in several cities. The Ford Motor Company has used the Dearborn, Michigan cable system as the local link in a teleconferencing network of eight locations in Dearborn, the company headquarters, and 46 field offices in the United States.

Manhattan Cable now has several miles of dedicated coaxial cable for business data. Customers have included banks, insurance companies, brokerage houses, and government services. Manhattan Cable also offers video and music services in addition to the private data communication network.

Not many cable companies have ventured into the business communication market. Perhaps justifiably, they are afraid of head-to-head competition with the telephone companies. Futhermore, it is an alien business for most cable operators. There is a somewhat different technology and certainly a different marketing and customer service function than providing basic and premium cable services to residences. Nonetheless, since many of the cable systems that might eventually enter the business data and teleconferencing market have already built an under-utilized institutional network, business communication services can be priced very favorably against competition. In some instances communication network operators such as MCI and Sprint may move into joint ventures with cable systems to develop the market and supply the necessary expertise.

Public service uses of the institutional networks will develop quite differently. Where the cable system must provide the communication channel free to public service users, the cable system has a *disincentive* to develop applications. There is no cable operator profit from the public service uses, and eventually the cable system could be able to claim the unused channels for revenue-producing business communication services.

Public service uses have developed on I-nets where aggressive public agencies have realized that cable can supplant costly telephone line service charges. The cost of connecting to the institutional network at each service point is relatively small against this saving. I-nets are used for voice lines, computer links between buildings, connecting remote monitoring devices in water, sewage and other publicly owned utilities, and traffic signal controls.

With several channels available, schools and government agencies may also find the institutional network a convenient, low-cost means of distributing live presentations or audiovisual material for in-service training.

Most institutions have a continuing need for in-service training, such as teachers, firefighters, police officers, social workers, and medical personnel. These people usually work at locations scattered throughout the city. It is costly and inconvenient to have them travel to central training or meeting facilities for all types of training. The institutional network is useful and effective for some of this training.

The network is also useful for distributing instructional materials to schools. Distribution via cable permits greater use of the materials because the time for physical delivery and pickup is eliminated. The instructional materials center maintains better control of the materials if they are handled only by staff people in a central location. Most audiovisual materials can be adapted to cable distribution—slides, films, and, of course, videotapes or cassettes.

The institutional network can also be useful for internal government agency announcements and briefings. This means of communication may be more effective than memos and newsletters and more efficient than passing information down an organizational hierarchy from one individual to another.

The institutional network may be a convenient link to the government and education access channels that reach all subscribing households in the cable system. It avoids physically delivering the tapes or other materials to the cable headend and permits live cablecasts from the institutional locations (such as city council and school board meetings).

Television teleconferencing in either public or private institutional channels is achieved by use of the institutional network to interconnect the conferencing parties. Each location must have at least minimal television studio equipment and a modulator. The most advanced systems are automated on multiple channels. Where a large number of conferees and more than one camera exist at each location, the system automatically switches to the camera that covers the microphone picking up the dominant voice. It is also possible to conduct teleconferences with a chairperson or an associate determining which location will be active at any time and switching manually to that location.

The cable system in Reading, Pennsylvania has had an operational tele-conferencing system for use, principally by the elderly, as an information resource, a means of exchanging views, and for entertainment. The project was initiated by the New York University Alternate Media Center under funding by the National Science Foundation.[1]

Teleconferencing need not be confined to a single cable system if two or more systems are interconnected by terrestrial or satellite links.

SUBSCRIBER NETWORK SERVICES AND MARKETS

The subscriber network, reaching all residential customers of the cable system, may develop several types of two-way service. These include pay-per-view television, home security, instruction, home shopping, automated utility meter reading, polling, videotex/teletext, and channel monitoring.

Pay-Per-View

Two-way is one of the means of accommodating PPV. It could be the first of the aggregated two-way services that make two-way cable commercially viable. PPV is discussed in Chapter 8.

Home Security

Several cable systems have offered alarm services. The idea of home alarm monitoring systems is appealing to homeowners and to some public safety people. The traditional security alarm business, using phone lines, has so far served mainly industry, business, and a few wealthy homeowners with a good deal to protect and the resources to do it—about 3 percent of the residential households.[2] If reasonably priced fire and burglar alarms, along with medical alerts or ambulance call buttons, could be offered to a broader base of homeowners, life and property would be protected at an efficiency previously unknown.

In general, the two-way cable alarm systems work like this. Smoke or heat detectors, and sensing devices at doors and windows, are usually hard-wired into the cable terminal although some systems are wireless. A medical alert button, probably located in a master bedroom, may also be wired into the cable. When the fire or burglar alarms are triggered, an audible alarm is heard in the household if it is occupied, and at the same time a signal is sent to alarm system operators or directly to police and fire department dispatchers. The signal to public safety officers may also include vital information about the household, called up from computer storage, such as location of handicapped persons, number of occupants, and placement of utility shutoffs. The medical alert is received by the proper medical authority or ambulance service. A medical history may be communicated along with the alert.

Home security systems are difficult for cable systems to market. A full-service installation cost may be as high as $2,000 while the monthly service fee averages around $20. This high cost puts cable security systems in competition with established alarm companies using other communication technologies and confines the market to upper income homeowners.

Cable does not now appear to have a cost advantage over telephone for home security, and it may be less reliable. If reliability and cost factors do not improve, cable alarm systems may not expand greatly beyond a few pre-wired, planned communities.

Instruction

There are many applications of cable television to education and training. In communicating, television may appear to be as efficient or more efficient than the conventional live classroom, but however stimulating the visuals and the presentation, students generally find it difficult to maintain interest and attention. Interactive instructional television has the promise of addressing this problem.

In an *automated* interactive instructional system, the learner logs in and out of the program and participates throughout it by receiving various types of feedback acknowledging the responses. This "registration" and active participation keep the learners *attentive*. Awareness that feedback will be given on each response and that an individual record of responses may be made serves to *motivate* people to create a good record. By the same token, the frequent positive feedback provides *reinforcement*, an important element in learning. If there is an opportunity to *compete* against a group, the interest value is enhanced. By making it necessary for most viewers to guess at an item, interactive items can be used to tease the learner into the next segment. *Drills* on material that must be learned by rote can be frequent and cumulative in interactive television. Because of the log-in and the availability of a computer printout tracking attendance and responses, the interactive instruction helps in the *administration* of an instructional program.

In a *live* interactive instructional system, the instructor can do everything just mentioned and, by looking at aggregated responses from the students, make instant adjustments in the instruction to rectify any difficulties observed. This is a highly desirable form of interactive instruction, but since the lesson must be created anew by the instructor for each session, it is costly. The automated interactive instruction, which sacrifices the adaptive option, makes it possible to rerun the instructional program over and over again without additional instructional effort. Both methods may have a place in education.[3]

Home Shopping

The high price of travel and the high value of time is causing consumers to supplement traditional retail distribution systems with new methods of choosing, buying and obtaining merchandise.[4]

Americans are depending more and more on mail orders and toll-free 800 phone numbers for buying. This is the fastest-growing element of retail sales. Shopping by cable is a logical extension of this trend. Two-way cable has the unique capability to demonstrate products and services, respond to inquiries, consummate the sale, transmit the invoice, and transfer funds for payment.[5] Furthermore, in *transaction services*, the vendor, not the customer, pays for access to the cable system.

A videotex system may be used by consumers to obtain product information including price. This information might be prepared "objectively" by organizations independent of manufacturers and sales agencies or by the marketer as part of the sales effort. In this kind of system, the consumer with a particular type of product in mind enters the index to find the category number and, by transmitting the number through the home terminal, calls to the home television screen all the product information in that category. When a choice has been made, the order can be placed by transmitting an item number and an authorization code (for security). Since the transmitting household code is already stored in the computer, along with a credit card number, no further information is needed at the time of purchase. One merchant has offered a buying service via cable, where the merchant seeks out the lowest price for the desired item. The inquiring cable subscriber can then take it or leave it.

Local consumer staples found in supermarkets and drugstores could be ordered through the same system and delivered on a route along with other orders from the neighborhood. This might well be efficient for households where all adult members are working or where transportation to the stores is more costly than the delivery. People without an automobile or adequate transportation might find the service desirable.

When the product is complex and has a high price tag, the full video might be important to the decision process. The seller may buy a large block of cable channel time. Some cable video shows have been used to liquidate merchandise in an auction style or sell contemporary items at a discount with orders by phone or two-way transmission. Real estate is also effectively displayed by video.

The success of home shopping networks, now relying on the telephone for ordering, may inspire a renewed interest in two-way communication.

Automated Meter Reading and Energy Conservation

Two-way cable systems can be used to read home electric, gas, and water utility meters. The difficulty in making this application of the two-way technology has been in the cost. If this were the only service of the two-way system, the cable companies would have to charge far more than the cost of persons going house to house, which may be less than $10 per year per household. Another problem is in the cost of interfacing the cable to current meters. However, there are other values of automated meter reading and connection to the cable system. A major advantage of automated metering is the immediate signal that is communicated if the meter is damaged or if theft is attempted. Large electric utilities lose millions annually through theft of service and unrecorded consumption due to faulty meters.

Energy costs and shortages have forced consideration of electric power conservation measures.[6] Some of these measures require specialized two-way communication systems for load control, rate incentives, and consumer information. Cable is one of the communication systems under consideration.

Polling

Polling has some interesting prospects. A television viewer can participate in the programming with a sense of the other viewers as he or she receives feedback on the aggregated responses. Certain kinds of television are, no doubt, improved by this option. The entertainment value of sports events is also enhanced where viewers are given the chance to predict the plays, suggest game strategies, or judge boxing matches, round by round. Some people may like to hit the gong, figuratively, in a talent show. Meetings where the home viewer enters opinions or "votes" probably are more interesting and hold attention better.

Polling is seldom representative of a meaningful group. The sample is distorted. First, it is limited to the people who have cable television and then to those who subscribe to the two-way service. Finally, the users of the two-way system in response to any program are only a small portion of that already narrow group. But public officials can probably safely assume that people in a cable audience, responding to issues raised at a public meeting, are among the most interested in the issues. And, for certain subgroups, it is possible to imagine a meeting via cable television, where polling or voting would take place among a quorum of members.

In addition to sampling problems in polling, there is the danger of inadvertent or deliberate *leading* questions that draw out a particular answer. A politician, for instance, intending to show dramatic support for certain positions, may lead out the desired response. For example, "Do you want the historic character and economic survival of the downtown area in your city destroyed by unfavorable tax policies?" It would be hard to answer "yes," whatever the merits of the tax policies.

Nonetheless, polling for public opinion and polling for entertainment enhancement have a value in overcoming passive viewing in television programming. This is not an insignificant achievement.

Teletext/Videotex

Cable has a huge capacity to provide videotex and teletext services. *Teletext*, which uses a portion of a video channel without disturbing the picture for users of the video, could be available on every single cable channel. Or a few channels of excess capacity in the cable system could be dedicated entirely to teletext communication. Although teletext is actually a one-way cable service technology, it appears to the user to be two-way. *Videotex*, the interactive text service, is two-way, and can be provided by cable for a limited number of subscribers, substituting cable for phone lines in the delivery. Videotex and teletext can be received by either a television receiver or personal computer.

A "system provider" in videotex or teletext provides a communication channel and the hardware for databases to be accessed by the terminal or a computer. The provider may itself be a source of information, but also provide "gateways" to other databases to round out the service.

The Dow Jones News Retrieval system distributed by the Associated Press/Tribune United Media Services, now serves thousands of cable system subscribers through several affiliates. It was the first operational cable videotex system. It includes the business news service, the complete text of the *Wall Street Journal* by 7 A.M., sports, and weather.

X-Press, described in Chapter 5, is a teletext service delivered exclusively through cable systems to computers, mainly in businesses. The databases are news, news features, financial market reports, sports, and weather.

Experimental videotex and teletext systems have been technically successful and interesting to consumers. It is yet to be determined, however, if the appeal is principally in the novelty. Will consumers pay the relatively high equipment and service charges over the long term? Another question is can videotex and teletext service reach down below the high income consumers to develop a broader market?

Service providers are looking for revenues, beyond the installation, service, and usage charges, in two areas. Some expect small commissions on high-volume sales from home shopping services to be a substantial source of revenue. Others expect that advertising, interspersed with the information, will provide significant revenue. By picking databases carefully, advertisers could have a highly targeted audience. Despite these revenue prospects, the two major consumer videotex business attempts, Knight-Ridder's Viewtron and Times-Mirror's Gateway, have failed.

Channel Monitoring

The capability of a two-way cable system to *monitor viewing* among its subscribers is an internal use of the two-way system that has an important application as cable systems come under pressure to provide audience data for advertisers (see Chapter 16). Of course, this system is also of value to cable programming and marketing people. The programmers know which programs and channels are most attractive to subscribers and which, if underutilized, need to be either better promoted or dropped. The marketing people learn the strengths of their product from which they can make general appeals.

CONCLUSION

The cable industry is cautiously developing the two-way services described here. Some of them are unique. Still, the utility of most services is probably marginal for most consumers. In economic terms, there are close substitutes for most of the two-way services. As a result each one of the services discussed attracts only a small percentage of the cable subscribers. After high start-up costs, there are major administration and marketing expenses.

Furthermore, in two-way service, cable is up against some much bigger industries, such as AT&T, with a tradition of large-scale research and development. These competitors have the resources to subsidize development over a long period and survive some failures. On the other hand, cable has its broadband capacity. Alone among communication industries, it can provide all of the two-way services described here. If each of these services were to prove effective and find a consumer market at about the same time, then the entire cable industry could quickly aggregate these services and build the administrative and marketing structure to fully exploit the technology.

NOTES

[1]Mitchell Moss, ed. *Two-Way Cable Television: An Evaluation of Community Uses in Reading,* Pennsylvania, April 1978.

[2]Brad Metz, "Cable Home Security Battles Restraints, Criticisms," *Cable Age,* June 29, 1981, p. 10.

[3]Studies of these approaches to interactive instruction are reported in: Thomas Baldwin, Bradley Greenberg, Martin Block, and Nick Stoyanoff, "Cognitive and Affective Outcomes of a Telecommunication Interaction System: The MSU/Rockford Two-

Way Cable Project," *Journal of Communication*, Spring 1978, pp. 180–194; William A. Lucas, "Spartanburg, S.C.: Testing the Effectiveness of Video, Voice and Data Feedback," *Journal of Communication*, Spring 1978, pp. 168–179; and Maxine Holmes Jones, *See, Hear, Interact: Beginning Developments in Two-Way Television*, Metuchen, N.J. The Scarecrow Press, Inc., 1985.

[4]Edward L. Niner, "Marketing in the 80's," address to the 1980 Annual Meeting, Food Distribution Research Society, October 14, 1980, p. 3.

[5]Ibid., p. 7.

[6]This section is based on a series of papers: Thomas Baldwin, Martin Block, and Robert Yadon, "Application of Two-Way Broadband Cable Communication Technology to Energy Conservation for Consumers and Electric Utilities," *Proceedings of the International Conference on Energy Use Management*, Los Angeles, California, 1979; Frank Whitney and Thomas Baldwin, "Using CATV for Load Research, Automatic Meter Reading and Load Management," Rate Committee Workshop, American Public Power Association, Washington, D.C., January 23, 1978; T.F. Baldwin, "Two-Way Broadband Cable Communication and Power System Load Management," report to the Load Management Roundtable, Engineering and Operations Workshop, American Public Power Association, San Francisco, California, March 2, 1978.

CHAPTER **10**

The Cable Subscriber

"Hours of TV Usage per Week by Household Characteristics (Monday-Sunday, 24-hour total)," A.C. Nielsen Company.

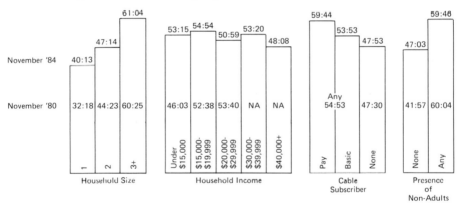

Knowledge of consumer interests and behavior is important to programming cable channels, marketing the services, and, more generally, assessing the social impact of cable communication. It is difficult to generalize about cable subscribers because the function of cable for its users is always determined by the television environment of a particular community. Cable television has different values for an individual in Liberal, Kansas, where the nearest television station is 60 miles away, than for an individual in Los Angeles where there are 21 stations in the market.

In communities with few good-quality broadcast signals available, almost all television households are cable subscribers. As the number of available, clear broadcast television signals *increases*, the number of cable subscribers *decreases*. In a community served fully by broadcast television—three network affiliates, one or more independents, and a PBS affiliate—cable television penetration hovers around 40 or 50 percent. In these places, subscription to cable television is clearly optional; a substantial proportion of the television households are electing not to subscribe.

SOURCES

Information on the size and demographic composition of cable audiences is supplied in individual market studies by Nielsen and other companies. Nielsen also surveys national cable network audiences monthly if the network is accessible to at least 15 percent of the U.S. television households (about 13 million cable subscribing households). The cable network research departments continually sample their audiences to determine usage, trends, satisfaction, and demographic characteristics.

DEMOGRAPHICS

Cable households are younger than the average noncable households, better educated, and with higher household income. Cable subscribers are more likely to have three or more persons in the household. These characteristics of all cable households are even more prominent in pay cable households. The pay cable households are still younger, better educated, and higher income, and they have more people than the basic cable subscribers.[1] Despite these general comparative characteristics of the body of cable and noncable households, it is important to bear in mind that there are many thousands, even millions, of cable subscribers in *all* of the major demographic categories: the older age groups, single and two-person households, as well as households with lower income and educational level.

DAILY CONSUMPTION

There are over 87 million television households in the United States. Among these television households, those without cable use television almost seven hours per day. *Basic* cable households use cable one hour more per day, nearly eight hours. The *pay* cable households use cable about three-quarters of an hour more than the basic cable households, over eight hours (Table 10–1). It should be noted that these data pertain to *household* use, not *individual* use of television. The average individual (men, women, and children 2 + , cable and noncable) watches four hours and 29 minutes daily.[2] Since there are more people in basic cable households than noncable and still more people in pay households than basic, these numbers may account for some of the additional viewing.

TABLE 10-1. Average Television Usage by Household.

	Nonsubscribers	Basic Cable	Pay Cable
Daily*	6:45	7:48	8:33
Weekly	47:17	54:35	59:51

*Averaged over 7 days. Usage higher on weekends, lower on weekdays.

Source: "1987 Nielsen Report on Television," A.C. Nielsen Company, 1987.

Viewing Shares

Broadcast network affiliate shares of the television audience are quite different among cable and noncable households. Over an average 24-hour day, broadcast network affiliates capture 79 percent of the *noncable* households. In *cable* households, the network affiliates get only 56 percent of all cable households viewing television, and independent broadcast stations (unaffiliated with a network) get 12 percent, totaling 68 percent for broadcasting. The advertiser-supported cable channels get a 24-percent share and the pay networks 10 percent. (In Table 10–2, totals exceed 100% because of multiset usage and rounding.) "All cable households" means all cable subscribing households, pay and basic.

TABLE 10-2. Television Viewing Shares of Services by All Cable Households.

Service	Viewing Share Total Day (24 hrs)
Cable satellite networks	15%
Cable superstations	9
Pay cable	10
Broadcast network affiliates	56
Independents	12

Source: "1987 Cable TV Facts," Cabletelevision Advertising Bureau.

Table 10–3 illustrates the viewing shares of *cable-only* services for all cable households and also for just those cable households that subscribe to pay channels. It may be seen across both categories of subscribers that cable-only services are relatively weak in prime time and relatively strong in the late fringe and on weekends. Pay cable households spend a greater share of their viewing time with cable services, as might be expected.

TABLE 10-3. Television Viewing Shares of *Cable-Only* Services by Cable Households.

Daypart	All Cable Households		
	Pay Channels (%)	Advertiser-Supported Channels (%)	Total (%)
Day: 10 A.M.–4:30 P.M.	6	24	30
Early Fringe: 4:30–7:30 P.M.	5	23	28
Prime: 8–11 P.M.	10	19	29
Late Fringe:			
11:30 P.M.–1 A.M.	16	20	36
Weekend: 1–7 P.M.	10	30	40
Total day	10	24	34
	Pay Cable Households		
Day: 10 A.M.–4:30 P.M.	11	23	34
Early fringe: 4:30–7:30 P.M.	9	24	33
Prime: 8–11 P.M.	18	17	35
Late fringe:			
11:30 P.M.–1 A.M.	25	20	45
Weekend: 1–7 P.M.	17	29	46
Total day	17	23	40

Source: "1987 Cable TV Facts," Cabletelevision Advertising Bureau.

It is important to note that television viewing shares are seasonal. Cable audience shares improve during the summer when broadcast networks are in reruns. See Table 10–4.

TABLE 10-4. Seasonal Viewing Shares in All Cable Households.

	Nov. 1985 (%)	July 1986 (%)
Ad-supported cable	22	24
Pay cable	9	13
Broadcast network affiliates	58	52
Independents	13	13

Source: "1987 Cable TV Facts," Cabletelevision Advertising Bureau.

Channel Availability

Except for the largest cities, households without cable are able to receive only a few broadcast channels and in some places none. Cable subscribers, except for those served by some old systems, are able to use many channels.

The number of channels available to cable subscribers is increasing as the older cable systems are rebuilt. Table 10–5 represents the situation in mid-1984. Nearly three-quarters of the subscribers have 22 channels or more. The subscribers with more than 22 channels are served by less than half of the cable *systems*. That is because the newer, large-capacity systems, although fewer in number, serve a great many subscribers. The most common cable systems are still 7–12 channels, but they serve only 12 percent of the cable subscribers.

TABLE 10–5. Television Channel Availability to U.S. Cable Households.

Number of Channels	Percentage of Cable Systems (%)	Percentage of Basic Subscribers (%)
0–6	0.9	0.1
7–12	35.1	11.8
13–21	21.7	14.4
22–35	29.2	48.2
36–53	9.7	16.5
54 +	3.5	9.0

Source: International Communications Research, Information Division (a subsidiary of International Thomson Business Press). In *CableVision,* November 19, 1984, p. 44. Figures as of May 31, 1984.

Not all cable households, of course, have access to all cable services. The percentages of all U.S. television households that are served by each of the cable networks are reported in Table 10–6. ESPN was available to 42 percent of the U.S. TV households in 1986, nearly all of the cable subscribers, while the Playboy channel was available in less than 1 percent.

TABLE 10–6. Percentage of all U.S. Television Households and Cable Households with Access to Cable Networks.

Service	Percentage of All U.S. TV HH	Percentage of Cable HH
WTBS-TV	42.89	95.11
ESPN	42.11	93.39
Cable News Network (CNN)	40.90	90.70
SIN Television Network	38.43	85.23
USA Network	37.66	83.52
CBN Cable Network	36.90	81.18
Music Television (MTV)	35.04	77.70
Nickelodeon	33.10	73.40
The Nashville Network	30.36	67.32
Lifetime	29.45	65.30
C-SPAN I	28.82	63.91
Financial News Network (FNN)	25.45	56.44
The Weather Channel	25.11	55.68
WGN-TV	24.40	54.10
CNN Headline News	23.94	53.09

TABLE 10-6. Continued

Service	Percentage of All U.S. TV HH	Percentage of Cable HH
Arts & Entertainment (A&E)	22.48	49.86
Home Box Office (HBO)	16.66	36.95
Video Hits One	16.09	35.69
Black Entertainment Television (BET)	15.98	35.43
PTL The Inspirational Network	14.84	32.90
Tempo Television	14.27	31.16
TelShop	11.41	25.31
Home Shopping Network I	11.16	24.75
The Discovery Channel	10.84	24.04
WOR-TV	9.89	21.94
The Learning Channel	7.99	17.72
Silent Network	7.76	17.21
Trinity Broadcasting Network	7.20	15.96
Country Music Television	7.08	15.69
National Jewish Television	6.36	14.10
Showtime	6.16	13.67
Home Shopping Network II	6.07	13.66
Cable Value Network	5.59	12.40
Eternal Word Television Network (EWTN)	5.48	12.15
ACTS Satellite Network	5.25	11.64
Cinemax	4.22	9.36
The Movie Channel	3.71	8.22
The Disney Channel	2.91	6.45
WPIX-TV	2.75	6.10
Hit Video USA	1.94	4.30
The Playboy Channel	.84	1.85

Sources: "1986-87 Universe Estimates Summary," *Arbitron Ratings: Television,* Arbitron Ratings Company, (estimate of television households, 87,614,900).

"Cable Stats," *CableVision,* November 17, 1986, p. 80.

"Cable Programming Status Report," *Broadcasting,* December 1, 1986.

"Cable Industry Growth Chart," *Cablevision,* December 1, 1986, p. 148 (estimate of cable television households, 39,510,000).

Cable Diversity

The diversity of programming available to cable subscribers is demonstrated in the study of an individual system in Columbus, Ohio by Becker and Creedon.[3] The system is typical of the urban cable operation. There are 34 channels including six broadcast stations. The data in Table 10–7 represent a week of programming in the prime time hour of 8 P.M. to 9 P.M. in 1985 comparing the programs on the six broadcast stations to the cable system of 28 cable-only channels *and* the six broadcast stations. Each unit in the table is one program hour.

Of the 26 program categories, the broadcast television service includes only 11. In this prime time hour cable has more news, comedy, drama, movies and sports and exclusively programs game shows, magazine formats, health shows, public affairs, education, local government and community news. A similar pattern of differentiation holds across other time periods analyzed: 5 A.M., 3 P.M., and 11 P.M.

TABLE 10-7. Content of 34 Cable Channels (with Broadcasting) and 6 Channels of Broadcast Programming in the 8 p.m. Time Period for One Week.

Program Type	Cable Hours	Broadcast Hours
Music	20	2
News	15	1
Discussion shows	8	6
Comedy	17	8
Westerns	3	0
Highlights/previews	1	0
Drama	28	15
Religion	3	2
Fantasy	1	0
Science	2	2
Game shows	5	0
Magazine	6	0
Documentary	3	1
Health	1	0
Movies	42	1
Weather	14	0
Outdoors	2	1
Public affairs	7	0
Sex news	2	0
Opera	1	0
Local government	4	0
Local education	3	0
Information	7	0
Community programs	5	0
Program guide	7	0
Sports	16	3
Miscellaneous fillers	15	0
Totals	238	42

Cable diversity has allowed the different tastes and interests of cable subscribers to emerge.

Table 10-8 provides an example. The top ten shows in broadcast network television, despite their general popularity among cable audiences, received substantially different ratings from cable and noncable homes. Number 10, "Magnum P.I.," was rated 21.1 in noncable households, 16.5 in all cable households, and 14.1 in pay cable households. "The Cosby Show" was 25.9 in noncable households, 29.7 in all cable households, and 29.0 in pay cable households. Cable and noncable households may be somewhat different to begin with and then also learn to use television differently as a result of the diversity of channels available. Furthermore, broadcast television networks are now deliberately programming to cable households against the more sophisticated content of the pay channels and to get a share of the upscale audiences.

TABLE 10-8. Comparative Ratings of Broadcast Network Shows in Noncable, All Cable, and Pay Cable Households.

	Program	Noncable HH Rating	All Cable HH Rating	Pay Cable HH Rating
1.	Dallas	26.4	24.4	22.3
2.	The Cosby Show	25.9	29.7	29.0
3.	Dynasty	25.2	25.1	25.7
4.	60 Minutes	25.2	23.8	20.1
5.	Simon & Simon	23.7	21.0	19.0
6.	Family Ties	23.6	26.1	28.0
7.	A Team	23.2	21.1	20.5
8.	Murder She Wrote	21.7	20.0	17.7
9.	Knots Landing	21.3	20.8	20.0
10.	Magnum P.I.	21.1	16.5	14.1
	Average	23.7	22.7	21.6

Source: A.C. Nielsen Company as reported in *Multichannel News,* November 11, 1985.

It is difficult to assess the popularity of the various cable program services. Some cable networks are carried by many more cable system affiliates and subscribers than others. The 1986 audiences for the larger cable networks are included in Appendix C. Some channels may have a very high *cumulative* audience, that, is the channel is used by a substantial proportion of the cable households over a specific period of time. Other cable services are valued intensely by a relatively few subscribers.

For the most part cable subscribers like what broadcast television viewers have always liked, although there are some cable subscribers who are attracted only by the nonbroadcast programming on cable. Most cable television viewers have a more complex set of options than broadcast television viewers. Some are only beginning to learn the breadth of what is available, partly because of inadequate programming guides and poor strategies in channel line-up. Cable television is different from broadcast television—subscribers are adapting to cable programming and cable programming services are adapting to a changing viewer.

Broadband cable broadens the functions of television for the user. The new functions are somewhat slow to emerge, and the cable household takes time to *integrate* new programs and services, that is, to first accept and then make regular or habitual use of the services.[4] Television becomes more of a utility that can be turned to for reference as well as for entertainment. This is true of the automated text channels, all-news channels, the Weather Channel, shopping services, and the like.

A five-year study by Sparkes and Kang, using panels of cable subscribers and nonsubscribers to determine changes in attitudes and behavior over time, gives the best insight into the "slow maturation of the public's understanding and use of this new communication technology."

Phase I. Cable is regarded as a simple extension of broadcast television, providing the same kinds of programming, but more of it. Those who adopt, however, can become disappointed and conclude that, as extension, cable is not worth the price. High churn in lower education and income brackets is the result.

Phase 2. Many more of cable's special channels are used and appreciated, although these channels are still closely related to the traditional fare of broadcast television. The subscriber population is now composed of persons who have either waited for these new services to develop, or who have, as a result of having subscribed to cable television, learned to value this increase in diversity.[4]

Background Use

Conventional broadcast television is used by many people as background while they are in and out of the room where the television set is on or as accompaniment for other activities (such as conversation, dressing, meals, housework, or homework). Attention to television comes into and out of focus depending on the primacy of other activities and the relative salience of the program, moment-by-moment. Cable programming enhances the value of television as a background medium with more programs that are meaningful in very short segments, such as the all-news channel and music videos. Conversely, many people who are accustomed to the use of television as background for intermittent short periods between other activities find it difficult to view the long, uninterrupted movies of pay cable. Some of these people do not subscribe to pay cable because they "don't have time" for that kind of viewing.

Viewing Styles

Because cable has so many channels, and often remote tuning, it has created a new kind of television viewer. Heeter, who has studied viewing behavior extensively, identified a number of dimensions to cable *orientation* and then several *viewing styles.*[6] In orientation, some viewers search channels in sequence, in numerical order. Others are more purposive, having their own order or sequence. The *search repertoire* may be *elaborate*, including all or most of the channels, or *limited* to a subset of specific channels. The search may be *exhaustive*, the viewer goes through all of the channels in the search repertoire and then makes a choice, or *terminating*, the viewer stops on the first acceptable option. Of course, still others may use a printed or electronic program guide, or their own memory, and go directly to a particular program or channel.

The styles of viewing in a two-way cable system where all households had remote tuners, and could be monitored minute-by-minute, were referred to by Heeter et al. as strings, ministretches, and stretches. *Strings* were defined as channel sampling, where channel changes are relatively frequent and no channel is viewed for more than four minutes. More than a half hour of the viewing day for most people was spent in string viewing. In a few homes almost all of the viewing was of the hopping around, string variety. Some homes make several hundred channel changes in a day.

Ministretches were viewing periods of between 5 and 14 minutes—more than a fleeting interest but not a real commitment to the program. Nearly an hour of the average household viewing day is viewing in ministretches averaging eight minutes. Certain of the cable channels lend themselves to watching in ministretches. For example, 91 percent of the changes involving MTV were in the short viewing category.

Viewing in periods longer than 15 minutes on a single channel was called *stretches*. This kind of viewing took up about three-quarters of the average household's viewing time, with the average stretch lasting 57 minutes.

So viewers in almost all households spent more than a half-hour flipping through channels with only a couple of minutes devoted to any given channel before switching. On average, an hour of the viewing day was given to eight-minute blocks of viewing. The rest of the time was in rather settled viewing of almost an hour at a time with a given channel. It is also clear from the data that some households may spend almost all of their viewing time in a rather manic pattern of strings and ministretches.

It was found that most of the channel changing behavior occurred on the hour and half-hour, conforming to television program schedules. But the researchers state that "channel changing is by no means confined to the time between shows." Throughout a one-hour cycle, changes occur regularly.

By channel type, the commercial broadcast networks got most of the cable viewing time and most of the channel changes. About six channel changes involved the pay movie channels, showing that people tune in and out of pay channels quite frequently. Pay movie viewing time averaged only one hour, not enough time to view an entire movie of 90 minutes or more.

Channel Use

Cable subscribers have more channels available than they use. Research on channel use indicates that, in the larger systems, viewers use only about half the channels available to them.[7] In part, this may be because people simply don't know the channels. In the large capacity systems, subscribers cannot name anywhere near all the channels available.[8] Increasing knowledge of channel offerings and consequently channel use are two major goals of cable system retention marketing. The idea is to get subscribers to take full advantage of cable, thereby making the service more valuable.

Guides

Since cable subscribers have more channels available, and use more channels than nonsubscribers, some have a problem of knowing what is available. As already noted, cable subscribers are likely to go through orientation searches (channel switching) to see what is on. They also rely on printed program guides, more than noncable subscribers before and during viewing. Pay subscribers are even more likely to use a guide than basic subscribers. The single most frequent reference to guides is to aid movie viewing decisions.[9]

Audience Diversion from Local Stations

The whole broadcast system in the United States is grounded in the concept of *local* service. Broadcast stations were originally assigned to as many different communities as technically practical. The bigger cities got multiple station assignments. The FCC has always protected a certain amount of the broadcast day from network domination and, at license renewal time, looked for evidence of local programming. Cable has the potential to destroy the effects of this carefully constructed *localism* policy by diverting audiences away from local broadcast stations to distant broadcast stations or to cable networks.

These suspicions have been confirmed by academic researchers and audience rating firms. Noncable subscribers spend almost all of their time with local stations, with only a very small fraction of the viewing time given to stations on the fringes of the market. Cable subscribers in the bigger markets with plenty of broadcast TV spend as little as 60 percent of their time with the local stations. In smaller markets, without a full complement of broadcast stations, cable subscribers may spend less than half their time with local stations.[10]

So cable *does* divert audiences from local broadcast stations. The smaller the market the greater the diversion. This may have some social significance in terms of the commonality of media experience as discussed in Chapter 18.

Diversion from local stations by cable may have an adverse effect on the business of broadcasting and could ultimately pull the rug out from under local information programming in smaller communities, but so far it does not seem to have affected local broadcast *news* audiences. The local broadcast news programs maintain their audiences with cable subscribers despite the entertainment programs and the substantial 24-hour national news and information programming offered as alternatives to cable subscribers.[11] According to Becker et al. "cable has been added onto the pattern of normal news media use rather than replaced it."[12] One study of a small town with two local broadcast stations found that of the subscribers who watched local news, nearly one-third chose local news originating from a distant city. These people, however, may *never* have watched television news from their own community.[13]

Another concern is that cable diverts audience from public television stations. There is a debate within public television about the net effect of cable. Cable does bring public television to many households that would not otherwise be able to receive it. Furthermore, cable converters put all channels on an equal footing; cable wipes out the *UHF handicap*, the handicap created by the need to switch over to the UHF dial and another antenna for UHF reception. But cable, particularly pay cable, also diverts audience from public television. Public television viewing shares are lower among pay cable subscribers.[14]

News Viewing

Cable households can have well over 200 hours per day of news and information available, counting information-oriented text channels. One study monitored the viewing of cable households, via a two-way system, to determine the extent of use of these news services. One-fifth of the cable household viewing time was given to news and information, a total of one hour and 20 minutes. Twenty-one of these news-viewing minutes was to cable-only channels. Since nearly half of the cable households did not use cable information channels at all, about half of the subscribing households were viewing cable-only news and information programming 40 minutes daily.

Surprisingly, 80 percent of the cable households ignored the all-news channels (CNN and a headline service). But the remaining 20 percent were heavy viewers, using these channels an average of 45 minutes daily. News viewing on cable channels is greatest in periods that are *not* traditionally broadcast news periods. Thus people did seem to be taking advantage of the 24-hour news availability on cable.[15]

Customer Satisfaction with Cable

Research has indicated—and "churn" figures for all systems confirm—that people do not always find what they want on cable. Cable subscribers complain of repeats, not enough quality programs on pay channels, and the failure of the content of basic programming to stimulate interest. But, surprisingly, according to a National Cable Television Association commissioned survey, only 25 percent of the disgruntled subscribers who quit cable do so for programming reasons. They are more likely to criticize the salesperson or the repair service.[16] A survey asking a national sample of 6,000 families how they rated the value they got for their money on a cross-section of consumer items resulted in a 41-percent response indicating "poor value" for cable. Cable ranked with used cars, health insurance, movies at theaters, and dentist's fees.[17]

Children and Cable

Because television is so important to children and because cable presents a different television environment for children than the heavily studied broadcast environment, researchers are beginning to look at the differences between children's television viewing behavior and other characteristics in cable and noncable homes.

The most comprehensive study to date compares cable and noncable children from the fifth grade and high school in Plymouth, Michigan with a full-service, 44-channel cable system.[18]

The cable children spent just as much time watching the eight available broadcast channels as their counterparts without cable, about three hours per day for fifth graders and two hours per day for high school people. But the cable fifth graders and high school students watched the cable channels too, adding to the total TV viewing time. The cable children could identify 20 channels on the system and said they regularly watched 12 of them. The children without cable watched six channels.

Cable children seemed to view in a less purposive fashion. They were less likely to check television guides or know what they wanted to watch before they sat down to television. The cable children were much more likely to continuously reevaluate what they were watching, checking other channels frequently and not watching whole shows. The cable children quite often would watch two programs at a time, switching back and forth. Although the fifth-grade cable children were no more likely to zap commercials than the noncable children, the high school cable people zap commercials significantly more often than their counterparts without cable.

There are fewer television viewing rules for cable children according to their own reports and the reports of their parents. The cable parents are less likely to recommend shows for children to watch or watch television with them. Cable children talk less with their parents while they watch television. Despite the presence of R-rated films in some cable homes, cable children are less often told not to watch some shows. They are not as frequently told they watch too much television. They can stay up later on school nights to watch television. As would be expected, the high school cable children have even less parental mediation than the cable fifth graders.

These are some basic facts about today's cable households. As cable grows as an advertising medium, it will generate more ratings and audience research from A.C. Nielsen, Arbitron, and others. As cable grows in social influence, it will generate more academic research. Between the two a better picture of cable consumption and its functions for users should emerge.

NOTES

[1]_____, "Cable Subscribers are Upscale," *Cable TV Facts, 1986,* Cabletelevision Advertising Bureau, p. 16.

[2]"1986 Nielsen Report on Television," A.C. Nielsen Company, 1986, p. 8.

[3]Lee B. Becker and Pamela J. Creedon, "Kabelfernsehen in den Vereinigten Staaten: Nutzung und Inhalte," *Rundfunk und Fernsehen,* 34(3), pp. 387–397.

[4]Dean M. Krugman, "Evaluating the Audiences of the News Media," *Journal of Advertising*, Vol. 14, No. 4, 1985, pp. 21–27.

[5]Vernone M. Sparkes and NamJun Kang, "Public Reactions to Cable Television: Time in the Diffusion Process," *Journal of Broadcasting & Electronic Media*, Spring 1986, p. 227.

[6]Carrie Heeter, Dave D'Alessio, Bradley Greenberg, and D. Stevens McVoy, "Cable and Viewing Styles," in Greenberg and Heeter, eds, *Cableviewing*, Ablex (in press).

Carrie Heeter, "Program Selection with Abundance of Choice: A Process Model," *Human Communication Research*, Fall 1985, pp. 126–152.

Bradley S. Greenberg, Roger Srigley, Thomas F. Baldwin, and Carrie Heeter, "Subscriber Responses to a System-Specific Cable Television Guide," Michigan State University, June 1985.

[7]Bradley S. Greenberg, Roger Srigley, Thomas F. Baldwin, and Carrie Heeter, "Subscriber Responses to a System-Specific Cable Television Guide," Continental Cablevision, Lansing, Michigan, June 1985.

[8]Carrie Heeter, "Program Selection with Abundance of Choice: A Process Model," Ibid.

[9]Bradley S. Greenberg, ibid. and Cecilia Capuzzi, "TCI 'Cabletime' Study: Subs Use Their Guides," *CableVision*, November 26, 1984, p. 40.

[10]Stuart J. Kaplan, "The Impact of Cable Television Services on the Use of Competing Media," *Journal of Broadcasting*, Spring 1978, pp. 155–165.

James G. Webster, "The Impact of Cable on Pay Cable Television on Local Station Audiences," *Journal of Broadcasting*, Spring 1983, pp. 119–126.

"The Changing Audience of the '80s," address by Dwight M. Cosner, A.C. Nielsen Company, to BEA Faculty/Industry Programming Seminar, Washington, D.C., November 7, 1980.

Gerald L. Grotta and Doug Newsom," How Does Cable Television in the Home Relate to Other Media Use Patterns?" *Journalism Quarterly*, Winter 1983, pp. 588–591.

[11]Joey Reagan, "Effects of Cable Television on News Use," University of Michigan (mimeo), August 1982.

Thomas F. Baldwin, Carrie Heeter, Kwadwo Anokwa and Cynthia Stanley, "News and the Cable Subscriber," in Greenberg and Heeter, eds., *Cableviewing*, Ablex (in press).

Morrie Gelman, "Cable News Fails to Harm Local Newscast Viewership," *Electronic Media*, October 6, 1983, p. 18.

[12]Lee B. Becker, Sharon Dunwoody and Sheizaf Rafaeli, "Cable's Impact on Use of Other News Media," *Journal of Broadcasting*, Spring 1983, p. 139.

[13]David B. Hill and James A. Dyer, "Extent of Diversion to Newscasts from Distant Stations by Cable Viewers," *Journalism Quarterly*, Winter 1981, pp. 552–555.

[14]James G. Webster and Donald E. Agostino, "Cable and Pay Cable Subscribers' Viewing of Public Television," Broadcast Research Center, Ohio University, April 1982.

[15]Baldwin, et al., Ibid.

[16]"A Summary of the 1984 ICR/NCTA Cable Satisfaction Study," International Communications Research, June 1984.

[17]_____, "Bottom Line," *Broadcasting*, December 16, 1985, p. 100.

[18]Bradley S. Greenberg, Thomas F. Baldwin, Carrie Heeter, David Atkin, Ronald Paugh, Michele Siemicki, Tom Birk, and Roger Srigley, "When Cable Television Comes Home," Michigan State University, July 1985.

CHAPTER **11**

Federal Policy

Federal Communications Commission Meeting. Photography courtesy of FCC.

Public policy governing cable television is formulated at several levels. The Congress sets communication policy through legislation and oversight of the Federal Communications Commission. The FCC acts for the Congress in administering federal policy. The FCC holds jurisdiction over microwave relays and satellite transponders used to bring distant signals to cable systems and over many years has established rules for the operation of cable systems. The executive branch of the federal government is represented by the National Telecommunications and Information Administration in the Department of Commerce. The White House may also have an advisor on communication matters. Local governments franchise cable systems because they use public rights-of-way in wiring a community. State governments may assume authority over cable in some matters and grant the franchising authority to local governments. The courts have frequently had to step into conflicts created by the development of this new medium in its relationships with government and competitive media.

Two subcommittees of the Congress concern themselves with cable: the Subcommittee on Telecommunications, Consumer Protection, and Finance of the House Committee on Energy and Commerce; and the Subcommittee on Communications of the Senate Committee on Commerce, Science, and Transportation. Communication policy resides in the Commerce Committees of both houses because communication is considered interstate and foreign commerce. Both subcommittees are staffed with specialists in communication.

The FCC is an agency of the Congress assigned to carry out the broad policies of the Communication Act of 1934, lending its accumulated technical expertise to the task.[1] Congressional input to the FCC is also made through oversight—the review of FCC actions and plans. The oversight hearings before each subcommittee, at least once per year, instruct the commission on Congressional intent and provide reaction to the Commission plans. One of the difficulties of this procedure is the mixed messages to the FCC that come from various personal and political interests of committee members and the President. Until the 1984 Cable Act, the FCC did not have direct guidance on cable matters in the form of legislation. It relied on a sense of what was expected by Congress, its own judgment, and decisions of the federal courts.

The FCC is composed of five commissioners appointed by the president with the approval of the Senate. No more than three may be members of the same political party. The commission administers (that is, enforces rules), legislates (sets technical standards, for example), and adjudicates (such as hearing a petition of a cable system to waive rules). Cable matters are handled in the Cable Branch of the Mass Media Bureau, one of several divisions of the FCC.

Through most of its history, the FCC has been staffed by attorneys and engineers. Only recently have economists been employed in greater numbers in influential positions. In an era of *deregulation*, there has been an attempt to structure the communications industries so that workable competition in the market substitutes for regulation. This trend has been abetted by new communication technologies, including cable, which have offered alternatives and have forced new thinking.

In this chapter, we discuss federal policy as it has developed in cable television and the contribution of the several interests that have shaped policy. Rather than a detailed history of government regulation in cable, much of which is not now applicable, we identify the main currents for purposes of perspective and concentrate on present policies. Emphasis is on the 1984 Cable Act, the First Amendment, and copyright.

Certain aspects of federal policy are covered more appropriately in other sections of this book. FCC technical performance requirements are in Chapter 3. FCC rules relating to local origination and the federal legislation on access channels are described in Chapter 6. The law on common carrier services is discussed in Chapter 9. Details of the functions of the franchising authorities, as defined by the federal law, are in Chapter 12. The federal limitations on cable system ownership are in Chapter 13. The law and rules on equal employment opportunity are covered in Chapter 14.

REGULATORY HISTORY

From the beginning, local governments granted franchises, licenses, and permits to community antenna television systems (CATV), although many systems emerged without any government authorization or attention. A few governments, themselves, organized or actually built CATV systems in the absence of entrepreneurial interest. It seemed purely a local matter with no significant bearing on television broadcasting. Most of the early systems were outside the defined television markets, and broadcasters were happy to have their signals extended. In 1959 the FCC ruled that it did not have jurisdiction over CATV. A Senate bill authorizing the FCC to license CATV systems failed.

Some of the cable systems were bringing in distant TV signals by microwave relay. The FCC had always licensed microwave relay systems and issued rules in 1965 governing certain functions of microwave relay used by CATV systems.[2] By this time, in addition to serving places without any broadcast TV, CATV was moving into communities with one or more broadcast stations in the vicinity. The broadcast industry became alarmed. Broadcasters felt, by bringing several more stations into a television market, audiences would be diluted and the price of advertising, held up by the scarcity of channels, would fall. In the smallest towns, barely able to sustain a single station by advertising sales, the stations might be so injured economically as to be forced off the air, leaving those communities with no local programming at all. This concern was expressed as early as 1959 when the FCC heard arguments by a television station protesting the award of microwave transmission licenses to bring distant television stations into its market.[3]

Furthermore, the FCC had just gotten fully behind UHF broadcasting. All TV sets built after 1964, by statute, would have built-in UHF-receiving capability.[4] The FCC at that time was also trying a "deintermixture" policy which would make some markets all UHF and some all VHF. The agency was not committed to UHF as a solution to the television station scarcity problem, but none of the UHF stations was financially healthy at this time. If audiences were siphoned off by cable in the bigger markets where UHF was beginning to gain a foothold, then the whole UHF policy might falter.

In 1966 the FCC extended its authority over all CATV systems, arguing that CATV, which at that time was only a community antenna service for delivering off-air broadcast signals, was "ancillary" to broadcasting. The FCC must therefore have jurisdiction over services derived from broadcasting and affecting the broadcast service. Anxious to protect the UHF stations, the FCC prohibited CATV in the top 100 broadcast markets if a UHF station, or even some group intending to build a UHF station, objected to CATV.[5] The Supreme Court upheld the FCC against several challenges of its authority to regulate cable as long as its regulation was "reasonably ancillary" to its regulation of television broadcasting.[6] Actually, it was suspected by economists at the time that cable would *help* UHF stations by equalizing signal quality and dial position (UHF stations were converted by the CATV system to a channel between 2 and 13) and extending the reach of UHF stations into distant communities. The real beneficiaries of the FCC policy were to be the big city VHF stations, and indeed the Association of Maximum Service Telecasters, representing these stations, was strongly behind the FCC position. The policies probably did delay cable competition with broadcasting in these metro areas.

In 1968 the FCC began to develop a comprehensive set of cable rules. Three years later an outline of the proposed rules was presented to Congress in a "letter of intent." The rules in the letter were modified on the basis of a consensus agreement, on distant signal importation and signal exclusivity, among broadcasters, the cable industry, and program producers. The rules went into effect March 31, 1972.[7] The 1972 rules required cable systems to obtain an FCC Certificate of Compliance and set out franchising standards, limited carriage of distant signals and required carriage of local signals, protected program exclusivity, required program origination and applied broadcast-type rules, required access channels, required 20 channels and two-way capacity, limited cross-ownership, and set technical standards. The rules were so detailed and complex that cable operators, franchising authorities, the public, and the FCC itself had difficulty comprehending them. After some experience with the rules, the FCC issued a "reconsideration"[8] and a "clarification."[9]

The FCC's feared "economic impact" of cable on local broadcasting service never did materialize. Cable did not enter many UHF markets in the 1960s and 1970s because it did not have a service to offer. These markets already had a full complement of stations available off-air with household antennas.

The small stations survived the competition with cable. Of 12 stations that petitioned the Congress for protection against cable in 1958, all were still in business in 1970.[10]

Under the growing political power of the cable industry, the awareness of consumer demand, better knowledge of economic impacts, and the deregulatory climate in Washington, in 1980 the FCC lifted the limits on distant signal carriage and syndicated program exclusivity.[11]

The FCC policy on program origination has a similar history. At first, looking at CATV as ancillary to broadcasting, the FCC was opposed to cable origination. Later origination seemed like a good idea, following out of the FCC's sustained interest in local program service. The Commission also realized that cable was earning the fruits of broadcast efforts without any of the obligations for service. So the 1972 cable rules required program origination by systems of over 3,500 subscribers, those thought to be big enough to support a studio and its operating expense. This rule was challenged in the federal courts on the grounds that it went beyond the FCC's authority. The FCC was upheld by the Supreme Court,[12] but by the time of the final verdict the Commission had decided to abandon the rule anyway.

With the development of satellite service for cable systems, it became clear that cable was no longer entirely ancillary to broadcasting. And, in the bigger cities, at least, with plenty of television available off-air, cable did not have a monopoly on the provision of television service requiring regulation to protect the consumer. Therefore, the federal and local policies that grew up under the early conditions of cable service did not fit the new industry. Federal legislation, the Cable Communications Policy Act of 1984, was written to accommodate the changes and clarify the relationships between the federal and local agencies in cable regulation.

THE 1984 CABLE ACT

The complete text of the 1984 Cable Act is in Appendix E. In this section we will describe its major provisions not covered elsewhere. The Act is a set of amendments to the Communications Act of 1934.

Access

The federal bill *permits* public, education, and government access channels. That is, the franchising authority may require provision of access *services, facilities,* and *equipment* either in the original franchise or at franchise renewal. Channel capacity on an institutional network may also be designated for public, educational, and governmental use.

FIGURE 11-1. Senator Barry Goldwater, credited with authorship of the 1984 Cable Act.

Commercial Use

All cable systems with 36 or more of the channels that can be made avail able to subscribers must make at least 10 percent of the channels available for "commercial use by persons unaffiliated with the operator." The more channels, the higher the percentage. The object is to assure the public of the widest possible diversity of information sources. It is intended to break the franchise monopoly on access to *cable* homes. "Commercial use"(leased access) means that the user, commercial or noncommercial, and the cable operator enter into a commercial agreement for the channel time. It could well be for a non-profit purpose, such as the public schools paying for an additional channel to supplement educational access channels. Commercial use in this context is for video programming not two-way or data communication.

The cable operators are protected from very direct competition on their own channels by the stipulation that the commercial use channels may be priced at a sufficient level to assure that the channels will not "adversely affect the operation, financial condition, or market development of the cable system." On the other hand, the price for commercial use channels must be "reasonable." This provision cannot be used to defeat the *diversity of source* purpose of the leased channels. The prospective user who finds the price unreasonable may have recourse to the federal courts. This avoids the need for rate regulation on these channels. The courts will serve this function but *only* if a dispute arises. Otherwise the price the commercial user pays for the channel use is negotiated between the user and the cable operator.

The cable operator is not to have any editorial control over the content of commercial use channels or even to consider the content except as necessary to assess a reasonable price. The franchising authority is given the responsibility of preventing a service that in their judgment is "obscene, or is in conflict with community standards in that it is lewd, lascivious, filthy, or indecent or is otherwise unprotected by the Constitution of the United States." We will return to this point in the next section on cable and the First Amendment.

It is too early to determine if or how these channels will be used. The provision seems to permit the cable operator to fully recover, by pricing, any loss to a competitive service. The losses might be in channel subscriptions where a "commercial use" channel cannibalizes system subscribers, or in advertising revenues, when a commercial use channel diverts audience from the cable system channels.

Regulation of Services

An important section of the bill attempts to take the franchise authority out of cable programming. In the request for proposal, the franchising authority cannot set "specific program services" or "broad categories of video programming or other information services" as a condition for obtaining a franchise. If the successful applicant for a new franchise or a renewal *offers* to provide broad categories of video programming or other services, the franchising authority can then build that into the franchise agreement. For example, the city's request for proposal cannot specify that Nickelodeon (specific children's program service) or children's programming (broad category) be provided. If the successful applicant *offers* Nickelodeon, the franchising authority cannot accept the offer. But, if the applicant offers children's programming, the franchising authority can accept it and later make sure that the operator provides some form of children's programming.

Programming services, facilities, and equipment required prior to December 29, 1984 are *grandfathered*. They may continue. Facilities and equipment in new and renewal franchises for a cable system must be *cable-related*. This would prevent future requirements of franchisees such as tree plantings, drug education programs, and capital improvements to libraries as they have in the past. This provision may allow states to preclude new franchises from requiring institutional networks since a cable system is defined as providing *video programming* to *multiple subscribers*.

The franchising authority and a cable operator may specify in the franchise or the renewal that services which are obscene, or otherwise unprotected by the Constitution, *not* be provided. The cable operator must provide by sale or lease, a device by which the subscriber can prohibit viewing of a particular service. This device is generally called a *lock box*. It is described in Chapter 4.

Privacy

At the initiation of service and at least every year thereafter, operators are to provide *notice* in the form of a separate, written statement informing the subscriber of:

1. the nature of personally identifiable information collected on the subscriber and the nature of its use,
2. the nature, frequency, and purpose of any disclosure that may be made of the information and to whom disclosure is made,
3. the period the information is kept,
4. the times and place at which the subscriber may have access to such information, and
5. the right of the subscriber to enforce some limitations on the use of information including the right to correct errors.

The notice to the subscriber about personally identifiable information is recirculated annually as a reminder.

The operator may not *collect* personally identifiable information without first obtaining written or electronic (in the case of a two-way system) permission, except when the information is necessary to offer the service (such as pay-per-view) and to detect unauthorized reception of cable communications. Cable operators may not *disclose* personally identifiable information without prior written or electronic permission except as necessary to conduct the business and provide the service, or under a court order. The operator may disclose names and addresses of subscribers to any cable service if the subscriber has had an opportunity to prohibit or limit the disclosure. For example, a cable company might give subscriber addresses to the Disney Channel for a direct mail piece. It is the subscriber's responsibility to exercise the option to limit this form of disclosure after having read the notice.

The cable operator may collect and use *aggregated* data which does not identify particular persons. This clause is important to operators who may desire to gather aggregated data on program audiences for advertising sales or programming purposes.

Privacy protection would be most important to the consumer in the circumstances where a cable system operated several two-way services that collected a significant body of data on each household. However, the provision applies to any information collected at the time the subscriber signs up, or at a later time, such a demographic description of the household for marketing purposes. It also pertains to viewing records that would be made for the purpose of billing PPV.

At this point, the privacy problem of cable companies is not as serious as the problem for telephone companies. The phone companies have access to whom and when calls are made and can actually record conversations as well as intercept data. The cable systems have access to far less information, at least for the present and near future.

Unauthorized Reception

The Cable Act makes it a federal offense to steal cable services either directly from the satellite transmissions or from the coaxial cable itself. Home dish owners and manufacturers and sellers of this equipment, for the purpose of making private use of the satellite signals in the home, are exempted as long as the signal is not scrambled or there is no marketing and licensing system for the signal (see Chapter 18). Unauthorized reception of satellite signals for other than private viewing (bars, hotels, multiple-dwelling units) is a violation.[13]

CABLE AND THE CONSTITUTION

Foundation

The Communications Act of 1934, of which the Cable Act is a part, draws on the authority given to the Congress by the Constitution to regulate interstate and foreign commerce. *Commerce* has been interpreted to include communications. As we have seen, when each cable system was a wired entity within a particular state, the authority of the Communication Act was extended by terming it "ancillary" to broadcasting, an extension of service already included in the Communication Act. Cable is now much more than ancillary to broadcasting, relying on satellite signals from out-of-state sources. A 1984 Supreme Court decision relies on the FCC's "*general authority* under the Communications Act to regulate cable television systems" (emphasis ours).[14] The 1984 amendment to the Communications Act (the Cable Act), of course, assumes this general authority.

First Amendment

The First Amendment to the Constitution states that "Congress shall make no law respecting an establishment of religion, or prohibiting the free exercise thereof; or abridging the freedom of speech, or of the press." Broadcasting and the press have been placed in different positions under this amendment. Because the scarcity of spectrum limits the number of broadcast frequencies available, the government chooses who will qualify and gives those who do a fiduciary relationship to the others who do not have licenses—the licensee must serve the public interest. To protect the consumer and preserve the consumers' right of access to information, certain rules intrude on the freedom of speech for the broadcaster. The sponsorship identification, political program restrictions, and other rules apply to origination cablecasting as well as to broadcasting so that it may be assumed that cable operators have some of the same fiduciary responsibilities as broadcasters. However, the ability of the government to directly regulate most *content*, as opposed to the rules under which that content is presented, is doubtful after *FCC v. League of Women Voters*.[15] In that case, the court declared unconstitutional the Congressional prohibition of editorials by public broadcasters who receive federal funds.

So ordinarily content is protected from government regulation. Obscenity, if it meets all three conditions of a test, is not protected. The test is severe. It is extremely difficult for a government to win a case when a defendant has a First Amendment claim. The "Miller test" in *Miller v. State of California* is a good example of the narrow grounds on which violations of the principle of free speech are permitted:

> (1) the average person applying contemporary community standards would find the work, taken as a whole, appeals to the prurient interest, (2) the work depicts or describes, in a patently offensive way sexual conduct specifically defined by the applicable state law, and (3) the work, taken as a whole, lacks serious literary, artistic, political or scientific value.[16]

It is important to note that *contemporary community standards* are to be applied, meaning that what is obscene in one community may not be in another. The sexual conduct that is not permitted must be specifically defined by state law.

In the past, broadcasting has had to meet a higher standard. The U.S. Criminal Code prohibits "obscene, indecent, or profane language by means of radio communication"[17] adding "indecent" and "profane." In *FCC v. the Pacifica Foundation*, the Supreme Court upheld the FCC position that *indecent* language, even if not obscene, may be regulated because broadcasting is "uniquely accessible to children." The case involved the broadcast of the George Carlin monologue, "Filthy Words," that couldn't be said on the public airwaves.[18] In this instance, Carlin was right. However, attempts to apply the indecency statute to cable have failed.[19] A Miami city ordinance and a state of Utah statute have been found in violation of the First Amendment.[20] The courts reasoned that the responsibility is on the viewers and parents, not the cable operator since cable households must affirmatively subscribe, and must make regular additional monthly payments for premium channels. Now under the Cable Act, subscribers may choose to lease or purchase lock boxes. The lock box is a "less restrictive alternative" than banning indecent speech.[21]

The Cable Act addresses both obscene and indecent programming. The cable operator is to have no editorial control over "commercial use" channels, but the franchising authority is to exercise its judgment to assure that content that is "obscene, or is in conflict with community standards in that it is lewd, lascivious, filthy, or indecent or otherwise unprotected by the Constitution of the United States" is not provided. This provision takes a very broad view of what constitutes unprotected speech, adding lewd, lascivious, filthy and indecent to obscenity. It does not fit the *Miller v. California* definition in its language, but is probably an attempt to put cable in a classification with broadcasting, the *Miami, Utah*, and *League of Women Voters* cases notwithstanding. The provision also seeks to *prevent* the undesired programming, an act of prior restraint that can be carried out only under very strict procedures that put the burden of proving the material is unprotected by the First Amendment on the censor, and if the material is to be restrained, the censor must go to court immediately.[22] Further, the government must show a compelling government interest, and that the interest is met *only* by the regulation, which must be drawn as narrowly as possible. This is a formula under which the government almost always loses.[23]

The section of the Cable Act on regulation of services, equipment, and facilities, referring to the cable system's own channels, permits the cable operator and the franchising authority to enter into a franchise agreement specifying that services that are obscene or otherwise unprotected by the Constitution not be provided. This simply restates other law that exempts obscenity from protection under the First Amendment.

Another section of the Act makes cable operators and programmers liable for obscenity (and libel, slander, incitement, invasions of privacy, false or misleading advertising) under federal, state, and local criminal and civil law but exempts them from responsibility for PEG access channels as discussed in Chapter 6. Finally the Act specifically makes obscenity by cable an offense punishable by not more than $10,000 and two years in jail.

The right of a franchise authority to select a particular franchisee has been challenged on First Amendment grounds. The U.S. 9th Circuit Court of Appeals *(Preferred v. Los Angeles)* ruled that a city cannot limit a franchise territory to just one franchisee when more than one operator could be accommodated on the utility poles.[24] In this context, the prospective operator is considered a First Amendment speaker and should not be denied that right any more than a person would be denied the right to publish a newspaper in the same area. The Supreme Court affirmed this decision:

> The activities in which respondent [the cable company] allegedly seeks to engage plainly implicate First Amendment interests. Through original programming or by exercising editorial discretion over which stations or programs to include in its repertoire, respondent seeks to communicate messages on a wide variety of topics and in a wide variety of formats.[25]

The Court did note that there may be circumstances under which First Amendment values must be balanced against competing societal interests and sent the case back to the District Court for resolution of disputed issues.

It was decided in the cases against the FCC's must-carry rules *(Quincy v. FCC* and *Turner v. FCC)* that forcing a cable operator to carry particular stations denied First Amendment rights. The object of requiring carriage of local broadcast stations was to protect those stations from economic injury resulting from loss of audience, which might eventually deprive a community of local broadcast service. The court determined that the government would have to make a very strong case for encroaching on First Amendment rights of the cable operator to preserve this social objective. The FCC had not proven the must-carry rules were necessary or that they were the least intrusive way of achieving the goals.[26] The FCC wrote new rules for must carry in an attempt to avoid the First Amendment violation by allowing cable systems to choose broadcast stations for carriage from among a specified pool of local stations. But in December 1987 the Court of Appeals again rejected the rules as a violation of the First Amendment.

In finding that the FCC exceeded its statutory jurisdiction in imposing the access rules on cable *(FCC v. Midwest Video)*, the Eighth Circuit Court of Appeals also concluded that the rules violated the First Amendment. Further, the Circuit Court found that the access rules violated the Fifth Amendment rights of the cable operator by taking property without due process[27] (an idea that was also noted in *Quincy v. FCC*). Upholding the circuit Court decision, voiding the federal access channel requirement, the Supreme Court relied on the statutory issue, but noted that the First Amendment question was serious.[28]

It has also been suggested that the cable franchising fee, to the extent that it exceeds the actual cost of regulation of cable, is a violation of the First Amendment. The Supreme Court, in *Minneapolis Star and Tribune Company v. Minnesota Commissioner of Revenue*, has determined that a special tax on a First Amendment speaker (in this case, a daily newspaper) is unconstitutional unless there is an overriding governmental interest. The latter would *not* include the raising of *general revenue*.[29] It is argued by Ciamporcero, Geller and Lampert, however, that the franchise fee funds in excess of the cost of regulation might be applied to development of the PEG access channels under the precept that such use *does* represent an overriding government interest as represented by the intent of the Congress in authorizing PEG channels.[30] It is now clear that the First Amendment will be a factor in determining the disposition of the franchise fee revenues.

Developments under the First Amendment in these cases, many outside the cable industry, have enlarged the free expression rights of cable system operators. This is consistent with the general approach of the Court that a free press "belongs to whoever owns one." Cable has been established as a First Amendment speaker, subject to some of the constraints of broadcasting (such as the rules governing local origination), but also with its own requirements (like the PEG access requirements), and is treated more like newspapers than broadcasters in other ways (that is, not being subject to indecency standards).[31] The 1984 Cable Act itself, which denies content control rights to local governments, is effectively a vindication of the First Amendment rights of system operators.

COPYRIGHT LAW

Background

Broadcasting and cable, of course, were not covered expressly by the 1909 Copyright Act, which was still in force at the time CATV systems began capturing and redistributing television and radio broadcast signals. Attempts by motion picture copyright owners[32] and broadcasters[33] to apply the Copyright Act to cable transmission of broadcast signals failed in the Supreme Court. According to the Court, CATV carried broadcast signals without alteration, thereby only enhancing the viewer's ability to use the signal. Technically, under the 1909 law, cable systems did not "perform" the copyrighted work.

Copyright Act of 1976

In the early 1970s, despite favorable treatment by the courts, much of the cable industry had conceded a copyright liability. The National Cable Television Association and the Motion Picture Producers Association of America worked out a compromise that was incorporated in the 1976 Copyright Act.[34] When a program copyright owner sells the rights to run a program to a broadcast television station, the program is priced to the station on the basis of a particular potential audience within that station's coverage area. If a cable operator takes that program signal off the air and for a fee offers it to subscribers outside the originating station's coverage area, the copyright owner is not compensated for the additional cable audience unless some mechanism is created to do so. That mechanism, a *compulsory license*, grants special privileges and creates responsibilities for the cable operator. The new law had several goals derived from the basic tenets of copyright law as applied to the television and cable technologies. They are stated in the purposes of the Copyright Royalty Tribunal, the body that collects royalty fees for cable system retransmissions of certain broadcast signals. The first purpose is "to maximize the availability of creative works to the public." The Copyright Act grants monopoly property rights in the production and publication of literary, musical, artistic, and dramatic works. It is assumed that these rights are necessary incentives to creators. But also, the ability to avail oneself of creative work is a public benefit that must be facilitated. Copyright law provides an administrative procedure by which original material may be shared.

The second purpose is "to afford the copyright owner a fair return for his creative work and the copyright user a fair income under existing economic conditions." The Copyright Act attempts to protect the livelihood of the copyright owner and at the same time offer fair incentives to the user of copyrighted materials (such as a cable company) so that they may be disseminated fully.

A third purpose is "to reflect the relative roles of the copyright user in the product made available to the public with respect to relative creative contribution, technological contribution, capital investment, cost, risk, and contribution to the opening of new markets for creative expression and media for their communications." This acknowledges the complexities of the modern creative process and the role and costs of communication technology in making original work available.

The final purpose is "to minimize any disruptive impact on the structure of the industries involved and on generally prevailing industry practices." The implementation of the copyright law is a delicate task. As new communication technologies emerge and royalty rates are established and adjusted, the Copyright Royalty Tribunal must be cautious not to upset or restructure communications industries to the detriment of copyright owners, disseminators, or the public.

The Compulsory License

The Copyright Act permits cable systems to obtain a compulsory license for retransmission of all television and radio signals that they carry. In return, the cable systems pay for the *distant, non-network* television programs carried. If the cable system is not licensed or does not report fully the distant, non-network programs carried, it is liable to those copyright owners. Since the blanket fee paid under the compulsory license is lower than what would be assessed by each copyright owner separately, and more convenient, it is in the best interests of the cable system to participate.

Only *distant* signals are covered because the retransmission of local signals is not expected to "injure" the copyright owner. Only *non-network programs* are covered because the network broadcasters pay for rights to a national audience. *Non-network programs* on a distant network affiliated station *are* included.

The compulsory license arrangement covers *secondary transmissions,* meaning the simultaneous retransmission of signals originating from another source. The definition is extended to include nonsimultaneous transmission in off-shore areas (such as Hawaii, Alaska, Puerto Rico) to cover the customary delayed transmission of programs due to time zone differentials. Since the compulsory license applies only to retransmission, cable systems have full liability for (the responsibility to pay for or get permission to use) secondary transmissions not intended for the public at large but directed to *controlled* groups. This would exclude from coverage by the compulsory license privilege subscription television stations, pay cable, background music services, closed-circuit broadcasts, and other programming intended solely for a *specially qualified audience* (qualified, usually, by paying for the service).

The secondary transmissions must not be altered in any way by the cable system. For example, the cable system could not substitute its own commercials for those broadcast except under specified conditions during market research. In the WGN case, the courts found that cable may not delete service such as teletext.[35]

The compulsory licensing arrangement also covers Canadian and Mexican broadcast stations that might be imported by a U.S. cable system near the borders.

Copyright Royalty Tribunal

The compulsory licensing system is administered by the Copyright Royalty Tribunal (CRT). The tribunal is composed of five commissioners appointed by the President for seven-year terms. The duties of the CRT are to (1) adjust royalty rates in accordance with inflation or deflation and changes in rates to subscribers and (2) to distribute royalty fees to claimant copyright owners after administrative costs have been deducted. Adjustments in the royalty rates are made every five years, starting in 1980. Copyright holders claiming distant, non-network, secondary transmission of their works may file individual claims with the CRT or participate in joint claims.

The computation of the royalty fee to be paid by cable systems semiannually is complex. A *statement of account* covering a six-month period is submitted. The statement lists the total number of subscribers and gross amounts paid to the system for basic service. The cable system must list all non-network programming carried from beyond the local service area. The CRT makes three distinctions in cable systems: Form 1 systems are those with semiannual gross receipts for secondary transmission of $55,500 or less. They pay a flat fee of $20 for the six months. Form 2 systems have semiannual gross receipts of more than $55,500 and less than $214,000. These systems use a formula based on a percentage of the gross receipts to determine the fee. Form 3 systems have semiannual basic gross revenues of $214,000 or more. Their fee is based on the number of *distant signal equivalents* (DSE) carried. A distant independent station is one DSE; each distant network affiliates and non-commercial station counts one-quarter DSE.

Until July 1986, gross revenue for most systems was the total of fees collected for basic services. This calculation of gross receipts has been brought into question by a District Court opinion that requires the Copyright Office to redefine "gross receipts" to exclude any revenues not attributable to broadcast signals when a tier of service carries broadcast signals along with non-broadcast services. Because almost all basic tiers include, at one price, broadcast stations as well as cable-originated nonbroadcast services, such as the cable networks, it is difficult to determine which portion of the monthly fee is for the broadcast stations carried.[36]

The Copyright Act states specifically that, if the FCC changes the signal carriage rules, then the CRT may make reasonable adjustments in the rates. When the FCC lifted its distant signal importation rule (and it was sustained by the *Malrite court* decision), the CRT imposed a flat 3.75 percent of gross basic subscriber receipts for every DSE added after the date of the Malrite decision. These are often referred to as the *post Malrite signals*.[37]

An alternative to the compulsory licensing procedure that has frequently been proposed is to eliminate the compulsory license altogether and require cable systems to obtain *retransmission consent* to carry broadcast signals. This would have the effect of placing cable systems in a position of negotiating the rights to use broadcast material with broadcasts or copyright owners. The cable industry fears both the cost and complexity of this plan. Proponents, however, argue that it would more realistically reflect the value of the signals carried; in other words, the market rather than the CRT would set the price for copyrighted works. An FCC notice of inquiry that included the retransmission consent proposal and an experiment that permitted a cable system to carry distant signals on a consent basis resulted in a rejection of the concept.[38] But in 1987 the FCC published a Notice of Inquiry which brings up the issue once again. The FCC is asking whether or not to recommend to Congress to eliminate the compulsory license.[39]

It should be noted that the compulsory licensing arrangement discussed here relates only to *secondary transmissions*. Cable system originations (fed into the system from the headend or elsewhere in the distribution system, whatever the source—original production, videotape, records, printed material, photographs, and so on) bear full copyright liability, meaning that use of copyrighted material is an infringement unless permission to use, or rights, have been negotiated with the owner. The law does permit limited *fair use* of copyrighted work for criticism, comment, news reporting, and teaching.

NOTES

[1]48 Stat. 1064, June 19, 1934.

[2]*First Report and Order in Docket Nos. 14895 and 15233*, 38 FCC 683, 4 RR 2d 1725 (1965).

[3]*Carter Mountain Transmission Corp.*, 32 FCC 459, 22 RR 193 (1962), affirmed *Carter Mountain Transmission Corp. v. FCC*, 321 F2d 359 (D.C. Cir. 1963), cert. denied 375 U.S. 951 (1963).

[4]76 Stat. 150, July 10, 1962.

[5]*Second report and Order in Docket Nos. 14895, 15233 and 15971*, 2 FCC 2d 725, 6 RR 1717 (1966).

[6]*United States v. Southwestern Cable Co.*, 392 U.S. 157 (1968).

[7]*Cable Television Report and Order*, 36 FCC 2d 143 (1972).

[8]*Memorandum Opinion and Order on Reconsideration of the Cable Television Report and Order*, 36 FCC 2d 326, 25 RR 2d 1501 (1972).

[9]*Clarification of Rules and Notice of Proposed Rule Making*, 46 FCC 2d 175, 29 RR 2d 1621 (1974).

[10]Ralph Lee Smith, *The Wired Nation* (New York: Harper & Row, 1970), p. 46.

[11]*Report and Order* adopted July 22, 1980, FCC 80–443, 45 *Fed. Reg.* 60186, September 11, 1980.

[12]*FCC v. Midwest Video Corp.*, 440 U.S. 689 (1979).

[13]For a complete analysis of the Cable Act, see Charles D. Ferris, Frank W. Lloyd, and Thomas J. Casey, *Cable Television Law: A Video Communications Practice Guide, Special Supplement, Cable Communications Policy Act of 1984, Enacted October 30, 1984* (New York: Mathew Bender, 1985). Daniel L. Brenner and Monroe E. Price, *Cable Television and Other Nonbroadcast Video* (New York: Clark Boardman Company, Ltd., 1986). For a broad view of U.S. communication law, see John R. Bittner, *Broadcast Law and Regulation*, Prentice-Hall, 1982, pp. 441. A survey of the relationships between cable companies and franchising authorities in the aftermath of the Cable Act of 1984 is found in Wenmouth Williams, Jr. and Kathleen Mahoney, "Perceived Impact of the Cable Policy Act of 1984," *Journal of Broadcasting and Electronic Media*, Spring 1987, pp. 193–205.

[14]*Capital Cities Cable, Inc. v. Crisp*, —U.S.—, 104 S. Ct. (1984).

[15]*FCC v. League of Women Voters*, 104 S. Ct. 3106.

[16]*Miller v. State of California*, 413 U.S. 15 (1973).

[17]18 U.S.C.A. 1464.

[18]*FCC v. Pacifica Foundation*, 438 U.S. 726 (1978).

[19]Indeed, the *Pacifica* case, decided by the narrowest margin, may not carry much precedental weight, particularly after *League of Women Voters*. The suggestion here is that the indecency standard may no longer apply even to broadcasting.

[20]*Cruz v. Ferre*, 755 F2d 1415 (11th Cir. 1985).

Community Television of Utah, Inc. v. Wilkinson, 611 F. Supp. 1099 (D. Utah, 1985).

HBO v. Wilkinson, 531 F. Supp. 987 (D. Utah, 1982).
Community Television of Utah v. Roy City, 555 F. Supp. 1164 (D. Utah, 1982).
Howard M. Kleinman, "Indecent Programming on Cable Television: Legal and Social Dimensions," *Journal of Broadcasting*, Summer 1986, pp. 275–294.

[21]*Carlin Communications, Inc. v. FCC*, No. 84–4086 (November 2, 1984).

[22]*Freedman v. State of Maryland*, 380 U.S. 51 (1965).

[23]*Vance v. Universal Amusement*, 445 U.S. 308 (1980).

Consolidated Edision v. Public Service Commission, 447 U.S. 530 (1980).

[24]*Preferred Communications Inc. v. City of Los Angeles*, 754 F 2d 1396 (9th Cir. 1985).

[25]*City of Los Angeles et al. v. Preferred Communications, Inc.*, No. 85–390, June 2, 1986.

[26]*Quincy Cable TV, Inc. v. FCC* (D.C. Cir. No. 83–1283).

Turner Broadcasting System, Inc. v. FCC (D.C. Cir. No. 83–2050).

[27]*FCC v. Midwest Video Corp.*, 571 F 2d 1025 (8th Cir. 1978).

[28]*FCC v. Midwest Video Corp.*, 440 U.S. 689 (1979).

[29]*Minnesota Star and Tribune Company v. Minnesota Commissioner of Revenue*, 460 U.S. 575, 75 L.Ed.2d 295, 103 S. Ct. 1365 (1983).

[30]Alan Ciamporcero, Henry Geller, and Donna Lampert, "The Cable Franchise Fee and the First Amendment," Washington Center for Public Policy Research, Duke University (mimeo, undated).

[31]See also: Edwin Diamond, Norman Sandler, and Milton Mueller, *Telecommunications in Crisis: The First Amendment, Technology, and Deregulation*, Washington, D.C.: Cato Institute, 1984, 164 pp. George H. Shapiro, Philip B. Kurland, and James P. Mercurio, *'CableSpeech': The Case for First Amendment Protection*, New York: Harcourt Brace Jovanovich, 1983. David Atkin, "The Cable Communication Act and Content Regulation: A Constitutional Collision Course," M.A. Thesis, Michigan State University, 1986. Patrick Parsons, *Cable Television and the First Amendment*, Lexington, MA: Lexington Books, 1987, 176 pp.

[32]*Fortnightly Corporation v. United Artists Television, Inc.*, 392 U.S. 390 (1968).

[33]*Teleprompter Corp. v. Columbia Broadcasting Systems, Inc.*, 415 U.S. 394 (1974).

[34]17 U.S.C.A. Sec. 1, *et seq.*

[35]*WGN Continental Broadcasting Co. v. United Video, Inc.*, 523 F Supp 403, revd, 685 F2d 218 (ND Ill 1981), rehg, 693 F2d 622 (1981).

[36]John Wolfe, "Ruling May Bring Down Cable's Copyright Costs," *Cablevision*, August 18, 1986, p. 57.

[37]*Malrite TV of New York, Inc. v. FCC*, 652 F 2d 1140 (2d Cir., 1981). Cert. denied.

[38]*Cable Television Report and Order*, 36 FCC 2d 143, 24 RR 2d 1501 (1972).

[39]"In the Matter of Compulsory Copyright License for Cable Retransmission," Notice of Inquiry, Docket 87–25, Federal Communications Commission, April 23, 1987.

CHAPTER **12**

Franchising

Public notice on cable license, City of Scottsdale, Arizona.

PUBLIC NOTICE

The City of Scottsdale, Arizona will be accepting applications for a license to construct, operate and maintain a cable communications system to serve city residents in recently annexed areas. Applications must be submitted by November 30, 1985. Copies of the formally issued Request for Proposals may be obtained by contacting:

**Mr. Doug Syfert
Manager, Office of Cable
Communications
City of Scottsdale
3939 Civic Center Plaza
Scottsdale, Arizona 86251
(602) 994-2796**

THE OWNERSHIP DECISION

Cable systems may be (1) owned and operated by the city, (2) owned by the city and operated by a cable company, (3) owned and operated by a public, nonprofit corporation, cooperative, or power company, (4) owned and operated by a private, profit-making corporation, or (5) owned and operated by a private company sharing a minority interest with the franchising authority. Therefore, prior to establishing a cable system, a community must decide the most appropriate ownership structure. Almost all cable systems are in the private, profit-making corporation category. But several systems are operated by cities or public corporations chartered to operate the system without profit. Most of these systems are in very small towns that did not initially attract private entrepreneurs.

Many larger cities (including Minneapolis, St. Paul, Cleveland and St. Louis), under quite different circumstances, have considered a publicly owned cable system. A variety of private-corporation/municipal-ownership plans have surfaced. For example, Rogers Cablesystems offered the city of St. Paul an option of buying back half the system at two-thirds its market value after five years.[1] Many cable franchises permit, and may give priority to, public or municipal acquisition at a fair price in case of revocation or nonrenewal. But none of the major cities have yet opted for public ownership.

Some features of cable suggest municipal or public ownership. Many of the newer services are *public* services or vital utilities—government, education, library, and public access channels; home security systems; automated meter reading and power-load management; and polling. A municipal system would be more likely to extend service to low-density and low-income neighborhoods. Such a system, funded by revenue bonds where permissible, would have a low debt service cost. These bonds, because of tax advantages, pay low rates of interest. Revenue bonds do not put the city or taxpayer at risk, since they may draw only on the revenues of the cable system.

On the other hand, there are some problems with municipal or public corporation ownership of a cable system. The tradition in the United States, of separating news media and other communication systems from government, is longstanding. It would be difficult to insulate a government communication system from political influence. It may also be difficult to build in management incentives for efficient and innovative operation. A greater possibility is that a municipal or publicly owned system would develop public or social services, at the expense of subscribers, that most subscribers do not want. A city ordinance for such a cable system structure would have to deal with these issues.

Since all but a handful of cable systems are franchised to privately owned, profit-making companies, the remainder of this chapter treats the franchising and regulation of such systems. Much of what is said, however, would apply to the selection of an operating company for a municipal system, for those municipal systems where that is appropriate, and to the monitoring of system performance, which would be much the same whatever the structure of ownership.

FRANCHISES

Local governments in most states grant franchises to cable systems under state-granted *home rule* powers. The franchise permits use of public land in wiring the community with aerial and buried cable. It provides a collective representation of the consumers. Franchises are usually granted "nonexclusively," but for practical purposes, under the present economics of the cable industry, the franchise has a monopoly on cable service. Only one applicant is accepted. Once a franchise is granted to a company, no other cable company would *overbuild* (build another system over the same territory) unless the system became obsolete. The reason is that there is probably not enough revenue to support the capital investment of two or more companies, and the service of one company could not be differentiated significantly from another. There are a few exceptions. In Allentown, Pennsylvania, for example, two overbuilt systems have existed for several years. In this mountain community, where there is a very high penetration of cable, the two companies split the subscribers and each has enough to survive.

The granting of cable franchises raises several issues. The possible violation of the First Amendment rights of an applicant who is denied a franchise is discussed in Chapter 11. There is also a possibility of municipal liability for antitrust violations in granting exclusive franchises or in selecting a single "nonexclusive" franchise. The Sherman Act prohibits unreasonable contracts, combinations, and conspiracies in restraint of trade and monopolization, attempts to monopolize, and conspiracies to monopolize.[2] The Louisiana Supreme Court ruled that a city is immune from such liability if the state has "a comprehensive and actively supported . . . policy permitting regulation to displace competition."[3] Does a state have such a regulatory system for cable, or is such power assumed by the municipality under privileges granted by the state?

The U.S. Supreme Court decided that the city of Boulder, Colorado is not necessarily exempt from antitrust liability on the strength of the state exemption.[4] This may open up competition for franchises or force states to adopt legislation specifically dealing with the franchising authority. Since the Boulder decision, some states have passed bills "clearly articulating" the franchising authority of the cities, presumably permitting the extension of the state antitrust exemption.

The Cable Act specifically permits a franchising authority to award one or more franchises within its jurisdiction, but the intent is not to give the franchising authorities automatic exemption from antitrust law. The grant of a franchise, whether or not expressly stated, gives the franchisee the right to construct the system over public rights-of-way and through easements already dedicated to utilities.

Cable services cannot be provided without a franchise. This prohibits telephone companies or other communications services from offering video programming on their facilities through their FCC authorization to provide services.

Cable is not to be regulated as a common carrier or public utility on the basis of its providing *cable service*, which is defined as a one-way transmission of video programming or other programming service and subscriber interaction required for the selection of video programming (such as PPV) or other programming service. The Act specifically permits regulation of the noncable services (such as data, two-way services).

The Cable Act gives the franchising authority the responsibility for assuring that subscription to cable services is not denied to subscribers because of low income (the practice of "redlining").

CABLE ORDINANCES

Most cable systems are governed, broadly, under a city ordinance specifically written for the purpose. The ordinance sets out the procedures for obtaining a franchise and may outline some basic and specific requirements of a cable franchisee. In a few jurisdictions, cable systems are not covered by a specific ordinance, but operate under the authority of a franchise agreement standing alone. Franchise agreements are discussed later in the chapter.

These are the common provisions of cable ordinances:

Definitions

An ordinance begins with definitions of key terms that are fundamental to the ordinance, such as "cable communications system," "city," "franchisee," "franchise agreement," "subscriber," public channels." Other terms may be defined when they appear. A number of concepts such as "two-way communication" and "institutional network" could have many different meanings and therefore should be described carefully in the ordinance or franchise agreement.

Procedure for Application

This section of the ordinance describes the process of soliciting applicants and, in general, the kind of information desired from applicants in the request for proposals.

Procedures for
Selection of Applicants and Franchising

Often the procedures for franchising are prescribed under state law or more general ordinances. These may be summarized in the cable ordinance. The ordinance may require public access to applications and other relevant documents, a public hearing on the applications, and a public hearing on the franchise agreement.

Certain criteria are important in the evaluation of a franchise, and these criteria are stated in the ordinance—among them financial capability, record of performance, character of the owner, and engineering design. Weights given to these criteria may also be stated.

How does a city determine its own interests in cable? An attempt can be made to systematically ascertain the community needs related to cable. There are several approaches. A convenient method is to call one or more meetings of groups of community leaders and government officials. The meeting might start with a discussion of general communication problems and later focus on the potential of cable in addressing the problems. A procedure and format for such meetings is included in Appendix F. An alternative is to ask the applicants for the franchise to ascertain community needs themselves.

A section of the ordinance indicates how applicants are to be evaluated. Because selection from among several applicants is a complex task, city council members are not likely to have time themselves for the basic work. A committee either of citizens or of citizens and government representatives may be organized to perform an advisory function. This committee can be charged to receive input from the entire community.

Monitoring Franchisee Performance

In granting the franchise, the city takes responsibility for the performance of the franchisee. Some mechanism is necessary to oversee construction of the system, hoping to hold it on schedule, to review service in accordance with the franchise agreement, and to arbitrate complaints from subscribers and system users that cannot be settled with the company. The Cable Act permits the franchising authority to require, as part of a franchise or franchise renewal, provisions for enforcement of customer service requirements (billing practices, notice of disconnect for nonpayment, methods of dealing with complaints, response time for orders and repairs, and the like) and construction schedules. If the city regulates rates in areas defined as noncompetitive by the FCC, rate increase requests must be evaluated. The ordinance may call for periodic reports on community programming, access channel, and institutional network use.

The ordinance creates the mechanism for performing these monitoring and, where needed, rate-making functions. If a commission of citizens and/or government officials acts for the city in this capacity, the ordinance will indicate how it is composed and its responsibilities.

Franchise Fee

The ordinance establishes a franchise fee within the limits set by the Cable Act. The percentage itself is stated, and the base against which it is to be applied. According to the Cable Act the franchise fee is not to exceed 5 percent of *gross revenues* from operation of the system. Thus the fee may be assessed against all of the revenues of the cable system, from all levels of basic and pay services, advertising sales, commercial use channels, and contract production. In some franchises, the fee is calculated as a percentage of only a portion of the revenues. In fact, the fee can always be less than the 5-percent limit and state governments are free to set a lower cap. It is not necessarily the best policy for the city to assess the maximum fee since it is usually passed through to subscribers.

The franchising authority may also assess a 5-percent fee on commercial users (lessors of channels or channel time).

Franchise fees may be assessed in advance but must take into account the time value of money. In other words, if the franchise authority is going to borrow from the franchisee, it must pay interest.

There is some controversy over the purpose of the franchise fee. The Cable Act does not specify how it is to be used, but some operators believe that any fee in excess of the actual cost of administering the franchise is improper and that the fees cannot be used as a tax on cable revenues to go into the city general fund. A Wisconsin case determined that any fee assessed beyond reimbursement for regulatory costs, as well as compensation or consideration for use of streets, would have to be justified by the city's *taxing* power.[5] Additionally, some cable operators allege that franchise fees in excess of the cost to regulate are unconstitutional as a violation of their First Amendment rights (see Chapter 11).

The operator is permitted to itemize the franchise fee on the subscriber's bill. Thus the cable operator does not have to hide the franchise fee from the subscriber. Increases in the franchise fee *may* be *passed through* (charged to subscribers) unless the rate structure in the franchise already includes the franchise fee. Decreases in the franchise fee *must* be passed through to subscribers.

Rates

In special cases where subscription rates may be regulated, the ordinance must set the procedures. Provision must be made for the city to obtain sufficient financial information to make sound judgments. To avoid a rate-making procedure just to cover inflationary needs, some communities permit automatic, periodic inflation adjustments, tied to the local or national consumer price index.

Access to Multiple Dwelling Units

For many years the right of cable companies to wire apartment buildings, whether or not the owner wants to have the operator, has been in question. After considering a provision for the Cable Act of 1984 to establish the right of access to multiple dwelling units, the Act was passed without such a mandate. This leaves the question of "access" to state law (see State Policy in this Chapter). The ordinance may include an access provision under direct authority of a state statute or implied authority. The ordinance may also settle such questions as to whether an access charge can be added to the regular rates.

Transfer

A city could go through a painstaking process of selecting a franchisee, only to have the system sold to another party, which might not meet the standards the city had set. To prevent this, the ordinance usually provides for a transfer of ownership process that is essentially identical to the original franchise award procedure except that other applicants may not be solicited. The prospective transferee must go through all steps in the evaluation process, including the appropriate public hearings.

Franchise Term

The franchise term is usually set at 10 or 15 years and is renewable for a period of 5, 7, or 10 years. Original franchise periods of less than 10 years are generally not acceptable to cable companies and their lending institutions. A shorter period would not be likely to attract applicants. Franchise periods of longer than 15 years constitute a commitment that most cities are unwilling to make.

Forfeiture

A section of the ordinance contains a listing of the sins for which a franchise may be revoked or the franchisee fined. Since revocation is a harsh penalty for minor failures, more recent ordinances have attempted to establish lesser penalties in the form of a schedule of fines.

Free Drops

Almost all cable ordinances specify free connections and at least free *basic* service to every fire station, school, and college in the service area. The obligation is usually to bring cable to a single designated room.

PEG Access

If a community wants PEG access channels, the ordinance or franchise agreement states the number and type. The ordinance may also say who is responsible for administering each channel, especially when one is shared; two colleges may be on the same channel, for example, or several public school districts and parochial schools on one or two channels.

Complaints

The ordinance may establish a procedure for handling complaints or unresolved disputes between the franchisee and users of the service. A cable commission may have authority to act for the city. New subscribers receive a notice of the complaint procedures.

Commission

The ordinance designates an official to act for the city in the administration of the franchise. The statute may establish a commission of citizens to set operational policies and oversee the city administrator/cable company/subscriber relationship. The commission may require regular reports from the cable operator and, in turn, report to the city council.

Other Provisions

Other routine items in the ordinance attempt to assure that the franchise will meet accepted engineering standards and follow safe procedures in construction, be prompt in connecting and disconnecting subscribers, be properly bonded and insured, not discriminate in the hiring of personnel and in the provision of service, and make the services of the system available in emergencies. If the franchise is not renewed or is revoked, the city may opt for buyback. The procedures for buyback are in the ordinance.

REQUEST FOR PROPOSAL

New cable franchises are still being granted, usually in new subdivisions in townships outside an earlier franchised area. The franchising authority prepares a *request for proposal* (RFP). The RFP itemizes all interests of the city and elicits conveniently comparable responses from all applicants. The RFP describes the city for the prospective applicant, such as square miles, street miles, number of households, household income, population, and major industries. The object is to give enough information to help in the initial decision on whether or not to apply; it need not be complete, since the applicants will make detailed analyses of the community on their own. A deadline must be stated. RFPs usually require a filing fee, which can be used to defray the cost of evaluating the applications.

For comparability among applicants, a rigid format is prescribed for the application, which forces the applicant to address the specific contents or, at the other extreme, a set of standard forms to be filled out.

Identification of the Applicant and Ownership Structure, Ownership Qualification and Character

The response to the RFP identifies all people, companies, and organizations with an interest in the applying company. These are the people who are investigated for "character" and other purposes that may be of interest to the franchising authority, such as racial and sex balance.

The form may ask the applicant to be explicit about ownership agreements. This addresses an historical problem in the franchising process referred to as "rent-a-citizen" or "rent-an-institution." Many cable companies have attempted to influence the franchise award by offering an ownership position in the company, often without significant financial investment, to prominent local citizens who would then lobby for the application. Local ownership could be highly desirable, if that ownership is active. People living in the franchise city will be more sensitive to the community needs and could exercise some control over the absentee multiple system owner. But, where the ownership is silent or to be bought out soon after the franchise is awarded, the purpose may be only to influence the process. The cost of buying this influence is passed through to the consumer.

In Omaha, Nebraska influential citizen-investors were solicited with this letter, quoted in part:

> Specifically [a company that eventually became one of the six finalists] is prepared to support up to 20 percent local investment on a carried interest basis; that is, these investors are required only to subscribe for their stock at a nominal par value, with the parent company advancing all necessary funds for the construction and operation of the franchise . . .
>
> Aside from financing, however, we view the local investors as full partners, particularly with regard to developing the strategy to obtain the favorable vote of the city for award of the franchise.
>
> Finally, the winning of a cable franchise is essentially a political campaign . . . . The ability of local investors to take the political temperature, make introductions and appointments on a timely basis,

and to lend their personal credibility to our formal business proposal is vitally important to the success of our proposal.[6]

The successful applicant in Omaha was Cox (not the letter writer), a major multiple system operator. Cox gave a 20-percent interest in the profits of the system to a group of eight investors who put up a total of $200 of the $36,879,000 investment. Cox estimated the 20-percent interest would have a value of $12 million in ten years, "roughly, a six million percent gain" on the investment. Or, if the profits were not as hoped, the eight investors would sell out in five years for a guaranteed $1 million.[7]

Once "rent-a-citizen" became exposed, "rent-an-institution" took its place. This plan worked for Warner Amex in Pittsburgh and again in Cincinnati while its competitors were being embarrassed by rent-a-citizen publicity. In this case, the local equity (usually 20 percent) is given, or sold at a token sum, to community institutions or groups presumed to have political power in the community. In Pittsburgh, it was minority groups; in Cincinnati, it was educational institutions.

The Fairfax County Council on the Arts accepted a 1-percent interest in the Storer cable subsidiary applying for the Fairfax County, Virginia franchise. In return, the council would advise on arts programming.[8]

These community institutions should have been evaluating all the applications and advising the city in their areas of interest. But, by becoming ownership participants, they had a conflict of interest.

The franchising authority may also ask for a record of current and past cable holdings so that these operations may be evaluated.

Financial Resources and Existing Capital Commitments

The RFP requests information on the capital capacity of the applicant. If the applicant is underfinanced, the system may never be built or be very slow in construction. A part of this response may be a financial pro forma that asks for itemization of revenues and expenses over a period, year by year. This can be an indicator of how realistic the applicant is in the business plan.

Service Area and Line Extension Policy

If the city does not actually prescribe the *service area*, the area to be wired initially, as would be the case in a community with greatly variable household density, then the applicant must be asked to indicate what areas would be initially served and to propose a *line extension* policy. The line extension policy specifies the conditions under which cable lines will be extended to new areas. In simple form, this might be to build into a new area when the housing density reaches 30 households per mile. The policy may also permit cost-sharing by households that do not qualify.

System Design, Facilities, and Equipment

The design of the physical plant and its capacities are quite permanent. Engineering consultants can help the city evaluate the design, as well as its capacity to provide quality service and meet future needs. The franchising authority can specify production facilities, equipment for PEG access use, channel capacity, addressability, distribution plant configuration, and headend design.

Programming and Service

The Cable Act prevents the franchising authority from specifying in the RFP program services or even broad categories of video programming or other information services. But the franchising authority can accept the applicant's *offer* to provide broad categories of video programming. Therefore, this section of the RFP may simply ask that applicants describe in general terms the programming and services to be offered.

Rates

Franchising authorities may wish to ask for the initial rate structure for system services.

RFP Announcement

Once the RFP has been prepared, it is circulated with a copy of the ordinance and a statement of community needs (if one has been prepared). It goes to all cable firms that have made inquiries. Announcements of the availability of the franchise are made through publicity and classified advertisements. Trade periodicals used commonly are *Cablevision, Multichannel News,* and *Cable Television Business.* A direct mail announcement to cable MSOs and companies serving the immediate region is also useful. The deadline is set at least three months from the date of the announcement so that interested parties have adequate time to prepare.

FRANCHISE COMPETITION

The value of many cable franchises is immense. Cable systems franchised before 1980 were built for $500 or less per subscriber. Those franchised after 1980 may cost $800 per subscriber. They are now being sold for as high as

$2,000 per subscriber. The difference between the construction price per subscriber and the market value is a rough index of the value of a franchise. Under these circumstances competition for franchises was intense. There were several applicants for each new franchise. The franchise bidding process forced companies into business and lobbying practices that were an embarrassment to the industry and to city officials.

Eventually, the acceleration in bidding went too far. Competitive zeal on the part of the bidders and greed on the part of the franchising authorities escalated the number of channels and services asked for and offered. *Pro formas* (financial projections) included revenues from high multipay penetrations and two-way services, neither of which materialized. Construction cost, much of it underground in urban areas, was underestimated. Dual cable systems with institutional networks were proposed. There was no immediate revenue potential for the second cable or the institutional network.

This was an awkward time for both the franchisee and the franchise authority. The only resolution was to *renegotiate* the franchises. City by city, almost all the urban franchises were renegotiated. Although each situation differed somewhat, the net result of the renegotiation in most places was: (1) to drop back to a single-cable system without an institutional network if the system construction was not underway, or to make the second cable and institutional network inactive if it was built; (2) to raise subscriber rates; (3) to cut down the number of access channels; and (4) to reduce or eliminate access support payments. Capital expenditures for access and franchise fees were also reduced in some places.

Renegotiation was not easy. Procedural questions arose when one applicant was chosen over others on the basis of the proposed system and then, after the others had been rejected, the original proposal was renegotiated. Some of the applicants who lost out originally asked that the process start all over again. It was also politically difficult for city council members to accept much less than they had originally bargained for. Renegotiation delayed some major city cable systems many months.

EVALUATION AND SELECTION

With few major markets uncabled, almost all franchising will be in suburban communities or subdivisions that grow up on the edge of places already cabled. The applicants may be limited to the franchisee already operating in the area. Since one or more of the companies making bids may have a local history, the study of that record is very important.

Consultants

Some cities hire consultants to help with the evaluation. The consultant can help to write the ordinance, prepare the RFP, and make the preliminary evaluation (leaving the final judgment to the client). Consultants like to be on board at the time the RFP is written because it provides the information they must work with during the evaluation.

The cable industry generally believes that consultants are underfunded for adequate analysis. A vice president of one MSO says, "The cities get what they pay for. The consulting reports today are quick and dirty."[9] The city should probably ask for intensive analysis *in the areas of major interest* and be prepared to pay for it. The consultant fee can be built into the applicaton charge.

The consultant only advises. It may or may not be "correct" advice or appropriate to the community. It must be weighed carefully with other information. After all the evaluations have been made—the reports from consultants, reviews by citizens, advisories of commissions—the city council must make a choice and negotiate a franchise agreement with the successful applicant.

The city has an obligation to the applicants, who have at this point made an investment in the application, to be careful and objective in the analysis. The city must protect itself from litigation which will prolong completion of the system.

FRANCHISE AGREEMENT

The franchise agreement sets the terms of the franchise; it is, in effect, a contract between the city and cable company. All the details of the *application* are written into the franchise agreement including the engineering design, facilities and equipment, broad categories of programming and other services, initial rates (in rate-regulated communities), access channel rules, location of public access studios and business office, construction schedule, line extension policy, and specification of periodic reports to the city.

The franchise agreement and the ordinance will be the principal documents against which the performance of the cable company will be judged. Consequently, the specifications for the cable plant and the description of services must be stated in the franchise agreement so that both parties have a mutual understanding of the meaning. Where the application is unclear, or there is insufficient detail, the clarification is provided in the franchise agreement.

MONITORING SYSTEM
PERFORMANCE AND DEVELOPMENT

Once the franchise agreement has been signed, the cable company and the franchising authority are partners. They share responsibility for, and interest in, the consumer welfare.

If the access channels and institutional services are to work, the city institutions must be imaginative and effective in the development. For the company there is a disincentive to develop most of the institutional side, which is not likely to be revenue producing. Therefore, the city and the users of these services must take the leadership in *development*. This is the major positive role of the franchising authority, and we deal with it first. The second major function is *regulatory*: to monitor the performance of the franchisee against the franchise agreement and ordinance.

Developmental Role

The city wastes the resources if it does not aggressively pursue the use of government and education access channels and the public institutional network channels, if any. Indeed, if they are unused, the cable operator has the right to use them for other purposes (see the later section, "Modification of Franchises").

Perhaps the best approach is to identify quickly the agencies that have the greatest interest in using cable and get them started as models. These groups will have emerged in needs assessment or in other ways during the franchising. If the fire department has expressed a strong interest in training by cable, the department of social services wants to prepare and evaluate foster parents by cable, and the director of the science resource center in the public schools sees merit in teleconferencing between the center and user schools, then these services should be encouraged.

A city employee responsible for coordination of cable uses, along with a coordinator for the cable company, can initially play a promotional role to initiate service. The enthusiasm and energy of these individuals in the initial promotional role are as important as knowledge.

In *some* cities, the government takes responsibility for the public access channel. Rules are established for orderly use and fairness in access. Funds available for access users may be administered by the government.

Regulatory Role

There are two main functions in the regulatory role of the franchising authority. One is enforcing the franchise agreement and applies to all cities. The other is rate regulation, which applies only to cities where there is no "effective competition."

During the construction phase, representatives of the city, usually the cable commission and city staffers, meet regularly with the cable company to monitor construction progress. In monitoring system performance after construction, it is helpful if the franchise agreement contains every obligation of the franchise. If so, it serves as a convenient *checklist* against which to evaluate performance.

If the cable company fails to perform in accordance with the city's interpretation of the ordinance and franchise agreement, the city has the burden of enforcement. The procedures will be spelled out in the ordinance. If fines can be levied by the city, the offenses and associated fines will be in the ordinance. Where there is no fine system, or the performance of the company is seriously inadequate, the enforcement action is *revocation*. This, of course, would be time-consuming and costly for the city and the company. There have been few such cases. Most revocations have been for failure to construct a system after the franchise has been awarded, where companies have been simply warehousing franchises awaiting the most opportune time in which to build or have had difficulty in financing. Since franchises are not exclusive, another approach to dealing with an inadequately performing system is to grant a second franchise.

The city sits in an appellate capacity for disputes between subscribers and users of the cable system and the company. Most of these issues should be settled without city intervention, but some may reach an impasse. When major issues come up, the action of the city establishes policy, and future cases are then settled more efficiently. In an active cable system, this function is likely to require a monthly meeting of the cable commission or smaller subcommittees that hear various categories of cases.

The Cable Act permits subscriber rate regulation only in places not subject to effective competition. It was left to the FCC to define "effective competition." The FCC assumed that, where several broadcast television stations are available, people can take cable or leave it. If they opt not to take cable, they will still have fairly good television service. But, where few channels are available from broadcast transmitters, the household is obligated to accept cable to receive adequate television service. In these places the Congress and the FCC felt it necessary to permit rate regulation so that the cable operator is not in a position to take advantage of the consumer. According to the FCC:

> A cable system will have effective competition whenever at least three unduplicated stations serve the cable community. The signals are to be counted if they place a Grade B contour over a substantial portion of the community, are significantly viewed within the cable community or are translator stations located within the cable community service area.[10]

Grade B contour is the predicted coverage area of a broadcast television signal, usually about 75 miles radius from the transmitter. If the cable system falls within that radius the signal is counted. (Actual coverage may differ from the "predicted" coverage. The FCC allows franchising authorities to make engineering studies to show that a signal does not actually cover the predicted area.) A station is *significantly viewed* in a cable community if that station has been deemed by the FCC to be significantly viewed for the county in which the cable community is located.[11] The FCC decision is based upon the station obtaining a specified share of the viewing hours and in reaching a specific percentage of the country population. However, it is possible that a station that has obtained significantly viewed status for a whole county does not in fact cover the particular cable community. Again, the franchising authority may utilize engineering studies and audience surveys to show that the signal is not truly available to the cable community.

Even in rate-regulated areas, franchising authorities may regulate only the *basic* service rates. "Basic" means the service tier that includes the retransmission of local television broadcast signals.

Cities that have authority to regulate basic subscriber rates under the FCC rule are not required to do so. If they choose to regulate rates, the FCC specifies the procedure for the regulation of rates: (1) give formal notice to the public, (2) provide an opportunity for interested parties to make their views known, at least through written submissions, and (3) make a formal statement (including summary explanation) when a decision on a rate matter is made.[12]

Rate regulation is a complex form of the regulatory function. How it is to be done in the cable industry, under the Cable Act, has not yet been settled. In the case of common carriers a "rate of return" regulation is applied. The carrier is allowed a fair return to equity and debt investors. But the Cable Act does not allow common carrier rate of return regulation: "Any cable system shall not be subject to regulation as a common carrier or utility by reason of providing any cable service." The legislative history of the Cable Act is more explicit: "A cable system would not for instance, be subject to rate of return regulation . . . to the extent that the cable system is providing cable services."[13]

If rate of return is prohibited, what meaningful regulation can be imposed? Communications Attorney Wesley R. Hepler believes:

> The only workable alternative is a "comparative" marketplace analysis: whereby the franchising authority may consider the number of services provided by the cable company and the quality of service provided— as compared to a similarly situated and geographically relevant cable system. This analysis should also include any relevant competitive alternatives in the local marketplace, such as thriving VCR rental business, the widespread existence of home earth stations, or other competitive factors that may indicate the cable system's rates are in fact subject to a competitive marketplace.[14]

The FCC, however, has chosen not to actually *impose* a "comparability" standard and simply states that "the means by which the appropriate regulated rate is determined is best decided consistent with the statute by the local franchising authority."[15]

The success of rate regulation in the United States—in energy, transportation, and communication—in terms of price and performance efficiency has been very limited. If the state and federal commissions, with professional staffs, are unable to regulate rates, can a small town city council or cable commission of volunteer citizens do so?

Nonetheless, in rate-regulating communities, the formal rate regulating procedure may serve to restrain rate increases. A rate hearing can also function, informally, to raise pertinent issues about the quality of the service. In the give and take of the rate proceeding, problems in service may come up, and corrections promised by the cable system.

MODIFICATION OF FRANCHISES

The franchise and the franchising authority may agree to change the franchise provisions at any time. In effect they would be negotiating a different contract that would supercede the previous agreement.

Under certain conditions, modification is sanctioned by the Cable Act. This provision strengthens the operator's negotiating position and, if negotiation fails, modification can be requested from the franchising authority under the federal law. *Services*, except PEG access channels, can be modified if the operator can demonstrate that the same "mix, quality and level" of services are preserved. *Facilities* and *equipment* (including PEG access facilities and equipment) may be modified if it can be demonstrated that those originally intended are "commercially impracticable" and the operator has a reasonable alternative. The franchising authority is to make a decision on such a request within 120 days.

The cable operator may unilaterally, on 30 days' notice to the franchising authority, rearrange, or replace channels, or remove a cable service that is no longer available or where the copyright royalty has changed substantially. And services can be changed from one tier to another. In a rate regulated system, all of the unregulated services, beyond basic, may be changed according to this provision. Franchises or renewals granted after the 1984 Cable Act must be much more flexible in the programming area (see "Regulation of Services," Chapter 11). The need for modification of newer franchises, therefore, would infrequently occur.

The Cable Act permits a franchising authority to set rules for permitting a cable operator to use PEG access channels for other purposes if they are not being used by the public, educational, or government institutions.

RENEWAL

One of the purposes of the Cable Act is "to establish an orderly process for franchise renewal which protects cable operators against unfair denials of renewal where the operator's past performance and proposal for future performance meet the standards established by this title."

A franchise may be renewed through *informal negotiation* The operator and the city work out the details of the new franchise through meetings and correspondence. At the end of this process and before the franchise renewal is granted, however, the public must be notified and given an opportunity to comment. While the informal negotiations are in process the cable operator should, as a formality at the appropriate time, notify the city of intent to seek renewal. Therefore, if informal negotiation is not successful, formal proceedings will have been initiated.

The *formal renewal process* follows two steps, and a third, if necessary.

Step 1

Between 36 and 30 months before the expiration of the franchise, the city begins a two-part *assessment (or ascertainment) procedure* on its own initiative or at the request of the operator. One part of this procedure is to review the performance of the cable operator under the current franchise. If this has been an ongoing activity, as suggested in a previous section, the city need only summarize the results of all the years of monitoring. The cable operator may *not* have complied entirely with the franchise, in which case it is necessary for the operator to establish that the city has waived compliance or effectively acquiesced to the nonperformance. Although the city does not have the right to review programming services, it can review customer service—the prompt handling of orders, complaints and repairs, as well as reasonable billing procedures.

The second part of this assessment step is to identify future cable-related community needs and interests. This should help the city to know what to look for in the next step, the renewal proposal. Determining future needs could be handled by a city cable commission following a community leader survey/public hearing procedure something like the one outlined in Appendix F, or a consultant may be hired to specify a procedure and administer it.

At this point it would certainly be prudent of the city to find out what cable services are available in other comparable communities, in case there are some locally "unrecognized" needs and interests.

Step 2

The operator seeking renewal submits a *proposal*. There may be an RFP for the renewal quite similar to the RFP for a new franchise described earlier, with the addition of a section asking the incumbent franchise to provide an overview of its current franchise period. The franchising authority would be limited in what it may ask and require by the Cable Act provisions described earlier in this chapter and in Chapter 11. The Act specifically permits the franchising authority to ask for an *upgrade* (new construction to bring the system up to the state of the art).

Upon receiving the proposal, the franchising authority makes a public notice of the proposal. The city may construe "public notice" according to its own judgment or in line with generally established procedures, which may mean publishing the proposal in certain ways and inviting public comment informally or at a public hearing. As a result of this step the franchising authority must either renew the franchise or begin an *administrative proceeding* (to be described later). The proposal step must take place within a four-month period beginning at the end of the assessment step. Although no period is specified for the assessment step, the proposal and proposal review must be handled promptly. The franchising authority needs to set a deadline within the four months that will permit adequate time for review.

If the franchise is renewed, the franchising authority and franchisee enter into a franchise agreement that has the same elements as an original franchise agreement.

Step 3 (if necessary)

A preliminary finding that the franchise should not be renewed triggers the administrative proceeding. This also requires prompt public notice. The administrative proceeding considers whether (1) the operator has complied with the terms of the existing franchise and applicable law, (2) the service has been reasonable in the light of community needs, (3) the proposal for renewal indicates that the operator has the financial, legal, and technical ability, and (4) the proposal reasonably meets the future cable-related community needs and interests, consistent with practical costs. The cable operator is given the right to full participation in the proceeding including the right to give evidence and the right to require the production of evidence. To deny the renewal requires an adverse finding on at least one of the preceding four factors. The cable operator who believes these procedures have not been followed, or who has been denied renewal, can appeal to a U.S. district court or a state court with jurisdiction.

It is important to note that a city cannot deny renewal of an existing franchise on the basis of superior promises made by a competitor for the franchise. Renewal must be denied before another application can be entertained. The city cannot read a competing application and then, *post hoc*, decide that the community's needs are items that are in that proposal.

The effect of the Cable Act's provisions on renewal has been to make it extremely difficult to displace an existing franchisee and, consequently, to remove the principal threat under which the city exercised authority over franchisees.

If a renewal of the franchise is *denied*, and the franchise authority acquires ownership or takes responsibility for a transfer of ownership, the original owner is to be compensated at *fair market value,* not including the value of the franchise itself. If a franchise is *revoked* and the franchise authority takes over, the owner must be paid an *equitable price.* This provision of the Cable Act applies only if the franchising authority chooses to buy back the system. It is not required to do so.

TRANSFER

Many cable systems are now changing hands as MSOs grow or consolidate geographically. Even if the city ordinance specifies an orderly procedure for transfer of ownership, the process is complex and, for a number of reasons, the city may be in an awkward position. Because purchase prices are now quite high, the new owner will have a substantially higher capital investment than the previous owner, and probably much more debt. Under these circumstances, although the transfer includes an agreement to assume all of the franchise obligations of the previous owner, it may be "commercially impracticable" to do so. The new owner, under the Cable Act, may modify the commitment to provide facilities and equipment, including those for public, educational, and government access.

In many recent franchise transfers, the new owner makes more limited franchise requirements a condition of the transfer. The franchise is renegotiated. Often the city is under pressure to be prompt in the negotiation because the buyer and seller have a deadline after which the deal is terminated. Even if the city does not accede to a tight deadline, it must act within a reasonable time.

An *acceptance fee,* paid by the franchisee, usually covers the city's costs for investigation, consultation, hearings, and deliberation.

STATE POLICY

Local governments carry most of the responsibility for granting franchises, construction of cable systems, and the administration of the franchises, under powers given them by their own states and the federal government. Some of the states are very actively involved in cable; others are not. This section describes some of the areas where states play a role.

Pole Attachment

Under the Communications Act Amendments of 1978, the FCC is required to regulate rates, terms, and conditions of cable television's use of poles, conduits, ducts, and rights-of-way in privately owned telephone and electric utilities. But the FCC only exercises this authority in states that have not assumed jurisdiction and actually asserted their authority. (See: "Pole Attachments," Appendix E and FCC Report and Order, note 10.) Several of the states have now done so.[16]

The object of pole attachment regulation is to prevent discrimination against cable companies and cable consumers in rates charged and the conditions of use. The legislation gives cable companies and utilities a forum for adjudication of conflicts.

Historically, delays in the granting of *pole rights* and *make-ready* (pole preparation for cable) and struggles over the charges have been the bane of cable operators. Utilities may have higher priorities related to their own interests or may overvalue the cost of pole use. Telephone companies may even have selfish interests in deterring a competing communication system.

The state may preempt the FCC pole attachment jurisdiction only if it shows that it "has the authority to consider and does consider the *interests of subscribers of cable television services,* as well as the interests of the consumer of the utility services" (emphasis added).

To determine the fairness of pole attachment rates, the FCC establishes a *zone of reasonableness* with the *minimum*, the additional costs to accommodate cable system users, such as special maintenance and administrative costs, and at the *maximum*, the cable system share of the fully allocated costs. The latter is based on the original cost of poles and the maintenance cost. The cable system share might be calculated by determining the portion of the usable space (the amount of pole above the required ground clearance) occupied by the cable.

The FCC has issued a number of rulings that have brought pole rates down in nonregulated states. For example, in Winter Haven, Florida, Tampa Electric Company wanted to raise the rate paid by Warner Amex from $4.50 to $9.24 per pole. The cable company challenged the rate before the FCC. The FCC ruled that a reasonable yearly rate would be $1.36 per pole.[17]

Generally, the states that regulate pole attachment rates have not been as favorable to the cable companies as the FCC. In some of those states, the cable industry has been successfully challenging the jurisdiction of public utility commissions to regulate pole rates under the existing state statutes.

The federal authority over pole attachment was brought into question late in 1985 in *Florida Power v. FCC* when the 11th Circuit Court of Appeals held the 1978 Pole Attachment Act to be unconstitutional on the grounds that it was a "taking" of property in violation of the Fifth Amendment. The Court concluded that the FCC was taking property and then, after having taken it, making a determination of just compensation. But the Supreme Court overruled the District Court and affirmed the FCC pole attachment rate regulation procedures.[18]

Theft of Service

Prior to federal law on theft of service, many states enacted statutes making it a crime to make unauthorized use of cable signals or to manufacture or sell devices to aid in the theft.[19] The operator, in prosecuting theft, can choose state or federal law. Local law enforcement officers may help cable operators enforce the state law on theft.

Noncable Services

As noted, noncable services, data transmission, telephone bypass, two-way, and nonvideo may all be regulated by state utility commissions. The FCC or the state utility commissions can require "informational tariffs" on the noncable services. These would provide the rates, terms, conditions of service and to whom the service is available. For those states requiring them, the informational tariffs would keep the utility commissions informed of what cable is doing and provide a basis for deciding whether or not to regulate. While "cable services" cannot be regulated as common carriers, most of the noncable services would qualify as common carriers and could be regulated by the states.

Access to Multiple Dwelling Units

Many states have written statutes to deal with the issues of "access" to cable services in apartment buildings and mobile home parks. Access may have two meanings in this context. The first is the right of access to cable services on the part of the resident. The second is the right of access by the cable operator to the building and the individual units within the building.

The Supreme Court, in *Loretto v. Teleprompter Manhattan CATV Corp.*, upheld a New York statute that prohibits interference by an apartment owner with the installation of cable in the apartment.[20] But, the decision, mindful of Fifth Amendment proscriptions against "taking" of private property, required "just compensation" from the cable operator to the owner. In New York this turned out to be only $1 per building. The case makes two important points: (1) that state laws mandating cable access to multiple dwelling units are constitutional, and (2) that there must be some formula for determining just compensation to the apartment owner for the intrusion on the private property.

Unless the state law mandates access, cable operators may encounter difficulties serving large multiple dwelling units. The owners of these buildings may seek to enter into exclusive contracts with private cable operators (satellite master antenna systems) who are willing to pay a royalty.

States may also be involved in construction requirements and regulation of access channels. Cable may be specifically identified in state obscenity laws. States may set limits on franchise fees (within the federal limits) and may take a portion of the fee for the state, reducing the amount permissable for the city. The Cable Act specifically permits state equal employment opportunity, theft of service, and privacy laws that are more strict than federal law.

NOTES

[1]"Municipal Cable Ownership Reconsidered in St. Paul," *Cablevision*, February 2, 1981, p. 37.

[2]15 U.S.C. 1 & 2.

[3]*City of Lafayette v. Louisiana Power and Light*, 435 U.S. 389 (1978).

[4]*Community Communications Co. v. City of Boulder*, et al., Case No. 80–1350 argued October 13, 1981, decided January 13, 1982. Certiorari to the U.S. Court of Appeals for the 10th Circuit.

[5]*Ripon Cable Co. v. City of Ripon, Wisc.* (82 C.V. 684, Wisc. Circuit Court, November 2, 1984).

[6]Warren Buffett, "How Cable Franchises Are Bought (Not Won)," *Access*, October 6, 1980, p. 1. The letter was received by the publisher of the Sun Newspapers in Omaha, of which Mr. Buffet is board chairman.

[7]Ibid.

[8]"The Emergence of Pay Cable Television," Volume IV, "The Urban Franchising Context," Technology and Economics, Inc., Cambridge, Massachusetts, prepared for the National Telecommunications and Information Administration, Washington, D.C., August 1980, p. 43.

[9]Hugh Panero and Fred Dawson, "The Cable Consultants: Are They Doing the Job?" *Cablevision*, December 8, 1980, p. 40.

[10]Report and Order in the Matter of Amendment of Parts 1, 63, and 76 of the Commission's Rules to Implement the Provisions of the Cable Communications Policy Act of 1984, FCC, April 11, 1985, Docket No. 84–1296. Translator stations are not counted if they are used to retransmit stations already providing Grade B contour or significantly viewed stations. At this writing the original FCC decision to count stations that put a signal over "any" portion of the community had been rejected by a federal court. The FCC was proposing a standard of 75-percent coverage of the cable community.

[11]47 C.F.R. 76.54. The FCC definition of "significantly viewed": "Viewed in other than cable households as follows; (1) For a full or partial network station—a share of viewing hours at least 3 percent (total week hours), and a net weekly circulation of at least 25 percent; and (2) for an independent station—a share of viewing hours of at least 2 percent (total week hours), and a net weekly circulation of at least 5 percent." To establish "significantly viewed" stations, the FCC used an American Research

Bureau report. (The significantly viewed stations are printed in rules or may be obtained from the Mass Media Bureau of the FCC.)

[12]Report and Order, Ibid.

[13]House Report 98–934.

[14]Letter to Thomas F. Baldwin from Wesley R. Hepler, Cole, Raywid & Braverman, dated June 6, 1986.

[15]Implementation of the Cable Act, 58 R.R.2d 1, 123 (1985).

[16]In 1985 fifteen states claimed the authority to regulate pole rates and therefore assert jurisdiction: Connecticut, the District of Columbia, Idaho, Illinois, Louisiana, Maine, Maryland, Massachusetts, Michigan, New Jersey, New York, Ohio, Oregon, Utah, Vermont, Washington.

[17]*Warner Amex Cable Communications Company v. Tampa Electric Company*, file No. PA–80–0019, released June 26, 1981.

[18]*Florida Power v. FCC*, 11 Cir. Ct. (Supreme Court reversal reported in J.L. Freeman, "High Court Upholds Pole Attachment Law in Victory for Cable," *Multichannel News*, March 2, 1987, p.l.)

[19]47 U.S.C.A. 224.

[20]*Loretto v. Teleprompter Manhattan CATV Corp.*, 458 U.S. 419 (1982).

CHAPTER **13**

Ownership

A few weeks of cable trade magazine headlines.

Acquisition-Minded Comcast Eyes
Boosting Sub Count to One Million

Jones acquires
nine Tribune
cable systems

Taft Vows to Fight
Any Takeover Attempt

Biggest MSOs
to continue
accumulations

Big Trade Sealed
200,000 subs change hands.

$230-million deal

Storer, Times Mirror swap systems

TCI Acquires Stake
In Metro Cable Corp.

Group W, Cap Cities, Tribune,
Gillcable Systems Change Hands
As Cable Rings in the New Year

Tab could be $30 million in cash

CBS selling Black Hawk system

Conference spotlights buying frenzy

TBS seeking
to sell up to
40 pct. of CNN

Continental Says It Will
Buy Back 6.5 Million Shares

Heritage gets franchise

Dallas transfer approved

Warner offered $450 million

Amex makes huyout bid

Viacom completes
Showtime/TMC acquisition

Sacramento Cable
Buys SMATV Operator

Five File Bids
For Franchises
In Chicago Suburbs

Several characteristics of private cable ownership are discussed in this chapter: local monopoly, concentration, vertical integration, cross-media ownership, and the ownership involvement of other communication technologies in cable. These characteristics of the industry structure are important to its performance—that is, to the operating relationships within the industry and to the service and price to consumers.

LOCAL MONOPOLY

A principal ownership issue in cable is local monopoly or exclusivity. Cable companies make application for a franchise. One company is selected and builds the system. It is unlikely that another operator will apply for the same territory and also be granted a franchise. Almost all U.S. cable franchises are at least *de facto*, exclusive franchises. The absence of cable competition, and in some cases broadcast television competition, *could* result in a lack of motivation for providing good signal quality, quick response to service calls, quality local origination, maximum program carriage, and state-of-the-art technology and services. Furthermore, the cable service *could* be overpriced.[1]

The distinction between the cable monopoly of a *distribution system* and a *market* is important in this context. The exclusivity of most cable franchises assures cable of control of the cable means of distributing television. But it does not give cable control over the market for television because television is also available, in many places, via broadcast and videocassettes. In a few places multipoint distribution services, satellite master antenna systems, and direct broadcast satellites are also available in the television market that includes cable. Cable cannot behave as a monopoly in a television market where *close substitutes* (alternatives) to cable exist.

In small towns or remote areas where there is "no effective competition" to cable, as defined in Chapter 12, cable pricing is the most serious concern. In these places franchising authorities are permitted to regulate basic cable prices.

In all communities there are other regulations designed to mitigate the affects of the cable franchise exclusivity. The commercial use, or leased access, channels (described in Chapter 11 and required by the Cable Act of 1984) are designed to permit others to introduce cable services that a cable operator has not considered or is unwilling to offer. Cable franchise authorities regulate customer service, making sure that service calls for connection, disconnection, and repair are prompt and billing problems are resolved fairly.

Aside from direct regulation, another technique for reducing the effect of monopoly in larger cable markets is to divide the city into several franchise territories. While the franchisees in each territory are not in direct competition, they are naturally compared to the adjacent systems. The quality of service, number of channels and technology of one would probably have to be matched by the others. The city of Chicago *publicly* rates each franchise district on several performance categories. In metropolitan markets, the existence of several different suburban franchisees also serves this purpose.

CONCENTRATION

MSOs

Characteristic of the cable industry is its horizontal integration or, more accurately, *concentration of cable ownership* across communities. The big MSOs have centralized management resources, which, if not necessary to operate cable systems, seem to have been necessary in bidding and getting the big franchises. The large operators fare better in the capital markets. Because of volume discounting, very large MSOs have the advantage of paying much lower rates for basic and pay programming than small MSOs or independents. A critical mass for maximizing programming fee discounts is considered by many industry people to be about 250,000 to 300,000 subscribers.

The number of broadcast stations permitted any group owner is limited by the FCC. Limitations on the concentration of ownership have been considered by the FCC. In 1982 an inquiry was terminated without action.[2] In 1986, after the must-carry rules had been vacated by the courts, a small cable operator raised the issue again.[3] The petition argued that an MSO would have the power to destroy a local broadcast "voice" if it owned a cluster of systems that included most of a station's signal coverage area and elected not to carry that station. The petition also expressed concern for the power of a concentrated MSO over advertising markets and the "stifling effect" very large MSOs can have on program development when their decision to carry or not carry a new program service can determine its economic survival. The petition was supported by the Motion Picture Producers Association, the National League of Cities, and the independent television stations. Yet the Justice Department argued that there was no need to set limits and that a blanket limitation might actually prevent acquisition or expansion that could be in the best interests of consumers.

The FCC must approve mergers that involve cable television relay services (CARS) that are licensed by the FCC.[4] These licenses cannot be transferred without approval. The Justice Department and the Federal Trade Commission also look at mergers under the antitrust laws.[5] The Justice Department filed suit against the ATC and Cox merger proposal in 1972, and the two parties withdrew. In addition to surviving FCC, FTC, and Justice Department scrutiny, merger plans must be cleared by all the individual franchising authorities that, by ordinance or franchise agreement, reserve the right of approval for transfer of ownership (see Chapter 12).

The biggest cable MSO, TCI, had about 16 percent of the cable house-holds and 8 percent of the U.S. television households in its systems in 1987.[6] The ten largest MSOs have about 45 percent of the cable subscribers, a pro-portion that has not changed too much since 1972 when it was 36 percent (the number of subscribers has increased substantially, of course, since that time).[7] Still, most MSOs have an aggressive *acquisition* policy where they are adding franchises and subscribers. In some places they are acquiring and swapping systems with other MSOs to geographically *cluster* operations. The clusters are large blocks of adjacent cable systems under one ownership. As a result of these acquisition programs, the concentration of ownership will undoubtedly increase in the next few years.

INDEPENDENTS

Single-system operators and small MSOs, sometimes called *independents* because they are not affiliated with large corporations and are exclusively in the cable operation business, are still important in the industry. These com-panies serve a major industry function in developing small population pockets that might have been neglected by the bigger operators. Once developed, the systems may be sold to a nearby MSO, but often the independents run systems on the 5,000- to 10,000-subscriber scale with the same services as most major MSOs, including some local origination.[8] Economies of scale in system con-struction and operation are not too important with the exception of program supplier fees. The principal capital expense is based on a per-mile cable cost and the household terminal or converter. These costs do not vary significantly as the scale increases. Therefore, the small-scale operator can continue to sur-vive in the milieu of the big MSOs.

VERTICAL INTEGRATION

Historical experience in the telephone industry, with a strong integra-tion of equipment manufacturing and operating companies, and in the film industry, with integration of producers, distributors, and exhibitors, has created some reservations about the *vertical integration* of the cable industry. Jerrold Electronics, once the dominant force in the supply of cable industry electronic components, serviced the systems and was also an MSO. Jerrold sold complete cable systems (*turnkey* construction) to other operators and threatened to over-build and operate its own systems in communities where other operators did not purchase Jerrold equipment.[9] Justice Department suits ended the service contracts and temporarily halted the acquisition of new operating systems. Eventually Jerrold sold the operating systems to Sammons.

Cable operators, program suppliers, and equipment manufacturers are also owners of signal distribution networks either in common carrier microwave systems or satellite resale common carriers. Although not currently at issue, some anticompetitive practices could evolve from these relationships.

The FCC has allowed the market to determine satellite rates, deliberately choosing not to regulate the rates. Although satellite transponders available to cable are increasing, additional satellites require additional earth stations so that transponders on an established satellite may be more valuable to program suppliers and to cable operators than on a newly launched satellite. Control of certain transponders is therefore an important factor in cable programming success.

Vertical integration in program supply and system operation is quite common. Almost every major cable network has an ownership affiliation with one or more of the cable operating companies. The operating companies may favor cable networks, in which they have a financial interest, in carriage, marketing, and promotional efforts. MSOs with ownership interests in a cable network refer to the network, facetiously, as a "corporate must carry." Consumers have sued one cable system for its carriage of company owned networks to the exclusion of others.

Program producers may be integrated with program distributors and cable operators. Time, Inc.—which has been associated with several Hollywood film production houses, several cable networks (HBO and Cinemax, BET, USA), and a large operating company (ATC)—is a good example of vertical ties of this type. Several other ties between producers and the cable distribution systems have been formalized. In made-for-cable productions and "prebuys," the cable industry finances, or helps to finance, specials and feature movies. Some sports franchises are under the same ownership as the cable networks that carry their games. These associations give the cable industry a measure of control over the production industry and could restrict competition.

CROSS-MEDIA OWNERSHIP

A pattern of *cross-media ownership* that has developed over the years was extended quickly to cable. Other media are early to recognize the potential of a new medium, have complementary professional and management expertise, and can add to business security by investing in new technologies as a hedge against possible restructuring of the media industries. Cable has attracted ownership interest from all of the other media.

Newspaper and magazine publishers own cable systems. They are now prone to say, "We are in the information distribution business, not the printing business." Many of the print services can be provided faster by cable through alphanumeric and graphic displays on a television screen or computer monitor and printer. Newspapers do not want to give up these functions or lose the business. They believe that they are experienced and staffed appropriately for some of the new cable services if they develop as anticipated. Furthermore, cable offers new opportunities for entry into the more conventional television services. Newspapers can use their resources to enter television news and public affairs programming. Cable offers the opportunity for flexibility and depth in television news that is perhaps more akin to the function of a newspaper than a television station.

Cable is a natural extension of a broadcasting business. The cable industry will need the acquired knowledge of broadcasters as advertising and program production become greater factors. Now that the future role of cable is clear, broadcasters can hedge their profits best by maintaining an ownership position in the cable industry.

Broadcasters have been urged to negotiate long-term leases on their local cable channels for program origination. Alan Bennett of Katz Programming, a unit of the broadcast television station sales representative firm, has proposed that stations take on two channels, programming one with classic films, presumably from the station library, and the other with news and sports. The news-sports channel, he suggests, should be programmed with 15-minute news blocks hourly to provide a local feed to an all-news service such as CNN. He also encouraged the stations to "aggressively go after" television rights, local professional sports franchises, and college and high school sports events for use on the channel.[10] Local television stations have a leg up on other competitors in the same service area because of existing production facilities and personnel and contact with television advertisers.

Several media conglomerates have extensive holdings in cable. Under the Time, Inc. umbrella are: HBO; Cinemax; USA Network; BET; Time-Life Books; Book-of-the-Month Club; Little, Brown and Company; New York Graphic Society; Pioneer Press; *Time, Life, People, Sports Illustrated, Fortune, Money, Discover;* a paper mill and over a million acres of forest land; American Television and Communication. The Times Mirror Corporation is another with television stations, several magazines and major newspapers, book publishers, forests and paper mills, and now several cable systems. The New York Times Company, in addition to the *Times,* owns several other newspapers, magazines, television stations, book publishers, educational resource companies, the New York Times Information Bank, and cable systems. Viacom International, Inc. produces programs for broadcast television networks, and it has syndication rights to such prominent television series as "The Cosby Show" and "All in the Family," an interest in programming and cable enterprises in Japan, Hong Kong, and Britain. It owns or holds an interest in the cable networks Showtime, The Movie Channel, MTV, VH-1, Nickelodeon, and Lifetime, owns five broadcast television stations, eight radio stations, and is a large cable MSO.

Some media conglomerates have no principal ownership of cable system operating companies but do have divisions to develop specialized programming to serve the cable industry. ABC, NBC, Hearst, *Playboy*, and Twentieth Century-Fox are examples.

The Cable Act and the FCC have restricted cross-media ownership. The television networks are not permitted to own cable systems except under FCC waivers to allow experimentation with programming in a small system.

The law also forbids broadcast station ownership of cable systems serving subscribers in the same local service area, defined by the broadcast station Grade B signal contour (about 75 miles in radius from the transmitter). Television translator stations and cable systems may not be co-owned in the same community. This has prevented formation of local cable-broadcast television combinations.

There are major concerns arising from cross-media ownership. Most critical is the possibility of a complete media monopoly in a community. The FCC has taken steps to ban or break up radio, television, and newspaper monopolies through its regulation of radio and television, especially where there is only a single entity of each type. Such a combination, should it also include cable, would give inordinate control of a community's information resources to a single owner. Apart from the media monopoly of a single owner, the FCC and others concerned with media policy value maximum diversity of media ownership in any community. According to this policy, it is better to have three media owners than two, better to have four than three, and so on. Finally, there is a fear of too much political, economic, and intellectual media power, regionally or nationally.

An issue related to ownership diversity is the special question of minority ownership of media. Most newspapers were founded, television and radio stations licensed, and many cable systems franchised before minorities had the economic and managerial resources to participate in ownership. Limiting the holdings of established owners to accommodate new owners is therefore prerequisite to including minorities among them. A few cities have made it clear that minority cable applicants would be welcome.

The Justice Department can look at media cross-ownership and antitrust issues on a case-by-case basis. The Congress or FCC can create policy that is more general, but the First Amendment rights of communication media stand in the way. If all media are looked upon as First Amendment speakers, then one's right to acquire media channels is protected, as media conglomerates have argued in court.

OTHER OWNERSHIP ISSUES

Alien Ownership

The FCC has refused to prevent alien ownership of cable systems. Therefore, the issue is open for franchising authorities to decide. U.S. cable companies have been concerned mainly with the success of Canadian applicants for U.S. franchises in some cities. The Canadian companies, principally Rogers and Maclean Hunter, are MSOs with fully matured systems at the upper limits of their growth potential in Canada. Expansion south into the United States was a logical option.

Telcos and Cable

The cable industry and the related media organizations (that is, newspapers) have an almost pathological fear of American Telephone & Telegraph (AT&T) and the large telephone operating companies. The companies have several overwhelming characteristics: (1) command of vast capital resources, (2) a presumed ability to cross-subsidize new ventures, (3) enormous, well disciplined, and well paid staffs, (4) unmatched communications research and development facilities, and (5) the political power of organizations with thousands of stockholders that serve millions of U.S. households. The technical possibility that someday a "single wire" may provide all of the home communication services hangs over the heads of cable operators who fear that the phone companies would most logically own the wire.

The Cable Act endorses the FCC rule that prohibits telephone companies from owning cable companies in their own exchange area.[11] The object of the rule is to "deter anticompetitive and discriminatory practices that evolved from the local telephone company's monopoly position in the community and its ownership of the utility poles, which usually carry the distribution cables."[12] The FCC considers waiver of this rule for low-population-density areas where the conditions make it more practical for telephone companies to operate cable systems than separate organizations.[13] Telephone companies can build cable distribution plants for lease to franchised cable operators.

In the meantime, AT&T is developing high-capacity fiber optic communication systems, which could be installed in competition with cable. The company has tested videotex in several locations for an electronic *Yellow Pages* and general information service.

Television Receiver Sales and Services

Some of the earliest CATV businesses were established by appliance dealers and radio repair shops as a means of getting into the television sales or service business. As the cable industry grew, some ordinances and franchises specifically prohibited the cable franchisee from entering the sales or repair business. This was to protect the existing businesses and prevent anticompetitive situations where the cable system might use its position to dominate the market. To avoid conflicts, the bigger cable operators have stayed out of sales and receiver service, even where not denied the opportunity.

Recently some cable operators have attempted to sell or rent backyard dishes, VCRs, and videocassettes, raising protests against monopolization and possible cross-subsidation from the existing concerns in those businesses.

SUMMARY

Despite local cable monopoly and increasing concentration within the cable industry and across media, diversity of ownership of media and the diversity of media content is increasing exponentially, principally as a *consequence* of cable. Under the present circumstances, we should not expect much change in ownership controls. According to Christopher Sterling,

> Further controls over cable system ownership do not appear likely as long as: (1) there is generally adequate channel capacity to allow for expansion of programming "voices" reaching the public; (2) there are sufficient separate firms providing program material to cable systems; (3) there are sufficient different means of cable networking (satellites, microwave lines, etc.) available; (4) examples of outright system abuse of the community monopoly position are absent; and (5) competing systems of pay cable are available for local carriage.[14]

NOTES

[1]A thorough discussion of the issues in exclusive franchising is found in Thomas W. Hazlett, "The Policy of Exclusive Franchising in Cable Television," *Journal of Broadcasting and Electronic Media*, Winter 1987, pp. 1–20. Robert T. Blau, "To Franchise or Not to Franchise: Is That Really the Question?" Ibid., pp. 95–97. Michael O. Wirth, "Comment on 'The Policy of Exclusive Franchising in Cable Television,' " Ibid., pp. 98–101.

[2]Report and Order in Docket 18891, July 15, 1982.

[3]Petition for Rulemaking in the Matter of the Commission's Rules and Regulations Relating to Multiple Ownership of Cable Television Systems, February 21, 1986.

[4]*Cable Television Report and Order*, 36 FCC 2d 143, 24 RR 2d 1501 (1972).

[5]15 U.S.C.A. 4, 18, 21.

[6]Using the 6.7 million subscribers reported in systems fully or partially owned by TCI: Virginia Munger Kuhn, "TCI Expansion Paying Off With Leaps in Revenues and Cash Flow Margins," *Cablevision*, March 2, 1987, p. 14.

[7]Herbert H. Howard, "Ownership Trends in Cable Television: 1985," National Association of Broadcasters, September 1986.

[8]Victoria Gits, "A Thriving Band of Small Independents is Keeping Doldrums and MSOs at Bay," *Cablevision*, May 4, 1981, pp. 36–41.

[9]William H. Johnson, "Structure and Ownership of the Cable TV Industry," in *The Cable/Broadband Communications Book*, p. 12.

[10]"Bennett Urges Broadcasters to Program Leased-Channels," *Cablevision*, March 30, 1981, pp. 117–119.

[11]*Cable Television Report and Order*, 36 FCC 2d 143, 24 RR 2d 1501 (1972).

[12]FCC, "Information Bulletin, Cable Television," 00867, October, 1980.

[13]*Report and Order in CC Docket 78-219*, FCC 79–775, 44 Red. Reg. 75156 (December 19, 1979).

[14]Christopher H. Sterling, "Cable and Pay Television," *Who Owns the Media? Concentration of Ownership in the Mass Communications Industry*, Benjamin M. Compaine, ed. (White Plains, N.Y.: Knowledge Industry Publications, 1979), p. 308.

CHAPTER **14**

Business Operations

Customer service representatives, Columbus, Ohio.

This chapter covers the financing of cable systems and the functions of the various departments in a cable system office: general manager, accounting, marketing, advertising, sales, customer service, community relations, technical operations, engineering, programming and personnel. The chapter concludes with a listing of professional resources in the cable industry; trade publications and industry associations.

FINANCE

As new systems are built and existing systems expanded or rebuilt, the cable industry needs a continuous capital supply. Diverse sources are now available. Whereas the cable industry was once characterized by frantic entrepreneurs, trying to make the cable business credible and scrambling for money at any price, the game now is to weigh the wider options and manage equity and debt carefully.

Equity

Equity financing, where capital is supplied in return for ownership interest, is important to the cable industry. Prior to the 1986 tax law, a principal source of equity financing was the *limited partnership*. In a public offering limited partnership interests were sold to qualified investors in packages, as low as $1,000 per unit, who could take advantage of the tax write-off incentive of this particular investment vehicle. An amount equal to up to 100 percent or more of the limited partner's original investment could be sheltered from federal income tax, depending on the specific structure of the partnership agreement. Losses generated during the early years of the cable system development were "passed through" by the general partner to the nonvoting, limited investors who participated in a fund used to acquire or build cable systems. Private limited partnerships were also used by the industry. The limited partnership financing method raised about $340 million for the cable industry in 1984.[1]

Tax reform modified investment tax credits and the accelerated cost recovery system extending "at risk" to all forms of limited partnership financing. These changes have reduced the attractiveness of limited partnership financing.[2]

A *private stock* offering is the other major source of equity funding for cable. Shares of stock in cable companies are put on the market through investment houses. Stock offerings are estimated to have produced $223 million for the cable industry in 1984.[3] *Venture* capital, where private investors put up money for high-risk, high-growth-potential businesses, is still used in the cable industry, but much less so as growth has leveled and risk diminished.

Debt

The most common method of financing for established cable companies is through borrowed money. Cable is said to be a heavily *leveraged* industry, where the *debt-to-equity* (worth) ratio is high. Robert Morris Associates reports the debt-to-worth ratio (calculated by dividing total liabilities by tangible net worth) at 14.7. This is an extremely high figure. The comparable ratio for television broadcasting stations is 2.1.[4]

Money is loaned for construction and operating costs with the system assets as security. Equity in new construction (which must be a proportion of the capital cost) comes from cash flow. In 1984, there were more than $9.4 billion in loans outstanding to the cable industry.[5] Banks are by far the largest lenders (Toronto-Dominion, Chase Manhattan, Continental Illinois, among others). The largest banks may have a portfolio of over $500 million in loans to cable. Insurance companies (such as, Aetna Life, John Hancock Mutual Life, Mutual Life Insurance of New York) are also major lenders to the cable industry. The larger companies may have nearly $200 million in loans to cable.[6]

The loans range from about 5 to 10 years with lending rates one or two points above the prime rate for established MSOs or borrowers. The leveraged cable industry works hard to finance and refinance loans for the most favorable conditions.

At least one cable company, Comcast, has used *industrial development bonds*. These are low-interest bonds issued by state or local governments to stimulate business. They might be used to encourage cable expansion into areas that are marginally attractive for cable investment.

Franchising authorities try to look carefully at the financial structure of cable applicants, to protect themselves from subsequent criticism if the franchisee were to delay construction or provide poor service because of under-capitalization.

Senior cable creditors usually get a first lien on the cable plant, most operating equipment, accounts receivable, *and the franchise rights* of the borrower. In default, the bank, or other creditor, would take over all tangible and intangible rights of the franchise.

The organization of a cable system varies greatly from small to large and from independently-owned to MSO systems. In the remainder of this chapter we will discuss the departments of a relatively large independent system. Functions and departments are likely to merge in a smaller system. In an MSO, some of the work of the departments is taken over by regional offices.[7]

GENERAL MANAGER

The general manager oversees all departments and answers to the system owners. The general manager creates and monitors the *operating budget* for the system and each department. The operating budget is an advance estimate

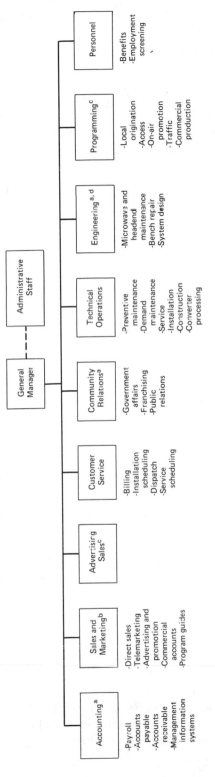

FIGURE 14-1. Organization chart for medium or large cable system.

Administrative Staff

General Manager

Accounting[a]
-Payroll
-Accounts payable
-Accounts receivable
-Management information systems

Sales and Marketing[b]
-Direct sales
-Telemarketing
-Advertising and promotion
-Commercial accounts
-Program guides

Advertising Sales[c]

Customer Service
-Billing
-Installation scheduling
-Dispatch
-Service scheduling

Community Relations[a]
-Government affairs
-Franchising
-Public relations

Technical Operations
-Preventive maintenance
-Demand maintenance
-Service
-Installation
-Construction
-Converter processing

Engineering[a, d]
-Microwave and headend maintenance
-Bench repair
-System design

Programming[c]
-Local origination
-Access
-On-air promotion
-Traffic
-Commercial production

Personnel
-Benefits
-Employment screening

[a] Often provided on regional or national basis by MSOs
[b] Often supplemented by regional or national offices by MSOs
[c] Combined with Sales and Marketing in smaller systems
[d] Combined with Technical Operations in smaller systems

229

for all expense accounts and revenue projections. Typical cable system costs and revenues are in Appendix G.

In addition to the responsibility for the smooth functioning and bottom line success of the cable system, the general manager represents the system in public and before the franchising authority. In this area, the general manager may be aided by the marketing department manager who is skilled in dealing with the public.

The general manager hires and fires higher level personnel, must keep people motivated, and be concerned for employee welfare.

ACCOUNTING

The accounting department keeps the books (usually for all departments), and it informs the general manager on the state of the operating budget. Cable systems are unique in several ways. The emphasis in this section is on those special features of the business.

Cable businesses are subject to several taxes, depending on the location. These may include franchise fees to the franchise authority and, in some cases, to the state; copyright fees; tangible personal property taxes on the cable plant, as well as on all other tangible personal property; intangible property taxes; and utility taxes in some jurisdictions. All the other normal business taxes and licenses are assessed against cable companies, such as federal and state income taxes. In addition, in some states cable service may be subject to sales tax, which must be collected from the customers and passed on to the state.

"Operating income" and "cash flow" are important terms in the cable industry. *Operating income* is operating revenues minus operating costs (before interest, depreciation and amortization). *Cash flow* is operating income less taxes.

It is appropriate to point out that terms such as "operating income" and "cash flow" do not have a similar definition in all industries. Furthermore, these terms are not used consistently by all cable companies in their published financial information since many are subsidiaries of larger corporations, some of them conglomerates with operations in various industries. Consequently, the financial information of these cable companies may be presented in conformity with the practices of other lines of business in which the parent company operates.

Operating income per subscriber is a figure that is useful in comparing one cable system with another because the figure is independent of capital structure (how the system is financed) and depreciation schedules (which are somewhat arbitrary accounting devices).

Cable is sometimes called a *cash flow business*. Limits on loans to established cable systems are often based on multiples of cash flow—up to about four and one-half times cash flow. Good cash flow is necesary for debt servicing in the leveraged cable industry, and the market value of a cable system may be indexed by its cash flow. The sales price of a system may be in the area of 12 times annual cash flow. Another more general rule of thumb takes into account current interest rates. Interest rates should also be reflected in the sales price. The rule states that the cash flow multiple when multiplied by the interest rate should equal 100. For example, if the interest rate is 9 percent, the cash flow multiple should be 11.1 (9 × 11.1 = 100). A system with an annual cash flow of $1,000,000 should sell for $11,100,000 (11.1 × $1,000,000) at a time when the expected interest rate is 9 percent.

Charges by Program Suppliers

Programming is a unique cost to cable businesses. Most of the basic program services charge a fee per subscriber. These fees are negotiated and agreed upon in an affiliation contract. A base per-subscriber fee is usually published by the supplier (1986 base fees are identified for basic satellite network in Appendix C). This fee is the discounted according to the number of subscribers the cable company brings to the basic program supplier. MSOs aggregate subscribers over all systems for purposes of this discount, although each headend is reported separately. The fee may be graduated up or down over the period of the affiliation contract, such as 7 cents per subscriber per month over the first year, 9 cents in the second year, and 11 cents in the third year. A network may give rebates on fees to operators proportional to the number of subscribers based on the advertising sales performance of the network. Shopping networks do not charge a program fee and pay a percentage of the revenues from the cable franchise area and, in at least one case, may give equity in the network company.

Operators ordinarily pay fees in advance based on the prior period's subscriber count. The reporting cycles and procedures vary among networks. The length of the contract also varies, although three years is common. Some MSOs have entered into long-term affiliation agreements with some program suppliers to help the supplier in financing operations and product acquisition in return for a favorable discount.

Cable operators are not of one mind on their willingness to accept basic service fees and fee increases. Those most opposed to the programming fees are operators with the least dependence on the basic satellite networks, that is, those in markets where they are providing clear broadcast signals that would not otherwise be available. Cable operators in urban areas, in range of many broadcast signals, need a strong satellite network system. They are less reluctant to pay the affiliation fees. Even so, there is a constant push and pull between operators and basic satellite networks over fees. The program suppliers are able to squeeze the operators by inreasing fees because the operator's only recourse is to drop the channel—a somewhat empty threat, where subscribers have come to accept and expect the channel. The one-of-a-kind networks are

FIGURE 14-2. HBO service charge card. Such rate card is subject to change at any time.

HBO® Service Charges and Applicable Volume and Performance Discounts Effective April 1, 1985:

Monthly HBO Service Charges

Supplementary Service Charges for Each HBO Subscriber:	$7.00 or less	$7.01 to 8.00	$8.01 to 9.00	$9.01 to 12.00	$12.01 to 13.00	More than $13.00
HBO Service Charges Effective April 1, 1985	$4.70	$4.70 plus 40% of charges over $7.00	$5.10 plus 20% of charges over $8.00	$5.30 plus 15% of charges over $9.00	$5.75 plus 20% of charges over $12.00	$5.95 plus 40% of charges over $13.00

Volume Discount Schedule and Application

Affiliates qualify for a reduction in their applicable HBO Service Charges in the event they serve the following number of HBO Subscribers:

Supplementary Service Charges:	$7.00	$8.00	$9.00	$10.00	$11.00	$12.00	$13.00

Volume Discount Schedule:

Number of HBO Subscribers:	HBO Service Charge Volume Discount:		HBO Service Charges With Volume Discount Application:					
10,000 to 24,999	2%	$4.61	$5.00	$5.19	$5.34	$5.49	$5.64	$5.83
25,000 to 39,999	3.5%	4.54	4.92	5.11	5.26	5.40	5.55	5.74
40,000 to 74,999	5%	4.47	4.85	5.04	5.18	5.32	5.46	5.65
75,000 to 99,999	15%	4.00	4.34	4.51	4.63	4.76	4.89	5.06
100,000 or more	20%	3.76	4.08	4.24	4.36	4.48	4.60	4.76

Performance Credit Schedule

Affiliates not receiving the HBO Service Charge reductions on the Volume Discount Schedule may elect to receive Performance Credits for each system that qualifies as determined by the following schedule:

HBO Subscribers as Percentage of Affiliate's Average Basic Cable Subscribers in Each System:	HBO Service Charges Before Applicable Credit:	
	Less than $5.20	$5.20 or More
40%—59%	—	2%
60%—79%	2%	4%
80% or more	3%	6%

Performance Credit Application:

Supplementary Service Charges:	$7.00	$8.00	$9.00	$10.00	$11.00	$12.00	$13.00

HBO Subscribers as Percentage of Affiliate's Average Basic Cable Subscribers in Each System:	HBO Service Charges With Performance Credit:						
40%—59%	$4.70	$5.10	$5.19	$5.34	$5.49	$5.64	$5.83
60%—79%	4.61	5.00	5.09	5.23	5.38	5.52	5.71
80% or more	4.56	4.95	4.98	5.12	5.26	5.41	5.59

Performance Credit General Terms: A. Affiliates receiving a Volume Discount are not eligible for a Performance Credit, and vice versa. B. A Performance Credit is only applicable on an individual system basis.

particularly strong in this respect. On the other hand, large multiple system operators that bring hundreds of thousands of subscribers to a network are critical to the network and its advertisers. These operators are in a good position to negotiate favorable fees with the networks.

Pay program suppliers charge much higher fees for programming. Some suppliers assess a flat fee and others tie the fee to the retail rate charged by the operator so that the supplier shares in the revenue in the systems with high rates. Pay services also discount fees on the basis of volume (number of aggregated subscribers), and in some cases the rates may be discounted on performance (for example, the higher the penetration of pay to basic for the pay supplier's service, the higher the discount).

Affiliation contracts for both pay and basic services may include making allowances for launch of the service and/or continuing promotion. A cable system that affiliates with more than one service from a single supplier may realize lower affiliation fees for the package than assessed for each channel separately.

Accounts Receivable

Cable companies maintain an extraordinarily large number of open accounts receivable compared with an average business, and none is a large account. This translates to high bookkeeping costs.

Several kinds of cable billing systems are in general use. Systems with under 1,000 subscribers may have a *manual* system. The subscriber is sent a book of coupons. One is returned every month, every two months, or each quarter with the payment.

Bigger systems may rely on *batch processing*, sending daily, or less often, by mail, subscriber information on specially provided forms to an organization which mails the bills and records collections. Most cable systems contract for this work with an independent data processing company that has service tailor-made for cable.

On-line, or *on-line shared-time* systems are used by the larger cable systems. In a shared-time system, the operator at the cable office enters information on a CRT, which is connected to the central processing unit (CPU) of the data processing company by a dedicated telephone line. In the on-line system, the computer is located at the cable system business office. The equipment itself is usually sold by a data management company which generally does not manufacture it, but which has developed programs (*software*) for this equipment and in turn licenses the programs and sells the equipment to the cable company. A monthly royalty is paid for the computer software, which is expanded and revised as necessary.

An operator enters all subscriber information, as soon as it is received, into the terminal—new orders of service, disconnects, upgrades, downgrades. Bills are printed by the computer. They are added to the appropriate stuffers or guides, and mailed. When the bill is paid, the credit is entered in the computer. If a customer calls to inquire about a statement, the operator need only enter the code into the terminal to call up the customer's history. The computer is also programmed to provide a series of summary reports and can be used for scheduling installations and routine maintenance, parts inventory, and so on.

The most advanced on-line systems combine data processing with converter addressability. One entry on the system accomplishes both order taking and converter channel authorization.

The handling of overdue accounts varies with the cable system and may be affected by provisions of the franchise that specify the conditions under which a subscriber can be disconnected. Most of these rules require formal notices before service may be cut off. The alert to overdue payments may be triggered by a prescribed number of days overdue. The first notice to the subscriber may be a printed notation of the overdue amount on the next month's bill or a special insert in the bill envelope with a mild admonition. If the overdue sum is not paid within a specified time, a special letter may be mailed taking a harder line and spelling out the consequences of nonpayment, including the charge to reconnect. The same notice accompanies the next billing. If this does not bring payment, a telephone call by a customer service representative or a territory manager may attempt to save the account. This failing, a disconnect notice is sent. The service is disconnected if the bill is not paid on a particular date (usually at about 65 days overdue). A technician collects the converter and may make a physical disconnect. In calling on the household, the technician may be instructed to make a last-ditch effort to collect the sum owed and save the account. To reconnect, the once disconnected subscriber may have to repay the previous sum owed and make a higher-than-usual prepayment.

Collection agencies may be called in to clear up accounts that have not paid through the normal system follow-up procedures. Generally, they split any amounts collected with the cable operator.

MARKETING

The marketing department produces most of the revenue for the system and, as noted, must keep the general manager informed so that revenue projections can be made and adjusted as necessary.

The functions of the marketing department are described in Chapter 15.

ADVERTISING SALES

Advertising sales may be a part of the marketing department in some companies, but is likely to be independent in the larger systems. Sales of advertising time to businesses is substantially different from cable subscription sales and retention. The functions of the advertising sales department are described in Chapter 16.

CUSTOMER SERVICE

As a service business, particularly one in which the service is not necessarily essential, the cable industry considers customer service a very significant function. The term *customer service* is used here to mean the functions of dealing with customer installation orders, billing questions, and requests by the telephone or at the business office for repairs. In small cable companies, one or a few persons may handle all these areas. In the larger company, different groups of people specialize in installation, billing, and repair with separate phone listings for each. A *uniform call distribution* telephone system distributes calls evenly among several lines in use.

Orders

The importance of customer service to the marketing function is emphasized in Chapter 15. Product knowledge and sales skill is necessary for the customer service representative (CSR) taking orders. To keep work current, the CSR writes the order while the customer is on the line. Manuals and scripted protocols help on programming, installation, and billing.

CSRs taking orders may earn a *spiff*, or small commission, on each sale. In some systems, a salesperson may get credit for orders from people who have been contacted. The customer may only be entitled to a special rate on installation if the salesperson's name or number is mentioned. A salesperson, failing to close, suggests that the individual think about the service and call in the order to take advantage of the special offer. The CSR records the salesperson's ID to give proper credit.

The CSR may schedule the installation at the time of the order and then make the credit check. If the customer does not clear a credit bureau check and the cable company's own records, then a letter is sent that may require a full deposit on the converter before installation. The CSRs in this department also handle requests for downgrades or the ultimate downgrade, a disconnect. The CSR may also earn a commission for each *save*, when a customer is dissuaded from downgrading.

Billing

Customers call the Billing Department because they do not understand the bill, the handling of partial months (*prorating*), and installation charges with debit and credit entries to indicate waived charges. In batch-processed billings, in particular, there is a lag between the receipt of a payment and posting. The customer may be concerned that a check mailed several days ago does not appear to have been received by the cable company. Then there are the inevitable mistakes on the bill. Inquiries come by phone, messages written on returned bills, or letters.

When the customer calls or writes, the CSR gets a microfiche record of the account or calls up the computer record. The CSR can then see all the transactions, current level or service, and account history.

If there are several billing cycles, or continuous billing throughout the month, the call load on billing CSRs is relatively even.

Repair

Dealing with technical faults in the cable system is now generally referred to as *repair* since "service" may also refer to the product offered. When the cable system receives a request for repair, the dispatcher first asks a series of questions about the exact nature of the problem reported. A clear understanding of the symptoms is the first step in determining the solution to the problem. A high percentage of the calls may be the result of an inability to tune the receiver, unplugging the set, or disconnecting the cable during a system outage. The customer is talked through a series of steps that may rectify the problem without an expensive service call. The customer is grateful, having been saved the embarrassment of having a repair person call just to plug in or fine-tune the receiver. About 25 percent of the calls can be "cleared" on the telephone by talking customers through front-panel adjustment and other checks. This requires dispatchers who can relate various problems to the proper adjustments and have an awareness of the different types of receivers.

Before dispatching a repair call, the dispatcher checks billing status. The customer may have been disconnected for nonpayment, in which case the bill must be paid before reconnection. If the account has not been disconnected but is in arrears, the dispatcher may ask for payment prior to repair work. The dispatcher then takes all appropriate data from the subscriber. It is important to record accurately the problem for the technician who must develop a priority for the calls. A loss of all signals is obviously a higher priority than a single impaired channel.

The dispatcher locates the repair call in an area or section of the system. This may be done by attaching an area or section designation code to each address in the subscriber data base. The code may have an electronic significance, such as all households served by a particular bridger amplifier. If several calls are received with the same code, the technician has a good idea of where the trouble lies. If the cable company has a status monitoring system, the dispatcher may run a check to pinpoint the trouble. The dispatcher then enters the repair request on the appropriate dispatch log. Dispatch logs list all issued repair calls for each technician, the time they are issued, and the time they are completed. The log may also list the nature of the problem and other data. Many cable companies have computerized this function. The computer can keep the dispatch clerk informed of the status of the log at any moment and also generate reports on service problems in each electronic section of the system, the efficiency of technicians, and so on. If the dispatch log for a technician has too many calls, exceeding an established quota, an adjacent technician with fewer calls may be transferred temporarily into the territory. The technicians are in contact with the office by two-way radio.

CSRs working on the telephone in the order, billing, and repair departments take pride in their ability to deal patiently with all types of customers and "turn" an irate caller into a satisfied customer.

Customer Service Audit

Many cable systems periodically, or continuously, perform a *customer service audit* to review customer service performance. *Every* customer service task is identified and, if necessary, broken down into the simplest units. For example, the task of "installation" would involve a CSR scheduling the installation with the customer over the telephone after the order has been received. Another unit of the task would be the physical installation at the subscriber's home.

Standards of performance for each of these tasks or task units are developed. In our example, the customer telephone contact should be friendly and businesslike. The CSR telephone contact might be monitored by a supervisor who would then rate the contact on the appropriate criteria. The rating should achieve a minimum score on each criterion. The work of the installer at the home should also be friendly and businesslike. Furthermore, the installation should satisfy the customer (for example, be neat and in the correct location). The installer may also be responsible for explaining the operation of the converter, a VCR, and other equipment. The supervisor would follow up installations asking the new subscribers to rate the performance of the installers in all these areas. Again, a minimum rating would be expected.

A final performance standard for these tasks might be the timeliness of the installation. The standard might be a completion of the installation within two working days after the order. To measure this standard, the supervisor would have to know the day of the order and the day of the installation.

All customer service tasks would be similarly evaluated against the performance standards. In areas where the standards are not met, the procedures must be studied. For instance, recruiting, selection, and training of personnel, as well as motivation, supervision, and compensation plans, might have to be revised. Support systems such as instruction manuals, office facilities, and communication systems may need improvement.

This systematic approach to customer service evaluation of at least a sampling of customer transactions is useful in focusing attention on effective performance and detecting problems.

COMMUNITY RELATIONS

One of the principal tasks of the community relations department is to maintain a positive relationship with the franchising authority. The operator makes periodic reports directly to the city council or through a department of telecommunication or cable commission. Community Relations is responsible for preparing for franchise renewal, as described in Chapter 12, and may seek new franchises in adjacent or nearby areas. The department must respond to complaints that come to the city. It represents the company, at least *ex officio*, on a number of city committees and civic groups. It is also the liaison with government and education access channel managers and users.

The department handles public relations. It may represent the company at community events, take on the company's share of work in various civic projects, work with student interns, respond to requests for information about cable, prepare company literature explaining cable, and release news to other media about company activities.

TECHNICAL OPERATIONS

A cable plant is maintained by the systems maintenance staff, consisting of the *chief technician* and *maintenance technicians* (who maintain the cable distribution system) and *service technicians* (who answer individual service complaints). In very small systems, there is usually no distinction between those categories, with a small number of technicians performing both tasks. Very large systems often have two or more *lead technicians* or *hub supervisors* who are responsible for the maintenance technicians within a particular hub or area of the system. Often only the highest-level technicians maintain the trunk line and less experienced people maintain the feeder and individual drop lines.

Chief Technician

The chief technician in a cable system is both the manager of the field technical staff and the person called upon to deal with technical problems that are above the ability of the maintenance technicians. Frequently, chief technicians come up through the ranks of the field technical staff. They may have additional training and, in some cases, college degrees. They are familiar with use of various types of sophisticated test equipment. Often the chief technician also supervises a construction crew, which makes system extensions and moves cable facilities.

Installers

The installers make the connection to the cable distribution plant, bringing the drop line from the feeder system, if this has not already been done, to the home. The installers must know how to attach the cable to the home, go through exterior walls, and get to receiver locations with minimal interior exposure.

The installers are not usually radio dispatched. They pick up their orders for the day at the business office.

An important function of the installer is to inform the subscriber about how the cable is connected and how the converter works. Use of the converter is demonstrated. The installer must make a good connection, and a good impression, to get the cable company and its customer off to the right start.

Installers also make converter changes, as orders for new services are received, and disconnect subscribers who move or cancel. The installer may be asked to make a last attempt to discourage disconnect either by collecting money from nonpays or reselling cable to those who may disconnect from dissatisfaction with the service.

Some installers are independent. They are hired on a commission basis by the cable operator.

Service Technicians

Service technicians are generally less technically qualified than maintenance technicians but must be presentable and courteous, since they are in constant contact with subscribers. Most subscriber complaints relate to problems with operation of their TV set or converter.

Cable service technicians must be capable of retuning television sets and of making minor adjustments to them. One of the service technician's main tools is a properly working TV set, which is often brought into a subscriber's home to demonstrate that the problem is with the subscriber's television set rather than with the cable service.

Service technicians also troubleshoot problems in individual subscriber drop lines. Sometimes the cable will have been damaged or water may have entered one of the cable fittings, causing corrosion. If a service technician traces the problem back to the cable plant, a maintenance technician is notified to correct the problem.

Service technicians are often used to move cable outlets, add additional outlets within subscribers' homes, and install new equipment for new service, instructing subscribers in the proper use.

Maintenance Technicians

Plant maintenance can be divided into two categories. First is *preventive maintenance*, which consists of routine inspection and testing of the system, particularly the trunks and supertrunks. In large cable systems, one or more people are often assigned solely to this task. A van is equipped with sophisticated test equipment, including frequency measuring equipment, spectrum analyzers, video monitors, and video test instruments. Usually, proof-of-performance testing is done by the preventive maintenance staff. Systems with status monitoring equipment require fewer preventive maintenance people, since many tests are performed automatically.

The second type of maintenance is *demand maintenance*, which is performed in response to subscriber complaints. Status monitoring is of significant help in localizing faults. It is often difficult to distinguish between individual service complaints and plant problems. Status monitoring automatically pinpoints the exact location of a plant fault or informs the dispatch personnel that there is no system fault so that a service technician can be dispatched.

Most routine plant maintenance, such as failure of amplifier modules, blown fuses, or physically damaged cable, can be repaired by the maintenance technicians. Occasionally, however, faults are difficult to isolate. Intermittent problems that come and go with different conditions or problems that are subtle in nature may require the services of a lead technician or even the chief technician.

Maintenance technicians usually drive small vans equipped with ladders, a stock of spare equipment, and a minimal amount of test equipment. Lead technicians and chief technicians have much more sophisticated test equipment and often have access to bucket trucks (vehicles with hoist equipment to lift a technician up to an amplifier or other piece of equipment to be serviced).

Maintenance technicians are assigned to a territory for long periods so that a particular area of the plant becomes their responsibility. Periodically the territory is changed to lend some variety to the work and to break up cliques of regional technicians who may begin to establish policies independent of the central office.

Technicians are issued tools for which they are responsible. At the termination of employment, they must turn in the tools or pay for them. Broken tools are replaced from inventory.

Parts inventory control is very rigid. The minimum levels must be set high enough to accommodate the worst periods of demand, usually after severe thunderstorms. Not all repairs are made in the field. Field technicians may bring electronic parts in for bench repair by another group of technicians. Once repaired, the parts are put back into inventory.

Cable company service hours vary greatly. Some companies provide around-the-clock maintenance staffs; others respond to a problem only during normal working hours or in emergency conditions. Usually, after working hours a maintenance technician is on duty with a lead technician or a chief technician on call in case of an emergency.

System Faults

Cable system faults, even in a well designed and well managed system, are fairly frequent. The most common problems are as follows:

1. *Electric power outage:* Often, electrical service will be out in one part of town causing the cable system to be out not only in that area but also farther down the trunk. Newer systems, with standby power, will keep operating, but systems with normal power supplies will be out of service. If a power outage lasts for a significant period, a portable generator can be connected to the cable system to provide electric power temporarily.

2. *Damage from lightning and electrical storms.* Even the most modern cable amplifiers are subject to damage from high-intensity electrical discharges from lightning. In some areas of the country, particularly the South and Midwest, lightning damage is the main source of cable outages. Usually the damage is to amplifier modules, but sometimes passive devices and even cable can be damaged.

3. *Physical damage:* Aerial cables can be damaged by automobile accidents, wind, and ice. Underground cables can be damaged by new underground construction, installation of fences, mailboxes, and even flower gardens or by accidental or intentional damage to pedestals. In most cases, when physical damage occurs, a section of the cable must be replaced.

4. *Failure of active equipment:* Electronic components have a finite useful life. Although cable equipment has become more reliable over the years, amplifiers are still subject to occasional failure.

5. *Connector problems:* Over the years, connectors in cable systems can fail, allowing water to enter the cable. Often, signal ingress or egress and *flashing* (the presence of intermittent flashes in subscribers' television pictures) may occur. Many times these problems are difficult to find since they come and go with different weather or wind conditions. Two-way systems in particular are vulnerable to this type of problem.

ENGINEERING

In small and medium-sized cable systems, the chief technician performs all the maintenance on the headend equipment and repairs of equipment that fail in the field. In larger systems, however, a separate engineering department often exists, headed by the chief engineer. This department has responsibility for maintenance of all headend equipment, video equipment, computer systems, as well as defective amplifiers, passive devices, and converters. The engineering department is also responsible for preconstruction activities such as design, pole permits, easements, and the walkout. These functions are described in Chapter 3.

The Chief Engineer

The chief engineer frequently has a college degree and may have an advanced degree. The chief engineer is usually familiar with television broadcast and production equipment in addition to cable equipment. The chief engineer must have computer experience as well.

Bench Repair

Under the supervision of the chief engineer is usually the bench repair staff. In this area, defective cable TV equipment is repaired. The bench repair technicians have diagnostic test instruments and simulations of the cable system, which allow them to repair the test cable equipment under conditions similar to those encountered in the field. Smaller cable companies may subcontract the repair of cable equipment and multiple system operators frequently centralize the repair of equipment.

Video, Microwave, Earth Stations, and Computer Equipment Maintenance

In small and medium-sized systems, the chief engineer is often personally responsible for this type of maintenance. In larger, more sophisticated systems, one or more engineers or technicians may be required for these functions. Recent major market franchises include elaborate studio facilities, microwave equipment, and earth stations together with complex computer systems. In these systems, several maintenance engineers and technicians may be required to keep all the equipment operating. The chief technician in smaller systems usually maintains the headend equipment. Larger systems, with more than one headend or several hubs, may require one or more people devoted entirely to headend maintenance and testing.

Technicians, installers, and dispatchers may have a weekly meeting. Part of the meeting is devoted to training: new procedures are outlined, new parts are described, and ways of meeting difficult problems are discussed. The meeting also serves to keep the technicians and installers in touch with the company, since they are otherwise occupied continuously in the field. Correspondence courses are encouraged by partial tuition reimbursement and pay increases on completion. Some investment of the employees' own money may be required to assure a commitment to continuing the self-study.

SYSTEM SECURITY

Technologies for dealing with the challenging problems of theft of service are discussed in Chapter 4. Theft of service may occur in several ways. Subscribers may tamper with junction boxes or pedestals in apartment complexes or underground construction. Dishonest employees or former employees may hook up their friends. And subscribers may actually climb poles to connect themselves, insert splitters for additional connections, buy their own converters, or tamper with converters supplied by the operator.

Many people also get free cable service through administrative errors on the part of the cable company. Even a very small error rate (such as the failure to disconnect when someone moves) will cause a large cumulative loss. Most cable systems conduct physical audits of their systems every three to five years. Typically, 2 to 5 percent illegal connections are discovered.

Theft of converters and descramblers is becoming serious because it is not feasible to ask for a deposit large enough to cover the cost and still sell the service at an attractive price. Cable companies have a variety of policies to protect themselves, including the following. Homeowners are given a converter with no deposit (saving the cost of a credit check). A credit check is run on the apartment dweller (at a cost of about $1). If the potential subscriber meets minimum credit standards, no deposit is required. If not, a substantial deposit is necessary. A customer with very poor credit will be refused the service entirely. Theft of addressable converters may be deterred by informing the subscribers that the converter will be deactivated if they fail to return it after discontinuing service.

The subscriber must sign an agreement form that usually states the following: (1) the converter (terminal, console) is the property of the cable operator; (2) if the service is terminated, the converter must be returned promptly—the service representative is to be allowed to enter the premises to retrieve it; (3) the value of the converter is stated and the subscriber is obligated for that amount if it is not returned; and (4) if there is physical damage, beyond reasonable wear and tear, the subscriber is obligated for reasonable costs of repair. The subscriber is informed of the law on theft of cable television service and the penalties for violation.

PROGRAMMING

Cable programming is discussed at length in Part II of this book. The programming department of a cable system is responsible for selecting from among available off-air signals and satellite-delivered channels. The more channel space that is available, the less difficult this task is. There are many more channels available than a 12- to 20-channel system can use. This is complicated by the fact that newly available channels, if carried, must displace existing channels that have already established some viewer loyalty. Sometimes a programming department will *cherry pick* from several available services, taking the most attractive programs or program dayparts from many different services and putting them into just a few channels.

Program services must be assigned to channels. It is the responsibility of the programming department to develop a strategy for the most favorable line-up. Cable systems usually attempt to keep local broadcast stations on their own channel number although this may not always be possible because of co-channel interference.

Most programming departments coordinate PEG access programming. This may mean only providing a cable liaison with channel users located in public buildings that feed the cable system. Or it may mean providing and maintaining company-owned production equipment and studios, training users, and scheduling. Programming departments procure or produce LO programs. They often produce local advertising and schedule all advertising spots.

The programming department usually takes responsibility for program promotion. This includes everything from tune-in advertising and publicity, to contests, to supplying program schedules to the various guide publishers.

Program Guides

The number of channels offered by cable systems makes program guides essential. Several program guide options are available to serve this consumer and cable system need. Cable channel electronic text guides are discussed in Chapter 6. This type of guide is always available, current and convenient to the viewer.

Most printed guides are custom-printed by specialized television guide publishers, of which there are several. The guide publishers will do as much or as little as the cable system desires. Some will take over the selling of advertising. Program descriptions for all broadcast stations and all basic and premium cable networks are collected by the publisher from the stations and networks. The cable system need only provide program information for local origination channels. Printed guides are sold to the cable system and then sold or given to subscribers. Some are sold directly to subscribers by the publisher, with the operator providing mailing lists.

Some guides, *generic guides*, cover only the major cable basic and premium networks and are published nationally with the same content except for time zone differences. Newspaper guides do the same thing for their circulation regions. Users of generic guides must cross-reference a channel line-up for their own system printed somewhere in the guide or on a card with the converter, or rely on their memory of key channels. *System-specific* guides include all, or almost all, of the programming of a single cable system including the proper numbers of that system's channel line-up.

Guides may be weekly, bimonthly, or monthly. Guides are produced in a variety of sizes, but the digest size, about 5 × 7½ inches, is preferred by most users. Many program suppliers provide monthly listings that can be used as bill stuffers.

Program logs list programs with times chronologically in three formats: (1) *title only*, for monthlies or bill stuffers; (2) *partial descriptions*, with titles for series and descriptions for movies, talk shows, and specials; and (3) *full description,* with every program titled and described. Generally, the more frequent the publication, the more descriptive the guide.

Guides may also have programs presented in a *grid* format, where the channels are listed down the grid on the left and the times across the top. Only titles or very brief descriptions describe the programs filling the grid.

Guides may have *breakouts,* separate listings of each day's programs in particular categories such as sports, movies, and specials. Some have sections for program highlights. And, of course, guides may include picture and text features on programs and performers. Most of the full-sized guides sell advertising space.

PERSONNEL

Personnel departments must keep the system in compliance with federal and state regulations on equal opportunity employment and other rules on hiring and firing. The department creates and implements benefits packages for employees. It helps employees file insurance claims.

Federal Equal Employment Opportunity

FCC rules enforce the 1984 Cable Act equal employment opportunity provisions. Operators with more than six full-time employees must report annually statistics on new hires, promotions, and recruitment sources. Random audits are conducted to verify the reports. Every fifth year operators must file more detailed reports for a more thorough FCC investigation. Cable systems with 6 to 10 full-time employees must employ a proportion of at least 50 percent of the proportion of minorities and females in the labor force overall and 25 percent in the top four job categories. Systems with 11 or more full-time employees must have 50 percent overall *and* in the top four job categories.

Training

Several cable MSOs have their own training facilities where technical people learn basic electronics and cable system design, installation procedures, and line technology. These schools require outdoor facilities such as "wall laboratories" or "install houses" where aluminum, wood, brick, and asbestos walls present different installation problems; "pole farms" or "pole orchards" for line work; and earth stations. The Society of Cable Television Engineers has a broadband cable technician/engineer training and certification program.

Training is a regular function of management people in all areas. In addition, large systems or regional offices within an MSO may have an individual with an exclusive training function working with other staff members to train CSRs, marketing personnel, and technicians.

Unions

Because the classic cable systems employed few people and were in small towns where the climate for unionization was not especially hospitable, few such cable employees are unionized. Big cable systems with hundreds of employees present a better opportunity for union organization. The most active unions in cable are the International Brotherhood of Electrical Workers (IBEW) and the Communications Workers of America (CWA).

Some managements have attempted to preempt union organization by offering regular cost-of-living increases and benefits packages that include profit sharing and stock options. These plans also help to prevent raiding of both technical and management personnel by other cable companies.

MSO ORGANIZATION

A typical MSO centralizes some of the functions of a cable system at a regional level. Personnel at the system level may have lower salaries and carry less responsibility than in an independent system. Regional MSO engineering and marketing people may handle an entire section of the country. An individual MSO system may have no accounting function other than billing. The personnel department may be entirely removed from the system level.

The general manager at an MSO system usually hires and fires personnel, but the regional office may have a *dotted line* relationship with counterparts at the regional level. For example, the system marketing department head is hired by the general manager and works under the general manager, but also works directly with the marketing director for the MSO region.

PROFESSIONAL RESOURCES

Two major professional associations serve the cable industry. The largest is the National Cable Television Association (NCTA) with about 2,000 members serving 70 percent of the cable subscribers. The association lobbies for the industry in Washington; advises members on business, public policy, and technical matters; sponsors an annual convention and other meetings on special topics; and attempts to educate the public and other businesses about the nature of the cable industry. A number of state cable associations are affiliated with NCTA.[8]

The Community Antenna Television Association (CATA) performs similar services for the smaller cable systems.[9]

The Cable Television Administration and Marketing Society (CTAM) is a forum for the exchange of ideas and information. The organization sponsors an annual meeting, regional and special-interest seminars on management topics.[10]

The Canadian industry association is Canadian Cable Television Association (CCTA).[11]

The work of the Cabletelevision Advertising Bureau (CAB) is discussed in Chapter 16.[12]

Public, government, education access users of cable and LO programmers are served by the National Federation of Local Cable Programmers (NFLCP).[13]

Cable programmers and program suppliers participate in the National Association of Television Program Executives, a professional association for programming interests and a marketplace for program wares.[14] Media, government, and community relations people may hold membership in the Cable Television Public Affairs Association.[15]

Women's interests are served by Women in Cable,[16] minorities by the National Association of Minorities in Cable, Inc.[17] Franchising authority interests in cable are represented by the National Association of Telecommunications Officers and Advisors, an affiliate of the National League of Cities.[18] Engineers are represented by the Society of Cable Television Engineers.[19]

The backyard dish trade group is the Satellite Broadcast and Communication Association.[20]

There are several periodicals serving the cable industry. They are *CableVision*,[21] *Cable Television Business*, *Cable Marketing*,[23] *Cable Strategies*,[24] *Broadcasting*,[25] *CED (Communications Engineering Digest)*,[26] and *Multichannel News*.[27] Many newsletters cover communications. Most specific to cable is a series by Paul Kagan Associates.[28] Directed to LO public access, government, and education users of cable is *Community Television Review*.[29] Three directories for the industry include listings of all cable systems, the MSOs, suppliers, and associations. They are *Cablefile*,[30] *Broadcasting Cablecasting Yearbook*,[31] and *Television Factbook*.[32] A text titled *Telecommunications Management* covers management principles in cable and broadcasting.[33]

NOTES

[1]Virginia Munger Kahn, "Fountainheads of Capital: At Peak Flow with No End in Sight," *CableVision*, February 9, 1985, p. 27.

[2]Interview with Glenn Friedly, Horizon Cablevision, January 23, 1986.

[3]Ibid.

[4]"Statement Studies," Robert Morris Associates, Philadelphia, PA, 1980.

[5]Kahn, p. 27.

[6]Kahn, p. 29.

[7]For a review of cable and other intertainment industry economics see Harold L. Vogel, *Entertainment Industry Economics: A Guide for Financial Analysis*, Cambridge University Press, 1986, pp. 457.

[8]National Cable Television Association, 1724 Massachusetts Avenue, Washington, D.C. 20036, (202) 775-3550.

[9]Community Antenna Television Association, 3977 Chain Bridge Rd., Fairfax, VA 22030, (703) 691-8875.

[10]CTAM, 1220 L St., N.W., Washington, D.C. 20005, (202) 371-0800.

[11]Canadian Cable Television Association, 85 Albert Street, Suite 405, Ottawa K1P6A4, (613) 232-2631.

[12]Cabletelevision Advertising Bureau, 767 Third Avenue, New York, NY 10017, (212) 751-7770.

[13]National Federation of Local Cable Programmers, 906 Pennsylvania Ave., Washington, D.C. 20003, (202) 544-7272.

[14]National Association of Television Program Executives, 342 Madison Avenue, Suite 933, New York, NY 10173, (212) 949-9890.

[15]Cable Television Public Affairs Association, c/o Dave Andersen, Director of Public Affairs, Cox Cable Communications, Inc., 1400 Lake Hearn Drive, Atlanta, GA 30319.

[16]Women in Cable, 2033 M Street, N.W., Suite 703, Washington, D.C. 20036, (202) 296-7245.

[17]National Association of Minorities in Cable, Inc., 1722 Lafayette St., Denver, CO 80218.

[18]National Association of Telecommunications Officers and Advisors, 1301 Pennsylvania Ave., NW, Washingotn, D.C. 20004, (202) 626-3250.

[19]Society of Cable Television Engineers, 1900 L Street, NW, Suite 614, Washington, D.C. 20036.

[20]Satellite Broadcast and Communication Association, 300 N. Washington Street, Suite 208, Alexandria, VA 22314.

[21]*Cablevision*, 600 Grant St., Suite 600, Denver, CO 80203.

[22]*Cable Television Business*, 6340 south Yosemite Street, Englewood, CO 80111.

[23]*Cable Marketing*, 352 Park Avenue South, New York, NY 10010.

[24]*Cable Strategies*, 12200 E. Briarwood Aven., Suite 250, Englewood, CO 80112.

[25]*Broadcasting*, 1735 DeSales Street, N.W., Washington, D.C. 20036.

[26]*CED*, 600 Grant Street, Suite 600, Denver, CO 80203.

[27]*Multichannel News*, 300 south Jackson, Denver, CO 80209-3194.

[28]Paul Kagan Associates, Inc., 126 Clock Tower Place, Carmel, CA 93923.

[29]*Community Television Review*, 906 pennsylvania Ave., S.E., Washington, D.C. 20003.

[30]*Cablefile*, 600 Grant St., Suite 600, Denver, CO 80203.

[31]*Broadcasting Cablecasting Yearbook*, 1735 DeSales Street, N.W., Washington, D.C. 20036.

[32]*Television Factbook*, 1836 Jefferson Place, N.W., Washington, D.C. 20036.

[33]Barry L. Sherman, *Telecommunications Management: The Broadcast and Cable Industries*, McGraw-Hill, 1987.

CHAPTER 15

Marketing

Disney Channel premiums.

The marketing of cable system services is of interest to the consumer and franchising authority as well as to the system operator. The cost of service to the consumer is a function of the number of subscribers who use the service. If the system is not marketed aggressively, the monthly charge to subscribers will be high, and the profits to the operator will be low. Furthermore, the market value of the cable system itself—the index that is often used to assess the value of the capital investment rather than the earnings—is usually stated in terms of a dollar figure for each subscriber. That value is currently $1,500 or more per subscriber. This clearly indicates the significance of signing and retaining a single subscriber.

This chapter describes and analyzes the marketing of cable television services to residential consumers. We will deal with the market response to cable, marketing techniques and their application to acquisition, upgrade and retention marketing, and finally marketing administration.

MARKET RESPONSE TO CABLE

There are positive values to cable now well known to marketing people. In approaching the market through sales literature, advertising, telemarketing, or personal sales, cable firms attempt to capitalize on these factors.

Some of the values are independent of specific programming:

1. A clear picture and balanced audio on all channels without adjustment of antenna or rabbit ears for each channel change are very important to almost all television viewers.
2. Television may already be a primary activity in the household; so even a marginal improvement in the satisfaction with that activity is highly valued.
3. Television may not be a dominant activity in the household and programs are viewed selectively. Therefore, an increase in program choice is valued. Several cable companies have used variations on the theme "The less you watch TV, the more you need cable."
4. Viewing may be individualized. There is always something for the individual taste. This is characterized by the advertising theme, "Not just more choice, but *your* choice."
5. Twenty-four-hour access to television may have specific value to people working odd hours or to insomniacs. It means that certain program types, such as news, may be available at times other than those few periods designated by broadcasters.
6. Although most people in the United States have become accustomed to commercial interruptions, many prefer movies and children's programs to be commercial-free. Most pay channels do not have commercials.

7. Since very little of the broadcast week is given to children's programming, cable can make a differance to households with young children. Having children's fare may allay the guilt feelings of parents who allow television to babysit their children. Age-specific children's programming is possible on cable (see Chapter 7).

8. Cable has significant instrumental value. It provides detail on the day's weather for all sections of the country, public announcements and calendars, stock and commodities prices, and discounted merchandise on shopping channels.

9. As discussed in Chapter 10 on the cable subscriber, cable is useful as a background medium. Much of the cable content does not require undivided attention. It can be meaningful in short bursts of attention.

10. A full-service cable system with a remote tuner provides a rich opportunity for channel hopping. People take pleasure in the orienting search that sweeps the channels or follows a patterned routine. People also have an easy opportunity to frequently reevaluate program choices. Cable permits viewing style not possible in broadcasting. This is perhaps too complex to communicate to someone who has never experienced cable, but it would be recognized by "former" subscribers who are being resold. Even those who have never subscribed can be interested in a remote tuner.

11. Repeats of major programs give the cable subscriber the opportunity to view programs originally scheduled at an inconvenient time. Over a viewing period, the most satisfactory time can be selected. Material on automated data channels, periodically recycled, is available to subscribers as needed.

12. Some cable channels supply stereo television sound, through FM connections. For this purpose the investment in a stereo TV receiver is not necessary.

13. Cable is inexpensive. If the average cable household watches 8 hours per day, basic cable is about 7 cents per hour. Pay movies compare favorably to theater ticket and consession costs. People of modest income rely on television for inexpensive entertainment. Those tied down with young families don't have much choice.

14. There is some social status in subscribing to cable. If subscribers have a great many television choices and learn to make use of cable to enhance the value of television viewing time, they may feel superior to non-subscribers.

Cable also offers specific programming values.

1. *Sports:* 24-hour sports, more local sports, minor sports (PFB arm wrestling, auto racing, lacrosse, horseshow jumping).

2. *Movies:* uncut, uninterrupted, earlier than the broadcast runs, cable television movies are a new entertainment medium for many people who no longer, for whatever reason, go to movie theaters.

3. *Music:* music videos, now in several forms; rock, middle-of-the-road, country, and so on.

4. *Information:* 24-hour, gavel-to-gavel, beginning-to-end coverage of many events.

5. *Culture:* PBS for many households outside the range of broadcast stations, Arts & Entertainment, Bravo.

6. *Sex:* "adult" entertainment, from R-rated movies to soft and hard porn.

7. *Business:* business news and business data.

8. *Nostalgia:* old movies—"Love Laughs at Andy Hardy," "God's Little Acre," "A Farewell to Arms," "Sweet Rosie O'Grady," "You Can't Take it With You;" old television series—"Leave it to Beaver," "Best of Groucho," "Rawhide," "Dragnet," "Flipper," "Ozzie and Harriet."

9. *Children and family viewing:* Disney, Nickelodeon, Discovery, G-rated movies.

10. *Foreign languages:* Spanish, French, Chinese, and other languages.

11. *Community programming:* local events, information, and people.

12. *Radio:* a much richer set of radio formats for people in small towns.

Any single, *intensely* desired program category or channel, such as sports for the sports fan, a business channel for an investor, or music videos for a teen, may be sufficient to justify a cable subscription.

While cable marketing programs play to the various values cable may have for people, they must also address some negative perceptions or points of resistance.

With some people these negatives are real and relatively firm; with others they arise from reticence and an unclear concept of how cable would fit into an established lifestyle. To acquire new subscribers, cable marketing programs must overcome this resistance and help the prospect form an imaginary picture of how cable could be used.

Cable marketing people are familiar with several negative reactions to cable.

1. *"I don't watch television."* This may be true, but often this response stems from the low social desirability of TV for some people. Cable can offer services that have a higher social acceptance than broadcast television: 24-hour news, performing arts, children's programming, unedited movies, financial and business information. For those people who don't actually watch broadcast television, it can be pointed out that cable is different and may provide programming and services that would be of utility.

2. *"I am satisfied with broadcasting."* Some people are indeed quite happy with broadcast television, its familiar programs, and its familiar format.

These people must be convinced that cable provides *more* of what they are already enjoying.

3. *"Cable costs too much money."* Paying for television, after years of "free" television, may be alien to some people. Because there is no direct charge for broadcast television, a monetary value for television must be established. The low cost per day or program viewed must be compared to the value of one's time. Cable may also be compared favorably to the costs of other forms of entertainment and information.

4. *"There is too much sex and violence."* People with this concern may not be aware of the levels of service in cable. Basic cable is no different from broadcasting in its standards. The pay services that are most likely to carry material perceived as objectionable (such as R-rated movies) must be ordered separately. Furthermore, basic cable and "family" packages of pay services offer wholesome entertainment that is an *alternative* to "sex and violence" on broadcasting. Lockout devices are available for adults interested in programming they would not want young children to see.

Former subscribers to cable, those who have disconnected, may have a different set of negatives, certainly better informed.

1. *"The programming is duplicated and repeated."* This may be overcome by addressing changes in programming strategies that are designed to differentiate services and by emphasizing the number of choices in any given hour compared to broadcasting. Multipay packages can be designed to minimize this objection, such as HBO-Disney, where the two pays do not duplicate each other.

2. *"There are not enough good programs or new programs."* New channels may have been added since the disconnection; certainly new programs have been added. The response to this objection requires a thorough knowledge of present and future programming and services to create a detailed picture of what a former subscriber is missing.

3. *"The service is poor."* Most cable companies continuously work on improvements in customer service. There is much less overpromising in the selling of cable. Previous difficulties and growing pains can be discussed frankly with former subscribers and the new customer service programs presented.

MARKETING TECHNIQUES

Cable is sold directly through advertising, by direct mail and door hangers, and by telemarketing and sales representatives calling on homes. Many companies use all of these techniques in a single system. Some favor particular methods. Sales approaches are different in acquisition, upgrade, and retention marketing and may also vary depending on whether the prospect is a former subscriber or has never subscribed. Marketing may also be tailored to specific demographic segments of the market. It is extremely important to differentiate between cable systems and markets in determining a marketing plan. For example, what is practical in the large system in a large media market may be impractical in a small system. We will first describe the general techniques and then discuss how they are applied in the different sets of circumstances.

Advertising

Publicity, advertising, direct mail, and door hangers are used to acquaint people with cable, cable services, and the company. Some of this may be done very early, before cable lines are actually in a neighborhood, to *tease* the market. Other advertising and direct marketing materials attempt to get the subscriber to order by phoning or returning a card or coupon. Usually these responses trigger a sales call or a telephone sales presentation so that the prospective subscriber is more fully informed.

Television is, of course, considered a good advertising medium for cable since it guarantees that those who are exposed are interested in television. As well as conventional 30- and 60-second spots some operators have purchased much larger blocks of broadcast television time to "preview" cable services. *Radio* has advantages in that it can be targeted by assessing formats and using rating book demographics. *Daily newspaper* advertising can be associated with television listings or any other section of the paper that would deliver the appropriate audience (such as the sports section for a sports channel, or the lifestyle section for the Lifetime network). A difficulty with all these media is that they almost always have a much wider circulation, geographically, than most cable franchises. A solution for the cable industry has been to form regional cooperatives to make media buys, dividing the costs in proportion to homes passed or number of nonsubscribers in each franchise. If all the companies and their areas served are not listed in the advertisements, then some central office that is listed must redirect leads to the appropriate franchise. Program suppliers advertise nationally using 800 phone numbers. They refer the orders or inquiries to system operators.

Some cable systems successfully *barter* or *trade* advertising with radio, television, and newspapers. Advertising time or space is exchanged dollar for dollar in value on the respective rate cards, but no money changes hands. In the case of radio, television, and cable, unsold commercial availabilities are used, and therefore no direct costs. For example, a broadcast television station takes local spots in CNN promoting its local news team in return for spots on the TV station offering discounted installation of cable.

Outdoor (billboards) and *weekly newspapers* do not have the waste circulation problem of television, radio, and daily newspapers. Billboards are limited to a simply visual and seven-word message (as a rule of thumb). Yet they can be useful in announcing the availability of cable service in a particular neighborhood as construction progresses, in promoting a daily program or cable network, and in reminding people who are interested in cable to subscribe. Weekly newspaper ads reach the target audience and help support publicity for the system.

Publicity

Publicity is helpful in several situations. News media cover the initial franchising process and supplement that with features about the cable technology and services. As each segment of the cable system is built and energized, the operator informs the news media that service will begin in that segment and then identifies the next area to be built. The operator also attempts to publicize any new channels or services prior to marketing the services.

Direct Mail, Door Hangers

Mailers and/or door hangers (where allowed)* may initiate a marketing campaign. These pieces, distributed to each nonsubscribing household in a campaign may contain a simple explanation of the cable technology, describe the hook-up for single receiver, multiple receivers, FM and VCRs, present the channel line-up and describe the major channels. All of this may be presented in a single brochure developed for the system or may be separate pieces of literature, some of it provided by program suppliers. The door hanger invites

FIGURE 15-1. A construction announcement door hanger teaser. Continental Cablevision of Michigan, Inc.

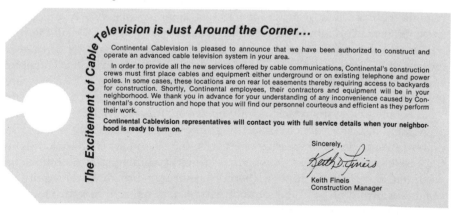

The Excitement of Cable Television is Just Around the Corner...

Continental Cablevision is pleased to announce that we have been authorized to construct and operate an advanced cable television system in your area.

In order to provide all the new services offered by cable communications, Continental's construction crews must first place cables and equipment either underground or on existing telephone and power poles. In some cases, these locations are on rear lot easements thereby requiring access to backyards for construction. Shortly, Continental employees, their contractors and equipment will be in your neighborhood. We thank you in advance for your understanding of any inconvenience caused by Continental's construction and hope that you will find our personnel courteous and efficient as they perform their work.

Continental Cablevision representatives will contact you with full service details when your neighborhood is ready to turn on.

Sincerely,

Keith Fineis
Construction Manager

*Door hangers are not allowed in some communities because, if not removed, they may signal to thieves that the occupant is not home.

the householder to contact the cable company and order service, but even if the household does not take this initiative, the materials prepare the household for a subsequent telemarketing or personal sales call. This is also an important function of advertising, publicity, and direct mail—*softening* the market.

Advertising and direct marketing methods are sometimes referred to as *cream skimming*. They get the easy sales, those people who are already anxious to get cable or are at least partially presold so that they need only to be triggered to action. These presold prospects (called "truck chasers" because they follow the construction or maintenance vehicles trying to sign up) or nearly presold prospects should be signed up by the most expeditious and inexpensive methods.

Telemarketing

Telemarketing is an important cable subscription sales technique. It is less costly (by about half) than door-to-door sales. It can be used as an immediate follow-up to advertising and direct mail, closing the sale, or used on its own. In either case, respondents do not need to act on their own initiative (that is, making a phone call, sending in a card, stopping at the cable office). The caller initiates the action.

Over impersonal sales methods, telemarketing has the advantage of permitting a friendly, personal approach with the opportunity to explain service or an offer, and to answer questions. The cable system can contract the telemarketing to a firm specializing in that service, thereby taking advantage of experienced personnel and the available telephone facilities, and help with the sales approach. This may be especially appropriate for a concentrated campaign. However, a relatively large cable company may have a telemarketing unit continually sweeping the system to sell new subscribers or upgrade existing subs.

Telemarketing people are scripted, but usually allowed to personalize the message somewhat so that it does not sound canned. A high-pressure pitch is avoided. The operator does not want a short-term subscriber who has been oversold and does not want to alienate the person who rejects the service now, but will be approached again. However, the caller must be persistent and also trained to have an appropriate response to a standard set of expected negative reactions. The telemarketing representative is also trained to pick up any positive notes and pursue them.

Telemarketing almost always involves a special offer—free trial, discounted installation, premium, and the like. This *hook*, of course, is mentioned immediately to justify the call and get the respondent interested.

Telemarketing is now a principal sales tool for most cable systems. Facilities and staff are available to use telemarketing for a large variety of purposes, although the tactic's value seems to be most effective with *current customers* rather than noncable subscribers.

Door-to-Door

Door-to-door sales are the most expensive and, many operators feel, the most effective way of selling cable. Here the salesperson has the opportunity to fully explain cable and cable services. Attractive and elaborate presentation (pitch) books may be used, and even portable VCRs to demonstrate. Here the salesperson has many more cues to the interests of the household. Income level is fairly obvious from the house and furnishings, children may be about, sports equipment may be in evidence, the television may be on, the channel tuned may suggest something cable has more of, and so on. Salespeople are trained to look for and respond to these things. The salesperson usually gets more time to make the presentation and ask questions in the door-to-door situation, once the foot is in the door.

Promotions/Premiums

In almost all marketing campaigns one or more *incentives* are used to encourage subscription. The most common incentive is an *installation discount*. The household is urged to sign up *now* to get a special installation price or have the installation charge waived entirely. Ostensibly this saves the subscriber $15 or more, although the operator does not plan on collecting many such fees because the installation fee is reduced in subsequent sales promotions as well as in the initial round.

A *free trial* is another workable device. This includes a full waiver of the installation price. Basic and selected, or all pay channels (except "adult" channels), are free for one month, sometimes two months. The period must be sufficient for the household to have time to explore a substantial part of the service. For this reason it is also very important to fully describe all the channels in sales presentations and literature and to promote all of the channels in cable advertising spots.

The free trial offer makes it clear that the household will not be under any obligation to continue the service. Indeed, a benefit of the free trial marketing technique is that the household should know at the end of the trial whether or not it really wants cable and which services. It is reported that there is less churn among subscribers who sign up after a free trial than among others, because the free trial household can make a decision based on experience with cable rather than conjecture.

There are two ways to make the determination if the household wants to continue cable. Sometimes one or more contacts are made during the trial period to elicit favorable reactions and deal with any objections. This builds a positive attitude towards the product *prior* to the decision point. At the end of the free trial, the people are called and asked if they want to continue. If they are interested, the details of subscription, various packages and prices are explained and the subscriber is put into the billing cycle at the end of the free trial. If the people do not choose to continue, an appointment is made to pick up the converter, at which point a "last ditch" sales presentation is made. The other method, used infrequently and only when a single service is offered free, is the *negative option*. Here the original agreement is that the person accepting the trial calls to cancel the service, otherwise it continues automatically after the trial on a billed basis.

Depending on whether and how thoroughly the area has already been marketed, the number of households accepting the free trial ranges from 50 percent to nearly 100 percent. Among those accepting the free trial, about 65 percent will subscribe after the trial,[1] and, as noted, the retention is above average.

But free trial marketing is expensive. At least two sales calls are made (by telephone or face to face), there is an installation (with drop) and later retrieval of the converter for those who do not subscribe, and in some cases the operator may be paying for pay channels during the trial period.

Some cable operators offer books of coupons entitling the new subscriber to discounts and premiums at cooperating merchants. The new subscriber can save money. The cooperating merchant increases store traffic and has a chance to win new customers. This is a particularly good technique for people who have just moved into a community (similar to the "Welcome Wagon" concept).

Premiums, or gifts, such as tote bags and sateen jackets, are also used with good effect to stimulate cable subscription sales. The premium often uses the cable system logo prominently, or the logo of a particular channel, further advertising cable. These offers help telemarketers keep people on the line and get door-to-door salespeople in the door.

Some cable marketers disdain any kind of incentives on the theory that the subscriber won by such devices may be persuaded by the incentive rather than by the merit of the service. Nevertheless, most operators think that it is necessary to provide an inducement to get people to *try* cable.

Promotions are used in cable marketing to generate excitement and give salespeople something to talk about. A promotional campaign may be built around a premium offer. Seasonal promotions can be tied to program content appropriate to the season. The promotion can involve a special two- or three-day *preview* where subscribers are given a pay channel at no charge. Between preview programs, on-channel live presentations urge people to call in orders. Often cable systems tie into other businesses for a joint promotion. In these promotions cable can place spots in unsold advertising availabilities and contribute its share to other advertising and promotional costs. Cable systems also tie into nonprofit organizations by offering to contribute money to the charity for every cable subscription order. Sometimes cable operators set up displays in shopping centers or at public events such as county fairs with program personalities making special appearances. Contests or sweepstakes are launched to promote special channels and program tune-in.

Cable Stores

A surprising number of people like to visit the cable company business office to inquire about service, and once a subscriber, to pay bills. Some companies have created a *cable store* for this purpose. The cable store can demonstrate the cable services, channel by channel, selling new subscribers and upgrading old ones. The store is a convenient place to exchange converters in a change of service transaction. If a subscriber orders a new service and brings the converter into the store, installation charges may be waived. The store may also retail VCRs, blank and recorded cassettes, and, for people outside the franchised area, satellite dishes. The cable store is also an ideal place to demonstrate and sell other peripheral video enhancements such as remote tuners, stereo adapters, and VCR hook-up kits.

Program Suppliers

Program suppliers to cable are wholesalers; they sell the product to operators but are dependent on the operator to retail the service. Since the suppliers' success depends on the efforts of the operator, they provide substantial marketing assistance.

On a major marketing campaign, or the launch of a new basic or pay channel, the network may be deeply involved from the planning stage to the

FIGURE 15-2. Order form for materials to promote the National Geographic Employer series, now on WTBS.

EXPLORER seen only on NICKELODEON
NATIONAL GEOGRAPHIC EXPLORER
Fulfillment Center
P.O. Box 20498
Columbus, OH 43220

For Assistance, Call EXPLORER
Fulfillment Center At 614-771-1876

OFFICE USE ONLY:
Order date _____
Order number _____
Ship via _____
Check amount _____
Check number _____
Customer number _____

Ordered by: **Please print clearly**

Name _____
c/o _____
Address _____
City _____
State _____ Zip code _____
Phone _____

Faster Delivery: We ship UPS unless you specify UPS 2nd Day; please add $15.00 for additional shipping. UPS cannot ship to a P.O. Box—only street addresses.

Ship to: (only if different from "ordered by")

Name _____
c/o _____
Address _____
City _____
State _____ Zip code _____
Phone _____

Signature of Person Authorizing Shipment

Item Code	Qty.	Description	Unit Cost	Total
EX101		EXPLORER billstuffer/packs of 100	$4/100	
EX102		EXPLORER national consumer TV spot*	$30	
EX103		EXPLORER Radio Ad—Reel	$3.25	
EX104		—Cassette	$3.25	
EX105		EXPLORER Ad Slicks**	free	

*You can tape this or other NICK promotional spots during the Satellite feed the first and third Monday of each month 6:30 a.m. ET.
**Also available from MTV Networks Inc. regional offices.

Subtotal		
Shipping/Handling	$3.00	
Add. Shipping Charge		
TOTAL		

**ALL ORDERS
MUST BE ACCOMPANIED BY A CHECK**

All checks made payable to the National Geographic Society
All Items Intended For Promotional Use Only

end. The major cable program suppliers have affiliate representatives whose principal responsibility is to assist the affiliate with marketing. They are armed with marketing plans, promotions, and materials developed by their network. They have extensive experience in marketing the product in a great variety of circumstances.

In addition, many program suppliers actually contribute funds to marketing their service, sharing in or covering the costs of advertising, promotional literature, direct mail, and perhaps even sales commissions. Promotional allowance, co-op money or discounts may be written into the affiliation agreement.

The catalogs of affiliate marketing materials supplied by the programmers are extensive, covering many areas:

Advertising materials:

- ad slicks with local imprint space, or ad copy
- television spots on cassettes, or delivered by satellite, room for operator message or identification
- radio spots, scripts, and cassettes with room for local ID
- outdoor billboard sheets

Premium and promotional materials:

- tote bags, backpacks, note pads, director's chairs, beach towels, bumper stickers, balloons, umbrellas, thermos bottles, coffee cups, windshield scrapers, watches, stuffed animals, caps, T-shirts, jackets. . .

Literature:

- bill stuffers
- brochures
- subscriber handbooks
- door hangers
- telemarketing scripts
- posters
- salesperson's presentation book
- banners
- four-color broadsides

Direct mail:

- sample copy and formats
- literature for enclosure
- self-mailers

Promotions:

- seasonal and special promotion plans with support materials
- incentive plans with support materials

Training:

- videotapes
- handbooks, kits
- theft of service kits
- monthly program highlights
- monthly newsletters or magazines

Manpower:

- toll-free (800) numbers where leads are compiled and forwarded to the operator for follow-up

Program guides:

- monthly guides
- materials for creation of local guides
- tune-in promotional materials

Public relations:

- news releases
- program listings
- photographs

While the assistance of the program suppliers is essential to the marketing of cable systems, the operator is careful to maintain the system identity in the process. All program supplier materials must fit into the overall system marketing strategy. The object of the marketing enterprise is to maximize *system* revenues and subscriber satisfaction. Program supplier marketing plans and materials may have a more narrow purpose.

Most of the cable marketing efforts are invested in television programming services but ancillary services (remote tuners, printed cable guides, FM connections, and additional outlets) also produce customer satisfaction and company revenue. These must be included as part of the service in literature, advertising, and sales presentations.

All of the marketing tools discussed here are applied in different ways for different objectives depending on whether the principal goal is acquisition, upgrade, or retention.

MARKETING STRATEGIES

Acquisition

There is always a universe of unsold homes passed. They may be people who have just moved in, newly constructed homes, former subscribers, and those who have never subscribed. A continuing effort, called *acquisition* marketing, is made to contact these homes—new arrivals as soon as possible, others at least once a year. Although not necessary, the addition of a new service or channel is an occasion for remarketing these people.

Cable systems frequently extend lines into new subdivisions, or obtain franchises for adjacent communities. And, of course, there are a few communities, and some major markets still to be built.

Cable systems are built a section at a time. As soon as a newly built section is tested, it is publicized as complete and cable subscriptions are sold. Thus the system begins to generate revenue as soon as the first section is built. Because only a small section of cable is being marketed at a time, television, radio, and daily newspaper advertising, if used at all at this point, is designed primarily to maintain anticipation in residents of unbuilt areas. Some combination of door hangers, direct mail, telemarketing, and door-to-door sales will be used for the new build areas.

It is necessary, in this section-by-section construction and marketing to be very well coordinated. Salespersons must follow construction crews and, in turn, be followed by installers very closely. If sales and installations don't follow construction, revenue is lost. If installers don't immediately follow sales people, the frustrated customer may be lost.

Remarketing of nonsubscribers may make a distinction between formers and nevers. The *formers* are households who have at one time subscribed but later disconnected. For a fairly large subset of households, cable is a marginal purchase. Their interest in cable runs hot and cold, or cable is the victim and beneficiary of highs and lows in the household economy. The cable operator loses significant revenue if these temporarily disconnected former subscribers are not periodically stimulated to consider cable again.

Ideally there would be a file on the formers listing the services that they previously subscribed whether or not they were *voluntary disconnects* or *nonpay disconnects*. In the event of voluntary disconnection, disconnect reasons may be put in the record for future remarketing use. Previous subscribers are very good prospects for reconnecting.

The marketing of former subscribers who were disconnected for nonpayment is a difficult problem for cable operators. Some of these nonpays could be reconnected and become good customers, and it makes sense for the operator to attempt to get them back on the system.

Though some portion of nonpays can be converted to good, paying subscribers, many cannot and will again become nonpays if reconnected. For this reason, some operators simply abandon nonpays in their marketing programs. One risk of this approach is that the nonpay subscriber may move out and the home becomes occupied by a new family. If that home is on the non-pay list, the new resident will never be marketed.

Nevers, the households that have resisted cable in all previous contacts, may also be sold on new services and special coming attractions, but it may be necessary to dispel misconceptions about cable. Common objections can be treated in advertising and direct mail. Through creative personal selling, enough questions may be asked to determine how cable could benefit a specific set of household interests. The free trial technique is sometimes used after several remarketing efforts, because it may be the only way for these people to learn how cable might fit in.

Illegals

A good source of paying subscribers is the body of people who are connected *illegally*. A cable operator periodically *audits* the system to identify those households that are not paying for all of the services received because they have connected themselves or the company has made an administrative error.

The audit team typically comprised of an installer and a salesperson, compare company records to actual pole hook-ups in the field. The installer notes discrepancies in a given neighborhood, turning over a "prospect list" to the salesperson for a personal visit. If the address represents a "first time" illegal, the salesperson may take the blame for the "error" and offer to initiate an account for the customer. Most first time illegals have become accustomed to the service and a substantial number actually become legal. Illegals who have tampered with the converter are more difficult to discover, but many reveal themselves by accident in service calls. In the case of recurring illegals, a criminal or civil complaint may be lodged with the court. In rare cases, arrest warrants are issued and the cable company may sue for damages.

Another technique for converting illegals is to mount an *amnesty* program where advertisements, and mailers ask illegals to identify themselves and turn in any illegal equipment, or become paying subscribers *with no penalty*. If they do not come forward, the implication is that they will be prosecuted.

Upgrades

It is the intent of marketing plans to have every subscribing household at the highest level of service appropriate for the household interests. Cable systems once used a *top down* approach to marketing, whereby a strong attempt was made to get the household to take every service available. Only

FIGURE 15-3. Mailer to basic subscribers encouraging upgrade to a satellite tier. Front and back pages shown. Inside pages describe the nine additional channels available at a penny a day per channel.

WHO
SAID
A

DOESN'T
BUY
YOU
A LOT!

NOW YOU AND YOUR FAMILY CAN ENJOY COUNTLESS HOURS OF TOP ENTERTAINMENT ... MUSIC ... SPORTS ... NEWS ... MOVIES ... KIDS SHOWS ... LOCAL COVERAGE ... PLUS MUCH, MUCH MORE.

THE COST FOR THIS ADDITIONAL 9 CHANNEL ENTERTAINMENT SERVICE IS ONLY $2.70 PER MONTH MORE THAN YOU ARE NOW PAYING.

ONLY AN ADDITIONAL
PENNY PER CHANNEL
PER DAY!

AND, WE'LL INSTALL THIS GREAT ENTER-TAINMENT SERVICE FOR JUST ONE PENNY. THAT'S RIGHT! AND TO MAKE THINGS EASY WE'VE EVEN PROVIDED THE PENNY!

TO START ENJOYING THESE
ADDITIONAL 9 CHANNELS
CALL

239-8500

USE THE PENNY ON THE FRONT
AND SAVE THE NORMALLY
$12.50 INSTALLATION FEE.

CALL **239-8500** TODAY. THIS OFFER EXPIRES AUGUST 31, 1985.

Coaxial
Communications
Technology and Service through Cable Television

after a strong pitch would salespeople back down to lower levels of service. This approach tended to irritate people who subscribed and resulted in high churn rates. Now it is considered a better strategy to find a combination of services that fits the household at the initial sale. But, since some households opt for a minimum level of service as a trial, and because the household needs may change or the cable system adds new channels, sales efforts to *upgrade* subscribers are always necessary.

Here internal mechanisms are effective and inexpensive. The unsold local advertising availabilities on LO and cable network channels are used to promote pay channels, pay programs, and upper-level tiers of service. Both electronic and print guides are used for the same purpose. Bill stuffers promoting a particular service may be inserted in bills for only those households not subscribing to that service.

Previews are important in upgrade marketing programs. Most suppliers permit several days of previews annually and have promotional and advertising materials designed to precede and follow the preview to maximize its value. Spots within the preview encourage phone-in orders. The preview may be immediately followed by a telemarketing campaign.

Telemarketing, with or without previews, is the most frequently used upgrade method. Selling upgrades is the most effective use of telemarketing.

Upgrade marketing usually involves some incentive. Internal cable promotional television spots, literature, and previews may also be used in the launch of a new pay channel, supplemented by radio, TV, print, outdoor advertising, and publicity.

Cable guides, printed or electronic, are considered to be instruments for promoting upgrade. If subscribers are constantly exposed to programming announcements on channels they do not have, they may eventually subscribe on their own initiative or become familiar with the service and therefore more receptive to a sales campaign.

Retention

The decision to subscribe to cable, or any pay channel or package of channels, may be *rescinded* at any time. A subscriber reviews the purchase decision each month when paying the cable bill. Monthly rates of churn for some channels may go to 6 percent or more. *Retention* or *maintenance* marketing are the terms given to the efforts to save these *disconnects* (discontinuing all service) or *downgrades* (dropping one or more services without disconnecting).

The economic benefit of saving a unit of service for one subscriber is significant. In a nonaddressable system, that loss means a costly service call, customer service representative time, and clerical/computer costs. Then acquiring a new subscriber, just to get even again, involves all of those associated marketing costs, another service call, and more clerical work. (The same costs are involved in a disconnect in an addressable system, but are somewhat less in a downgrade because of the elimination of a service call.)

Of course, there are costs connected with retention marketing, yet they are difficult to assess because the result of a particular retention program cannot be easily measured. For example, if a bill stuffer promoting next month's program highlights causes a subscriber to keep the service, the cable operator would never know. The problem of measuring the results of retention marketing will be discussed under the marketing administration heading below.

The entire operations unit of the cable system is involved in retention. Since a major reason for disconnect is customer service, the customer service procedures and training must be designed to maximize retention. This requires a friendly manner by customer service representatives (CSRs), sometimes enforced and fine tuned by *mystery shoppers*, members of the supervisory staff or outsiders who call with various problems to see how the CRSs respond. Supervisors may also listen in to CSRs periodically to make corrections in technique if necessary. CSRs will have a *save-a-sub* routine to use when a subscriber calls to downgrade or disconnect, attempting to dissuade the subscriber. Service technicians who also have contact with subscribers must be continuously aware of their responsibility to help keep customers satisfied. Installers must make a good impression and fully explain the use of the system. Sometimes they leave a handbook introducing the subscriber to how the system works. This is considered part of the retention program.

Sales presentations are developed with retention in mind. The subscriber is not supposed to be pressured into taking cable or any service that is not appropriate. Since the salespersons work on commission, certain *disincentives* to overselling must be built into the compensation plan. Commissions for disconnected or downgraded subscribers in the first two or three months of service may be deducted from a sum set aside from commissions earned. Or a more *positive* approach is a *retention bonus*, often paid 60 or 90 days after the sale for each unit sold that is still active.

A new subscriber to cable or an upgrade may receive a *follow-up* letter, phone call, or personal visit to welcome them to the service and to answer any questions they may have about the service. If there are any dissatisfactions at this point, they may be rectified. Since churn is usually greatest in the earliest days of subscription, operators usually find these follow-ups effective.

It is considered important to retention, to have subscribers well informed of programming. Several approaches are used. Tune-in ads run in the electronic and printed cable program guides, newspaper guides, and *TV Guide* promoting individual programs. *Cross-channel promotional spots* also serve this purpose. These are advertisements for individual programs or cable channels that are inserted in the unsold availabilities on cable networks and LO channels. Since the system operator has not been able to sell the spots, they are useless as a revenue source. Services are available that collect and package cross-promotional spots from all of the cable networks for operators. The value of *cross* promotion is that regular viewers of one channel are exposed to the programming and services of another channel that they might not ordinarily think about viewing. This expands the individual subscriber's use of the cable service.

Broadcast television, radio, newspapers, and outdoor is used for retention. External tune-in or program promotion advertisements may serve the dual purpose of attracting new subscribers and retaining old ones.

FIGURE 15-4. Tune-in ad slick. Operator fills in the ID and program time.

Bill stuffers are used as tune-in advertising and also to reinforce cable satisfaction. They are inserted with the bill by the operator or the operator's billing service for a few pennies per bill. When used for retention purposes bill stuffers usually go to all subscribers to keep them aware of the cable benefits. It is a bonus if they also sell upgrades.

Both tune-in ads and program listings in *cable guides* are good for retention. The guides can keep the subscriber fully informed of all programming and serves as a regular reminder of the full array of cable offerings. Editorial features and news columns can be used to build knowledge of cable and promote elements of the service. One field experiment with cable guides, found that the sample of households that received the guides had less churn for *all* types of subscriber. The reduction of churn, however, was offset by the cost of providing free guides only for multipay subscribers. These were the people, incidentally, who make the most use of the guides.[2]

Channel placement is considered critical in getting maximum exposure to cable-only programming. A channel line-up that mixes the major cable networks with the popular ABC, CBS, and NBC affiliates produces substantially better ratings for the cable networks than a line-up that buries the major cable networks in the higher channel numbers among channels that are not viewed frequently. Some of the cable networks give affiliates an incentive, in the form of per-subscriber fee discounts, to give their network favorable "shelf space." Some entire metropolitan areas have standardized on channel line-ups to aid in cooperative advertising and promotional efforts.

Subscriber *games* may be effective in retention marketing. Each month the subscriber may get rub-off cards, stamps, or some other device to qualify them for prizes or take deductions in the next month's bill. Because the games continue over several months, downgrades or disconnects are discouraged. Trivia games and contests are used to get people to watch the lesser known channels so that they can answer questions and qualify to win. *Contests* are used with cable system employees to provide incentives to save subscribers. Individual employees or systems of an MSO compete with each other in reducing churn rates to earn prizes.

Many subscribers are lost from a cable system every year because of nonpayment. Part of every retention marketing program, then, must be to keep these subscribers by encouraging them to pay. Early reminder of a delinquent account is important so that the unpaid balance does not accumulate and get too forbidding. Before disconnect, subscribers are contacted by phone, or in person, in a last effort to get them to pay and retain them as subscribers. Nonpays may be forgiven a portion of the bill. This is commonly done when the operator feels that the nonpayment may be due to poor service or other problems created by the cable company, but the operator runs the risk of encouraging subscribers to become delinquent in the hope of getting an adjustment to their bills. Another approach is to work out a *payment schedule* with the nonpay subscriber. Frequently, nonpays are reconnected with a promise to pay the delinquent balance over several months.[3]

Marketing Two-Way Services

Not too much is known about the marketing of two-way services. At this point, two-way services are almost always sold separately from the one-way entertainment and information packages. Although the basic and pay salespersons may offer the two-way services and make uncomplicated sales, more likely they get leads for follow-up by specialists. Incentives are seldom used because the sale is not triggered by impulse but by a well developed, rational argument on the merits.

Direct mail, advertising, and telephone are used to produce leads for face-to-face sales calls. Most of the two-way services require demonstration—if not on-line, then either through simulation with the equipment or on videotape.

Time payment plans may be necessary for the purchase of necessary hardware for security, videotex, and games.

In selling per-program pay services or any other usage-billed service, the cable operator is obligated to make a very careful credit check before making the installation, since the user is in effect given unlimited credit.

There has been a trend toward *third-party* sales of two-way services. A burglar and fire alarm company, for example, may take over the marketing and operation of alarm systems. Modest goals are set for the initial marketing of most two-way services, generally 5 percent or fewer of the homes passed.

MARKETING ADMINISTRATION

The whole marketing effort of a cable system, to be effective, requires systematic administration. There is much detail work in keeping track of the marketing program in acquisition, upgrade, and retention phases. The marketing department must know the status of every home passed by the system. Progress is tracked by keeping several critical statistics. Every person in the company inside and outside the marketing department, who has any contact with customers, must be trained and coordinated. Since marketing is perhaps the largest single variable expense in cable, cost analysis is extremely important.

Packaging and Pricing

The way cable services are put together, *packaging*, is an essential part of marketing strategy. The simplest package is a single-tier service with no distinction between basic and pay channels. For this plan no scrambling is necessary, solving some of the incompatibility problems with television sets and VCRs. Much more common is to offer a basic service, to which any number of pay channels may be added *a la carte*, at undiscounted prices. Or the system may have two or more levels of basic that may be ordered on a tiered basis, where the subscriber must take the lowest level of service to get to the next higher level. Rather than make a la carte pay channels independent, a system may want to discount each additional pay service (for example, the second pay channel less $1 from its listed price, the third pay channel less $2 from its price, and so on).

A cable system may also *group* pay services into packages that have some inherent relationship, and discount the package over the total cost of each of the channels in the package, *a la carte*. For example, a system may offer a "family" package that includes a G- and PG-only movie channel and The Disney Channel. Each of the channels may be priced at $7.95 separately, but at $14 in the package. A "movie" package may include HBO, Showtime, and The Nostalgia channel at a discounted price for the package. This kind of packaging scheme is intended to help the subscriber understand the services by putting clearly related channels in a meaningful group. Without a small meaningful set of packages, the subscriber is faced with an infinite combination of services.

Another packaging strategy is to allow the subscribers to put together their own packages by offering options within a service. For example, the "silver" service may include HBO and the choice of Disney or Playboy. The "gold" service may be HBO, Showtime, and the choice of two channels from among Disney, Playboy, Home Theater Network, and American Movie Classics.

Usually, a cable system will have a special, heavily discounted price for full service, everything the cable system offers.

Still another packaging strategy is to put together several specialty pay channels at a low price. The package might include Disney, The Nostalgia Channel, Bravo, and a regional sports channel, all for $14.95. The subscriber may not have an intense enough interest in any one of the channels to subscribe, but enough interest to take them all at a low enough price.

Integral to the packaging plan are the ancillary services. These services are available individually at a relatively high retail rate, but are heavily discounted or offered free with successively higher-level packages. For example, a printed cable guide, retailing at $1.00 monthly may be listed as "free" in a one-pay package, and the remote tuner, at $3.95, may be added free along with the guide in a two-pay package. For a three-pay or larger package, these two items may be put in free along with a $4.00 additional outlet and a $3.00 FM hook-up. When the full cost of each pay service, *a la carte*, is added to the retail value of the free ancillary services, the total "value" for each package is much higher than the actual package price, showing consumers substantial "savings." Ancillary services are referred to, internally within the cable industry, as *glue*, to cement the package and keep it stuck.

Packages, along with the ancillary services incentives, are designed mainly to develop the multipay market. The object is to (1) maximize revenues by increasing total dollar volume despite the discounts (incremental margin is sacrificed for cumulative margin), (2) reduce churn by structuring the prices and packages so that there is a disincentive to drop services, and (3) to provide identifiable, meaningful combinations of services so that the potential subscriber is not confused by the number of choices.

Some operators are concerned that packages may force people into services artificially, leading to dissatisfaction and churn. Packaging schemes may also be quite transparent to the intelligent consumer who may not see the *a la carte* and ancillary service prices as realistic. This household may feel it is paying a high penalty for not succumbing to a package deal, even if none of the packages precisely suits the household. The packages may not work well for operators either. As the discounting gets deeper in the more advanced packages, the margins get thinner and, with operating costs included, may be negative. Multipay packages may become less important marketing and pricing tools with increasing emphasis on basic sales, at higher prices, and pay-per-view.

There is still indecision, as noted earlier, about how to package some services. A regional sports channel or nostalgia movie channel may be used as part of the basic tier to *lift* subscriptions, and perhaps increase the basic price. A channel supplier may charge a much lower monthly fee per sub when the channel is on basic than with it is marketed as a pay channel.

The relationship of price and demand for cable is not altogether clear since, in the past, most cable prices have been regulated. In other words, operators have not been free to test demand by experimenting with pricing. It is generally assumed that basic cable prices are relatively *inelastic*, meaning that demand does not change much if prices go up or down. This may also be true of the first pay channel, but the second, third, and beyond channels are thought to be quite price *elastic*. The perceived value of each succeeding pay channel, and therefore the price one is willing to pay for each succeeding pay channel, drops considerably after the first.

Having prices deregulated for most cable systems, and with elastic demand for at least some services, optimum pricing becomes a complex problem. For each unit of service, the operator must estimate the number of subscribers, *in his or her market*, at each possible price.

Monthly Price	Estimated Number of Subscribers	Monthly Revenue
$5.00	600	$3,000
6.00	575	3,450
7.00	525	3,675
8.00	475	3,800
9.00	400	3,600

In this example the simple optimum price is $8.00 because it yields the greatest revenue. (Actually it is a bit more complicated. The $9.00 price might be optimum. If expenses were considered, the lower cost of servicing 400 subscribers versus 475 would probably mean a greater profit at the $9.00 price.)

The same estimates of subscribers and revenues must be considered for any packages that are offered. When packages are involved, the effect of the packages on *a la carte* subscriptions must also be evaluated.

It is not feasible to experiment extensively in pricing within a single system without confusing and alienating subscribers. The ability to experiment is further restricted by the pricing policies of other operators within the region. Although not competing, individual operators within a region cannot afford to get too far out of line with conventional pricing of the other operators. MSOs can compare results of different pricing plans across systems. This must be done with caution, however, because of the unique characteristics of each market.

For many cable systems installations produce significant income, even if priced below cost. The installation charge may be looked at by operators as earnest money. The installation fee is collected along with the first month's service charge, in advance, at the time of installation. With this initial *up-front* outlay, sometimes called *front money*, the new subscriber must take the subscription seriously and give it a fair trial. In other cable systems installation is almost always discounted or provided at no cost to the subscriber as a marketing incentive. Installation or service charges for an upgrade are usually low and quite often waived as an incentive to subscribe.

Cable operators are reevaluating the charges for second-set hook-ups. Some are eliminating or lowering these fees in an attempt to increase customer satisfaction.

The operator may or may not charge for swapping services, dropping one and adding another. This is called *spin*. Spin is both a problem and an opportunity. A subscriber may spin through the services to settle on the appropriate set of services for the household. Or subscribers may tire of one service and substitute another on a fairly regular basis. These things can be considered positive uses of spin for both the subscriber and the cable company. Indeed, some addressable cable systems are permitting change without cost. On the other hand, if subscribers, in a nonaddressable system spin frequently and frivolously, it could be a major cost for the operator. There may be a charge for each change to assure that the subscriber has given it some thought and to cover some of the cost.

Compensation

Salespersons are paid on a commission, with sales management personnel usually receiving an override commission or bonus. The commission plan is structured to provide several incentives. The most basic is, of course, to motivate sales effort. Commission plans are also designed to pay a higher rate for sales over a particular number within any week to keep people working hard through the week. Compensation plans may factor in penetration so that the commission is not only based on an absolute number of sales but on the *percentage* of sales within the territory—the higher the penetration, the higher the commission. This provides an incentive to deal with the tougher customers.

If the salesperson does not close the sale at the time of contact, the commission is still paid if the household orders the service within a specified time period. Door-to-door sales are compensated at a higher rate than are phone sales. Remarketing commissions may be higher than commissions in original marketing because the sales are more difficult and prospects scattered over a wider area. As mentioned in the section on retention, salespersons may be compensated only for subscribers who are *retained* for a couple of months.

Training

Training is a continuing part of cable marketing because of the high turn-over and frequent introduction of new services on which the sales and customer service staffs must be retrained. The first objective of the training is to acquaint the salesperson with the cable service. This is no simple matter in a modern, multitiered system. Any one of the channels may be an important attraction; therefore, the salesperson must be able to answer questions about each. The salesperson may be equipped with separate program guides for each of the channels.

The second objective is to teach sales technique. Most cable companies concerned with the company image and long-term retention of subscribers now use a low-keyed, soft-sell approach. The training explains the use of sales aids such as a flip chart or full-color brochure. Sales aids may also graphically explain the pricing scheme, which is complicated in the multitier, packaged system. The trainee must also learn to *close* a sale, to conclude a presentation with a businesslike request for the subscription.

The training is usually done in role-playing situations with the trainee attempting to sell a supervisor or trainer who knows all objections and techniques for avoiding a commitment. New salespersons may accompany experienced people on sales calls or listen to telemarketing presentations.

Territorial Management Plans

Now many cable systems have *territorial management* plans where marketing department persons are assigned to specific territories in which their responsibilities are quite broad. The duties include sales, follow-up calls, contacts with nonpays, attempts to dissuade disconnects and downgrades, picking up converters no longer in use, audits, amnesty programs, and so on—all within the territory. They are involved in or informed of virtually all company activity in the area. This gives the salesperson more responsibility and a more challenging diversity of tasks. It permits them to develop a knowledge of the territory and builds in them a long-range interest in the welfare of that territory. They are evaluated, and sometimes compensated, on the net gain or loss in subscribers and revenue in the territory.

Multiple Dwelling Units

Some urban cable franchises have well over 50 percent of the residents in multiple dwelling units (MDUs). Because of the higher rate of turnover, the greater piracy problem, and the reluctance of some building managers or owners to coopera e with cable systems, MDUs require special attention by cable operators. Territory management programs are very helpful. The representative can make contacts and attempt to establish a continuing relationship with building personnel. He or she can keep in touch with who is coming and going. They can watch for illegal connections within the buildings.

While most cable marketing techniques apply to MDUs, two unique approaches are used. *Resident manager programs* involve building superintendents in the marketing process through literature and special presentations intending to inform them of fundamental facts about the cable service. They are asked to help sell the cable service as a feature of building occupancy and keep the company informed of move-ins and move-outs. For these services they earn cash or credit toward gift catalog merchandise.

Group marketing in MDUs may be desirable because of residents' unwillingness to answer the door or building restrictions on solicitations. The resident manager may assist with the arrangements and room. An advance direct mail piece announces the meeting. Attendees may be offered a premium for being there. The programming can be demonstrated if the room has been equipped in advance with a cable outlet. Registration for door prizes serves to provide a check-off for those in the building who attended so that the remaining residents can be contacted by other means.[4]

Market Segmentation

Cable marketing people have had some success with *market segmentation*, breaking the market down into its demographic components and then targeting the marketing approach specifically to each segment. Several companies (such as Saxe Marketing, Inc. and Donnelley Marketing) can provide 30 or more demographic characteristics on households with the desired addresses. These can be combined to form categories of households that are relatively homogeneous. The categories may be senior citizens, households with children, wealthies, young professionals, and the like. The segment information may be overlaid on the system account history for each home passed (such as nevers, former basic, former pay, former multipay, basic subs, one-pay subs, multipay subs). From marketing research and cable sales experience, cable systems know the basic appeals to make to each segment. The marketing *strategy* can then be adjusted to the segments. Some cable companies, for example, have found that direct mail works well in upper income but not in lower income areas. The marketing *materials* may also be tailored to the segments. Direct mail pieces, literature, premiums, and sales presentations can all be individualized for each segment.

"Young professional nevers" might receive appeals emphasizing movie channels, diversity, and information programming. The same message to "young professional formers" might be refined to take into account the previous subscription, playing up the things they have been missing and any new features of the service related to movie channels, diversity, and information channels. A final statement could welcome these formers back with a special premium. Current "young professional basic subscribers" might not receive the diversi-

FIGURE 15-5. Mailer aimed directly at the young professional (yuppie) segment.

ty and information programming pitch, but would be approached with a pay channel emphasis on movies and specials that communicates the importance and convenience of using cable to stay on top of the entertainment world while maintaining a busy schedule.

Each other segment, in turn, gets its treatment related to the appropriate appeals and the cable account history.

Children are an important market segment. Households with children are more likely to have cable. In one study of the child market it was found that, in almost all households with children, the children participated in the discussion on whether or not to get cable. The majority of children urged parents to buy cable and were influential in the purchase decision. Quite a few parents said they would not have cable if it weren't for the children and, importantly for retention, three quarters said that children would protest if parents were to threaten to drop cable or a particular cable channel.[5]

Assessing Marketing Results

A number of figures are used as indices of sales productivity and for marketing cost accounting. The *universe* of potential subscribers is *homes passed*; all homes that have access to cable, that is, where feeder lines actually go past the home either overhead or underground. *Basic penetration* is the percentage of households sold to household passed by cable:

$$\text{Basic penetration } \% = \frac{\text{Number of basic subscribers}}{\text{Number of homes passed}} \times 100$$

Pay penetration usually refers to *all* pay units as a percentage of *basic*. This may also be referred to as *pay-to-basic* penetration. The percentage in a multipay system may be well over 100 percent.

$$\text{Pay penetration } \% = \frac{\text{Total number of pay units regardless of service}}{\text{Number of basic households}} \times 100$$

The penetration for any particular pay channel or package is calculated:

$$\frac{\text{Penetration of}}{\text{service (such as HBO)}} = \frac{\text{Number of subscribers to the service}}{\text{Number of basic subscribers}} \times 100$$

In judging the potential upgrade market, it is also useful to determine:

$$\text{Homes using pay } \% = \frac{\text{Number of subscribers with at least one-pay service}}{\text{Number of basic subscribers}} \times 100$$

Pay-per-view sales are indexed by the *buy rate*:

$$\text{PPV buy rate } = \frac{\text{Number of households taking program}}{\text{Number of subscribers}} \times 100$$

In this calculation, the "number of subscribers" in the denominator may have to be the "number of addressable subscribers" for those systems where only a portion of the subscribers are addressable and have access to PPV. The buy rate may also be calculated for a period by cumulating all the program purchases for the period in the numerator:

$$\text{Cumulative PPV buy rate } = \frac{\text{Number of program purchases in period (e.g., month)}}{\text{Number of subscribers}} \times 100$$

In this case the buy rate may exceed 100 percent. The PPV operators are also interested in the homes using PPV, usually over a month's time:

$$\text{PPV penetration } \% = \frac{\text{Number of homes using PPV at least once in a month}}{\text{Number of subscribers}} \times 100$$

Here each home using PPV during the month is counted only once, no matter how many different PPV programs they use during the month. Again, the "number of subscribers" may be the "number of addressable subscribers."

Since basic penetration varies significantly according to the location of the community and the value of cable as an antenna service, penetration of pay may be expressed as penetration of all pay units or individual channels using *homes passed* as the base:

$$\text{Pay-to-homes passed \%} = \frac{\text{Total number of pay units regardless of service}}{\text{Number of homes passed}} \times 100$$

$$\begin{array}{c}\text{Pay service (e.g.,}\\\text{HBO)-to-homes}\\\text{passed \%}\end{array} = \frac{\text{Number of subscribers to the service}}{\text{Number of homes passed}} \times 100$$

Churn represents a loss to cable companies and is the principal figure in assessing retention marketing efforts. It is usually calculated monthly and is also annualized:

$$\text{Percent churn} = \frac{\text{Current month's (or year's) disconnects}}{\begin{array}{c}\text{Average number of subscribers for the}\\\text{Month (or year)}\end{array}} \times 100$$

Sometimes a cable system will use the previous period's end subscriber count as the base instead of the average number of subscribers.

Retention rate is also important in marketing. *Retention rate* is simply the percentage of subscribers staying with the service for a particular period after a marketing campaign. The period used is often three months, the most critical period:

$$\text{Retention rate} = \frac{\begin{array}{c}\text{Number of new units after a marketing}\\\text{campaign retaining the service for 3 months}\end{array}}{\text{Number of new units after a marketing campaign}} \times 100$$

Acquisition and upgrade marketing costs are relatively easy to calculate. For marketing campaigns, all marketing costs—advertising, promotional materials, sales commissions, and the like—are totalled. This cost figure is divided by the number of new units sold to provide a *cost per unit sale*. To determine a more useful figure, the total cost of the campaign and follow-up activities associated with the campaign should be divided by the number of new units retained after three months:

$$\begin{array}{c}\text{Cost of sales}\\\text{per retained unit}\end{array} = \frac{\text{Total cost of campaign and campaign follow-up}}{\begin{array}{c}\text{Number of new subscribers retained after}\\\text{three months}\end{array}}$$

Close monitoring of the cost of sales per retained unit is particularly important in remarketing *mature* cable systems where an aggressive subscription sales campaign could *supersaturate* the system that subsequently will experience many downgrades or disconnects. Another useful figure is the *marketing cost per net unit*. This is the total cost of marketing for a period divided by the net gain in units during the period. The result is reflective of all marketing efforts for the period including retention marketing:

$$\text{Marketing cost per net unit} = \frac{\text{Total marketing costs per period}}{\substack{\text{Number of units at the end of} \\ \text{the period minus the number of} \\ \text{units at the beginning of the} \\ \text{period}}}$$

Of course, a net *loss* in subscribers is possible. In this case the total marketing cost would be assessed against the loss in subscribers for the period. Some judgment would be made to determine whether more or less of an expenditure was appropriate under the conditions.

Determining the *optimum* for specific retention marketing expenditures is difficult. The success rate on save-a-sub procedures can be determined along with the costs, but the results of other retention marketing techniques are never too clear. A cable operator must keep track of all retention marketing costs and look at those costs against the system churn rate. By comparing one period to another comparable period (keeping in mind that there are seasonal churn variations), the operator gets some idea of the appropriate overall expense. The object is to spend no more for retention than is necessary to keep churn to the expected minimum. When churn gets down to the acceptable minimum, then retention marketing expenses are gradually reduced to see if that low level of churn will tolerate less expense.

Certain retention marketing techniques can be assessed by experiments. A cable system newsletter with programming highlights may be sent to half of the subscribers for several months and churn rates compared for those who got the newsletter and those who didn't. If there is a difference favoring the newsletter subscribers, the newsletter cost may be compared with the cost saving resulting from the reduced churn. Also, retention marketing techniques that appear to be cost-effective in reducing churn in one system may be applied with reasonable expectation of the same result in a comparable system.

Looking at marketing costs in this way is essential as a means of cost control. Particularly in mature cable systems, where nonsubscribers and subscribers have been remarketed regularly, new marketing campaigns and retention efforts may cost more than they are worth in net subscribers gained or churn reduced. Or marketing methods may be used that do not compare favorably with other methods.

Cost analysis of direct mail and coupon advertising is done by coding the response cards or coupons so that the results of a particular mailing or advertisement can be properly attributed to the source. For example, for a direct mail piece that is sent to young professionals, the return address on the return postcard may include Department DMYP (or simply have that code printed in a corner of the card). The number of orders attributed to that mailing may then be divided into the cost of the mailing.

For individual salespersons, the cable company may calculate the conversion rate (sales) per contact or the conversion per presentation:

$$\text{Conversion rate} = \frac{\text{Number of conversions (sales)}}{\begin{array}{c}\text{Number of contacts (or}\\\text{presentations)}\end{array}} \times 100$$

To complete the picture, cable systems are interested in revenue produced per subscribing household and per home passed:

$$\text{Revenue per subscriber} = \frac{\text{Total subscriber revenues}}{\text{Number of basic subscribers}}$$

$$\text{Revenue per homes passed} = \frac{\text{Total subscriber revenue}}{\text{Number of homes passed}}$$

Revenue per subscriber is important in comparing the business performance from one period to another or in comparing comparable cable systems in different communities. Since the capital investment in cable relates to all homes passed and because homes passed is the universe of potential subscribers, the *revenue per home passed* figure is also valuable.

Revenue per household for particular services may also be calculated, as in:

$$\text{PPV revenue per HH} = \frac{\text{Total monthly PPV revenue}}{\begin{array}{c}\text{Basic subscribers (or addressable}\\\text{subs, if not all subs are addressable)}\end{array}}$$

The Cable Television Administration and Marketing Society (CTAM) regularly provides baseline data from a cross-sectional representation of cable systems tracking churn, upgrades, and downgrades. The CTAM Database, developed in cooperation with A. C. Nielsen, may be used to detect trends and as a standard of comparison for individual system and MSO data.

Substantial cable system resources are put into marketing. Almost every employee outside of the marketing department has some marketing responsibility as well. It is the area of cable system operation for which there is least precedent in other fields. The learning experience is only beginning.

NOTES

[1]Carol Siewert Mackey, "An Evaluation of Nonsubscribers' Perceptions of Cable Television to be Applied to the Development of Target Marketing in Three Suburban

Michigan Communities," M. A. Thesis, Michigan State University, 1985. Ronald Paugh, "An Analysis of Participants in a Free-Trial Cable Television Offer: Implications for Marketing Strategy," Michigan State University and Horizon Cablevision, January 1986.

[2]Bradley S. Greenberg, Roger Srigley, Thomas F. Baldwin, and Carrie Heeter, "Subscriber Responses to a System-Specific Cable Television Guide," Michigan State University/Continental Cablevision, Inc., June 1985.

[3]A thorough discussion of retention marketing techniques may be found in: Kenneth L. Bernhardt, "Designing and Measuring Effective Retention Marketing Techniques," Cable Television Administrative and Marketing Society, August 1984.

[4]A description of marketing procedures for MDUs is found in: "Marketing to Multiple Dwelling Units," HBO (undated).

[5]Ronald Paugh and Thomas F. Baldwin, "Children as an Influence in the Decision to Purchase and Retain Cable Television," paper presented at the Midwest Association for Public Opinion Research, Chicago, Illinois, November 1985.

CHAPTER **16**

Advertising Sales

Interconnect map for Chicago area.

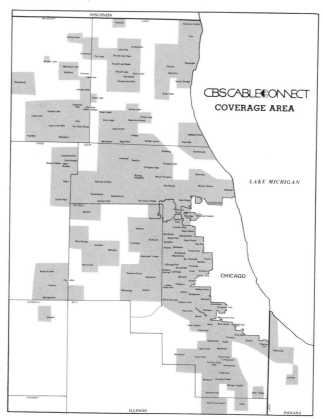

Advertising came to cable as a means of sustaining programming that could not be supported by subscription fees alone. Early CATV systems began selling advertising to help defray the costs of cablecasting local sports and other events. This was generally done on an ad hoc basis; when an event came up, the cable system staff would seek out one or two businesses that would put up some of the money, mainly as a community service contribution rather than as a marketing expenditure.[1]

Cable advertising may be the only opportunity for television advertising in some businesses. Broadcast radio and television stations serve such large areas that they are too expensive and wasteful for localized merchants. Other major television advertisers, because cable attracts significant audiences that are diverted from other media, cannot ignore cable. Broadcast television program ratings, as we have seen in Chapter 10, may not be as high in cable television homes.

Most cable networks were designed on the commercial broadcasting network model. Once the assumption was that they could eventually be entirely advertiser-supported. Per-subscriber fees for the networks were gradually reduced, in some cases to zero, but the trend has now reversed as programming costs and competition have increased. Nevertheless, advertising is a very important element in cable network revenues.

TYPES OF CABLE ADVERTISING

Network

Network advertising is sold by the cable network and inserted in the programming so that it reaches the entire universe of affiliated cable systems. There are about 12 minutes of advertising time interspersed throughout an hour of programming. Usually ten of these minutes are sold by the network; the other two, called *local availabilities*, are reserved for affiliate sales. Since not all affiliates sell advertising time these two minutes must also be filled by the network with program promotion spots, public service announcements, or bonus commercials for major buyers of network time.

Each cable network has its own sales staff calling on advertising agencies and advertisers. Almost all of the network time is purchased by major agencies for nationally distributed products. Cable network sales generate about 80 percent of the total cable advertising revenues.[2]

The *transaction cost* in buying network time is minimal. The advertising agency makes *one* buy (usually for several commercials at different times) to get the whole cable network. A single commercial or set of commercials is delivered to the network. Actual performance on the contract is fairly easy to monitor. A single bill is paid. Compared to other types of cable advertising (to be described), this is an administratively simple process.

Spot

Spot advertising may be national, regional, or local. *National spot* advertising is the purchase of commercials in an ad hoc network of cable affiliates selected to fit a product's market or marketing plan. The spots purchased come from the *local availabilities* that have been reserved for the affiliates use. Because advertising agencies must deal with many, many affiliates to put together a campaign with any impact, relatively little national spot advertising is sold in cable. However, a mechanism, *interconnection*, has been implemented to meet this problem. Many cable systems are regionally interconnected in a *hard* (wire or microwave) or a *soft* (nonphysical) interconnect. In either case, the order is placed with a single office for the whole group of systems or any combination of systems within the interconnect. In the hard interconnect, a single commercial tape is inserted in the proper programming for all of the affiliates within the interconnect. In the soft interconnect, commercial video tapes are distributed by the interconnect.

The interconnect aggregates enough cable subscribers to make the advertising buy of consequence to a major advertiser. The interconnect is represented by salespersons *(representatives* or *reps)* in the major cities where advertising agencies are located. It is not practical, of course, for a single cable system, whatever its size, to represent itself for advertising time sales in all of these cities because of the low dollar value of any contract in relation to the sales cost.

Cable interconnects have now been developed in most major television markets. This should stimulate growth of national spot sales.

Even without using interconnects, national or regional spot advertising in cable may be efficient for some advertisers, allowing them to approach the ideal in market segmentation. For example, an advertiser with a big price ticket product could buy time in local availabilities in the appropriate cable networks only in upscale suburban cable systems in exactly the geographic areas desired. The costs of identifying and contracting with these systems individually would be high, but this cost would be more than offset by the value of going directly to the right market segment.

Local spot advertising is sold to local advertisers in *local availabilities* of national cable networks and in local origination (LO) programming. Since many local advertisers do not serve as broad a market as a broadcast television station, one or more cable systems can be purchased to match the market without *waste circulation*. For bigger advertisers within the market, cable may be used to supplement broadcast television or to target a particular segment of their market. So far, these bigger advertisers, already in broadcast television, have been the major supporters of local spot advertising on cable. A local interconnect, if one exists, is a value to these bigger local advertisers because it reduces transaction costs, as it does in national spot.

Sponsorship

Spot advertisers are buying *participations* in network or local programming along with other advertisers. Another approach to advertising is to *sponsor* programming. Most broadcast television programs are too expensive, even for the largest advertisers. But cable networks and local origination are very much available for sponsorship. The sponsor gets exclusive identification with the program and does not have to be concerned about being lost in the clutter of spots for other advertisers. The audience for the program may be fairly loyal permitting certain assumptions about knowledge and recognition that the advertiser is building with the audience. Whatever credibility or affection is generated by the program can rub off on the sponsor. The sponsor can select a program that is a good environment for the product or service advertised. In some cases it is possible for the advertiser to produce and supply the program with the advertising already in place. This was once the way broadcast network radio and television programs were developed, but high costs have almost entirely eliminated the practice.

Sponsorship may be especially valuable to *institutional* advertisers who are seeking to create a substantial, community-spirited image. Financial institutions, major employers, and hospitals might do institutional advertising at least part of the time. A number of local origination programs could fit these needs.

A form of sponsorship is *barter programming*. In barter, an advertiser supplies a program with about half of the commercial spots filled, paying nothing for the advertising time. The cable network or cable system gets the program for nothing and has the other half of the spots to sell.

Another concept somewhat similar to sponsorship, used for institutional advertising, is *underwriting*. This idea, developed for public broadcasting stations, gives a company or institution an *enhanced identification* with a program in return for a contribution to its costs. For example, the ID might show a bank's logo with an audio message saying, "This program is made possible by a grant from the Liberty National Bank with full-service branches in seven locations." Cable systems, city governments, and the schools may accept underwriting for PEG access channels as well as the cable system's LO.

Direct Response

Merchandise or services sold directly to a consumer from catalogs or advertising is a fast-growing form of marketing, and as a result many consumers are no longer wary of *direct response* advertising. Because of cable's upscale subscribers, inexpensive spot rates, and the potential to select audiences, cable has been a natural medium for direct marketing. Since each sale or inquiry is recorded by medium, the effectiveness of the advertising is easily assessed.

Direct response advertising may be sold by cable networks or operators in the same way any other advertising time is sold, or on a *per inquiry* basis where the advertiser gives a percentage of each sale to the cable medium. If only inquiries are solicited, the medium gets a fee for each inquiry.

With the increasing popularity of buying direct, major magazines and national newspapers, computers, international cruise lines, insurance, and other high-priced items are sold through cable as well as the Slim Whitman records and bamboo steamers. In accepting direct response advertising, the cable operator and network must take some responsibility for the credibility of the seller, since long delays in delivery or dissatisfactions with the product may bring complaints directly to the cable network or cable system.

Co-op

Manufacturers give advertising allowances to their dealers or retailers to encourage local advertising tie-ins to national advertising and to take advantages of local discounts earned by large-volume dealer advertising contracts that include advertising for other than the manufacturer's product. Although all dealers may be entitled to the allowances, they are used mainly by the large volume retailers. *Co-op* means that the manufacturer and the advertiser share in the cost of the advertising using a predetermined percentage split. Usually the retailer accrues co-op advertising credits based on the amount of merchandise purchased. When the advertiser presents the cable operator's bill along with the claim documentation,[3] the advertiser is reimbursed.

The cable operator may assume that any branded merchandise advertised locally by a retailer in any medium is under some sort of co-op plan and can follow up to attempt to include cable in the merchant's plans. (Because cable is a relatively new advertising medium, the advertiser may first have to get permission to use cable.) Often there will be several merchants selling the same product. They may pool their co-op advertising dollars in a single contract with the cable operator or interconnect, billing each their agreed-upon share and including each in the dealer listing ID. Lists of co-op advertising plans that permit cable are provided by the Cabletelevision Advertising Bureau (CAB) regularly.

Co-op is convenient for cable operators and advertisers because well produced television spots are available from the manufacturer.

Infomercials

More or less unique to cable is the *infomercial*. The infomercial is a commercial that is longer than 60 seconds. It may be two or three minutes or more. Cable networks accept infomercials, but it is unlikely that a cable operator has the opportunity to sell infomercials in local availabilities because the availabilities permitted by the network may be only one minute at a time. Operators can, however, make time for infomercials in LO programming.

Infomercials are very valuable to advertisers who have a relatively complex message that cannot be adequately confined to the conventional 30- or 60-second spot (such as presentation of a new electronic banking system). A major event may be promoted by an infomercial to include all its dimensions and emphasize the importance (like the introduction of a dealer's entire new car line, or the seasonal sale at a shopping center involving several merchants). The infomercial may also serve as a miniprogram within which a product plays a role (such as the two-minute General Foods "Short Cuts" series in which cooking tips were given using the company's food items).

Program Length Commercials

Program length commercials or *shopping programs* use standard length programs to promote a single product or service or to combine several products within a shopping format. These programs are designed to be informative, not a hard sell for a product from beginning to end. They must have some utility for the consumer—interesting products, bargains, new sources. Real estate brokers have used such programs for providing a quick visual tour of several homes. Long-form, self-help venture programs also appear frequently on cable. "The Cable Shop," although it did not succeed in its attempt to syndicate nationally, found strong audience interest in a program format that was essentially several back-to-back infomercials for a variety of products. The success of the shopping networks in direct sales is further testimony to the cable subscribers' interest in television retailing.

Political Advertising

Cable gives most local political advertisers a chance to match their advertising to their constituency of voters. Other media, daily newspapers, radio, and television, generally have too much waste circulation for drain commissioners, township supervisors, treasurers, and the like, as well as some Congressional districts. Cable can also more conveniently provide political office seekers an infomercial in LO programming, permitting more time for exposition of an issue.

Other Cable Advertising

Not all cable advertising is in network programming or LO. The automated channels almost always devote a portion of the screen to advertising messages. Videotex services may be sponsored with advertising messages interspersed. In both cases computer-generated graphics can be used to attract attention and enhance the message. Audio advertising or audio messages from television commercials can be placed in cable radio services. National and local advertisers may also buy space in printed cable guides. The editorial matter and certain listings in the guide may be uniform and circulate nationally to hundreds of thousands of cable subscribers, a useful circulation for some national advertisers. Local advertisers can take space in the system-specific listings to reach an individual system's subscribers.

CABLE ADVERTISING STRATEGIES

The only channels on cable that attract a consistently large share of a system audience are the off-air broadcast stations, and pay channels, none of which are available for cable advertising. An advertiser-supported cable network very seldom attracts a large audience and averages about .5 to 2 percent of the cable television households. LO channels have even lower numbers, suggesting cable advertising strategies that are somewhat different from broadcast television.

The cable advertising vehicle (channel and program) is selected to meet precise *demographic* and *geographic* parameters, making the relatively small audience more meaningful. This advertising buying procedure is called *qualitative* because of the selection process (and also because cable, indeed, has little quantitative audience data available). Large numbers of spots are purchased so that the small audiences, aggregated over different time periods or channels, are of significant size. Large volume purchases are also necessary to get the desired *frequency* (the average number of times a target prospect is exposed to a message). Finally, volume purchases are essential to keep administrative (transaction) costs to a reasonable percentage of the total cost of the buy. In other words, if only a few spots are purchased, the administrative costs of selecting and buying the time, providing the commercial, verifying the runs, and paying the bill may be much too large in relation to the value of time.

The cable advertising buyer is interested in several factors. *Reach* is the number of different (unduplicated) people or households that are exposed at least once to an advertisement. It represents the breadth of the audience. Because cable network programming and local origination are often narrowly targeted, it may take advertising on several channels to produce adequate reach for a mass market advertiser.

As noted, the cable buyer is also concerned about frequency. A single exposure is usually not enough to make a significant impact on the prospect. The advertiser seeks sufficient frequency for an *optimum* impact without wasting money on diminishing returns (*commercial wearout*). To get desired frequency an advertiser would have to purchase a high volume of spots since cable audiences tend to be quite low. However, if all the spots are purchased within a single specialty network or programming type, frequency might be quite high at low volume because of audience loyalty.

Continuity of advertising is another concept of interest to advertisers. The spots may be *continuous*, maintaining the same volume in the same channels and periods consistently over time. Or a *flighting* strategy can be employed, where the advertiser advertises for a period of time and then rests (avoiding *wearout*). This continues in a planned off-and-on use of advertising. A *pulsing* strategy mixes both, maintaining a base level of advertising continuously, but pulsing periodically to greater levels.

All of this suggests the importance of planning in advertising. Ideally, it is not advertising spotted in a variety of places in a random or haphazard manner, but rather a carefully worked out schedule designed to deliver the desired reach, frequency, and continuity.

Advertising time is sold on a *fixed placement* basis or *run of schedule* (ROS). In *fixed placement* the advertiser selects the channel and the time period, or program, and pays a premium for this privilege. In *run of schedule*, the advertiser allows the cable operator to place the spots in a particular channel or across all channels in which advertising time is sold. Or the cable network or system will guarantee a percentage of the spots being in each of several categories of time from the least desirable to the most desirable, based on assumptions about the size of audience expected. In this latter case the various possibilities are called *bulk plans* or *packages*. The advertiser receives substantial discounts for allowing the cable operator to place the advertising or for participating in plans.

Most broadcast television advertising time is priced in terms of audience size. A rating point has a generally accepted value in a particular market. These audience figures are also related to the prices charged for time in terms of *cost per thousand* (CPM), where the cost of a spot is divided by the number of the audience in thousands. Cost per thousand[4] is calculated:

$$\text{CPM} = \frac{\text{Cost of the spot}}{\text{Audience size}} \times 1{,}000$$

If the spot on cable cost $30 and the audience at the time the spot was run were 2,200, the CPM would be $13.64—$30 divided by 2,200 and multiplied by 1,000.

Theoretically, the pricing of cable or broadcast spots should be structured so that the cost per thousand is identical in all time slots and programs. That is, if the $30 spot in the example is in prime time, the programming in the afternoon draws exactly half the audience, 1,100, then the spot in the afternoon should be priced at $15. The cost per thousand is still $13.64. All package plans should be worked out so that the cost per thousand of one package is the equivalent to the cost per thousand of any other package, however they differ in their make-up according to time slots or programming. Discounts are then given on the basis of volume (the more spots purchased, the lower the price per spot) and the amount of freedom the cable system has in scheduling the spots (the more freedom, the greater the discount).

But the cable advertiser may very likely desire a particular demographic group, and in fact may be coming to cable because its programming will select that particular group. This adds a new dimension to the calculation of the CPM:

$$\text{CPM} = \frac{\text{Cost of the spot}}{\substack{\text{Audience size in} \\ \text{(specific demographic category)}}} \times 1,000$$

In our example, if the $30 spot has a total audience of 2,200, but the advertiser is interested in women 18–49, and this audience includes 1,700 women 18–49, then the CPM for women 18–49 is $17.65. This figure may vary considerably across the time periods and programming because of the varying availability of demographic groups and the differing attractiveness of programs to demographic groups.

The cost analysis might be further refined to the CPM in the advertiser's *trading area*. If 30 percent of the population of television households in the broadcast television market is in the advertiser's primary trading area, then only 30 percent of the broadcast television audience should be included in calculating its CPM. The same is done for cable audiences in relation to the trading area. Only there a very large portion of the cable system audience, perhaps 90 or 100 percent, would be in the trading area because cable system service is so much more confined.

Ideally, this is how cable advertising time should be priced and how it should be purchased. It works for some of the cable networks. But this theory breaks down for the other cable networks and most cable systems because they do not have the audience data on which to base the rates, nor do they have audience demographics for time slots and programs. Therefore, time sales and purchases are made primarily on logic—the set of qualitative assumptions about the likely audience for individual programs or time slots and the probable impact of the advertising. This means that the cable advertising representative and the time buyer should make careful judgments about the market for the product or service to be advertised and then select the individual programs or time slots most appropriate. Package plans should fit the same logic, a much more difficult task because of the wide variety of time slots and programming usually covered. Run of schedule, while it suits the convenience of the cable operator and fills the spots in the least desired time periods, is extremely risky for the advertiser because it is almost impossible to apply the necessary qualitative logic. The risk is only worthwhile if the ROS prices are extremely low relative to fixed placement in specific times and channels.

ADVERTISING SALES PROCEDURES

Inventory

Local cable advertising sales are made from an *inventory* of available spots and programs for sponsorship. The inventory consists of all the spots in cable networks that are given to affiliates to sell—*local availabilities*. The system may choose not to sell local advertising on *all* cable networks carried even though there are local availabilities. Too much inventory of local "avails" may be confusing for advertisers and require too much administrative effort and equipment on the part of the cable operator. Most cable systems select three to six of the most popular cable networks. The list may include one or two other specialized networks that target a particular audience difficult to reach by any other medium.

The cable system may or may not sell advertising in local origination programming. If LO programs attract a fairly large audience or are known to have intensely loyal followings, they may be useful to advertisers who might want participating spots or sponsorship. If audiences are unknown or known to be quite low in relation to all the other competing programming, and there is no clear indication of intense audience interest, the operator is not likely to offer LO advertising. There is an expense in sales and insertion costs, and the results for advertisers would be disappointing.

If LO advertising is sold, then the spots available in LO are added to the inventory. The inventory is updated at least daily as spots are sold and taken off the inventory. Sponsorship is permitted in almost all LO programs. If an advertiser commits to sponsorship, it is usually for a relatively long period. All the spots in that program for the period of the contract are taken off the inventory.

Rate Cards

Without definitive local system audience data available for each channel, cable advertising time prices are often based on the total subscriber count for the system, much like newspapers and magazines. Cable spots (30 seconds) are sold at roughly $1 per 1,000 subscribing households.

The cost of local cable advertising is presented in the system *rate card*. Cable rate cards come in many varieties, but most of them list spot costs under successively greater purchase volumes in different program or time categories. The price of an individual spot always varies by its *placement* in the schedule and by the number of spots purchased in a particular contract (*volume*).

Volume discounting reflects the cable system's attempt to provide an incentive to purchase a greater number of spots and the actual saving in sales and administrative costs when a single advertiser buys in volume.

Placement of advertising by time period or program is priced to reflect the audience expected. The rate card will list the various dayparts as time classes.

Prime time in cable is usually defined differently from prime time in broadcast television, 8–11 P.M., because cable does relatively well in the early and late fringes of broadcast prime time. Therefore, cable prime time may be 6 P.M. to 12 Midnight. The afternoon period may be listed on the rate card discounted from prime time. Other dayparts may also appear, or the lowest rate may be listed for run of schedule (ROS) where all available time slots are used (see Table 16–1).

TABLE 16–1. Volume Discounts.

Spots	Rate Per Spot ($)		
	Prime	12N–6p*	ROS
1 time	60	40	30
10 times	55	36	27
30 times	48	29	22
60 times	39	20	18

*N = noon; M = midnight; p = P.M. Prime = 6p to M.

When advertisers purchase several spots in a program or period, the one-time spot rate is discounted. The larger the volume of spots purchased, the greater the discount is.

In an effort to sell advertising in volume across all dayparts, cable operators may put together *packages* or *bulk plans* that guarantee a proportion of spots in each time period. Discounts are given for the package purchase and increased as volume increases.

TABLE 16–2. Bulk Plans Based on Rates in Table 16–1.

	Total Cost ($)		
	10 Spots	30 Spots	60 Spots
Plan 1	389*	930	1,420
1/3 prime			
1/3 12p–6p			
1/3 ROS			
Plan 2	390	990	1,590
1/2 prime			
1/2 ROS			

*Table 16–1 rates in each category are discounted $2 in the bulk plan.[5]

This rate structure may be adequate in systems selling advertising in cable networks with nearly equal audience levels such as CNN, ESPN, MTV, and USA. Indeed, there is substantial merit in attempting to keep the rate structure very simple as in these illustrations. But if audiences vary considerably across channels on which advertising time is made available by the local operator, a more complicated structure may be necessary. One solution is a classification system in which a mixture of channels, program types, and time periods are grouped together on the assumption of *roughly equal audience size or audience value.* The advertiser's value for an audience may depend on demographic characteristics. For example, the "upscale" audience for the Arts and Entertainment network, with its buying power, may have a greater value than the audience for the childrens' channel, Nickelodeon. The classification scheme may be based on actual local audience data, local audience projections from national audience data, and the logic of program and time period attractiveness to various audience types.

Each of the classifications is given a name, number, or letter designation. A price is worked out for each classification representative of its value relative to all other classifications.

Table 16–3 is a sample rate card for a system that has CNN, ESPN, MTV, USA, Arts and Entertainment (A&E), and a local origination (LO) channel. Volume and bulk plan packages are then built on discounts from these rates.

TABLE 16–3. Audience Size/Value Classification System.

Classification A—$60
   6 p–12 M—CNN, ESPN, MTV, and USA
Classification B–$40
   12 N–6 p—CNN, ESPN, MTV, and USA
   6 p–12 M—A&E
Classification C—$30
   ROS—CNN, ESPN, MTV, and USA
   6 p–12 M—LO
Classification D—$15
   ROS—A&E, LO

Sales/Orders

Salespersons call on cable advertising prospects, or their advertising agencies, to learn the nature of the prospects' business and their marketing and advertising needs. In most cases it is necessary to acquaint the advertiser with the unique aspects of cable advertising—the number of channels available, the rate structure, and some general principles about using cable advertising effectively.

Once the salesperson is acquainted with the prospective advertiser, a specific presentation is developed. This may be a formatted, one-page proposal suggesting a plan using time periods and/or programs in one or more channels that are appropriate for the advertiser. Or the presentation may be much more elaborate, with background information on the cable networks and the programming suggested, as well as expected demographic characteristics of the intended audiences, all related to the advertiser's marketing needs.

The cable advertising sales representative and the advertiser work over the proposal, revising it to suit the interests and the budget of the advertiser. The rates charged or the price of the whole package may be negotiated. In large purchases of time that is not in high demand by other advertisers, when the advertiser's negotiating power is the greatest, the cable operator tries not to deviate from the published rates, but may include some *bonus* spots that get the advertiser to the number of spots desired or some extra exposure while maintaining the appearance of a firm rate card.

The sales-buying process is not always such a rational one. The cable advertising salesperson makes the best case for cable in the absence of concrete knowledge of audience and impact. The buyer has even less information and, often as not among cable advertisers, is new to television. The lure of television and the desire to believe the cable advertising salesperson's promises may overcome healthy skepticism.

The order itself is a contract: a form with blanks to fill in the number of spots in each time period or program on each channel involved, the dates, cost of production (if appropriate), and details about payment.

Once the contract is received, the spots purchased are taken off the inventory, as already mentioned. The spots are scheduled in the master programming log.

Sales for interconnects are handled locally in the same way as for an individual system. Each system in the interconnect receives a portion of the revenues after expenses in proportion to its number of subscribers.

The cable systems and interconnects sell directly to advertisers and to *advertising agencies*. Sometimes a major advertiser will have an *in-house agency*, a separate unit within the company. Advertising agencies get 15 percent of the advertising sales. This *commission* is actually a *discount* on the advertising time rate. For example, if cable advertising purchased for an automobile dealer through the dealer's advertising agency is priced by the cable company at $1,000, the advertising agency discounts the $1,000 by 15 percent and pays the cable company $850. The dealer pays the agency the full $1,000. If the dealer were to buy the advertising time directly from the cable operator the price would be $1,000 (not discounted). Thus, the dealer (advertising agency client) does not pay directly for advertising agency services. When the advertiser buys directly from the cable operator, the operator provides some of the services of the advertising agency.

The agency commission system does not always work well in cable because of the low cost of cable spots. Unless the advertiser buys a tremendous volume of spots, the agency commission may not cover its costs. If this commission does not even cover costs, the agency may choose to ignore cable, or assess a special fee to the advertiser for cable advertising. Either way, cable advertising sales suffer.

Production

Many spots run on cable have already been produced for local broadcast stations and are ready to use, as is, on cable. Some spots are supplied by home offices of franchises (real estate, fast food, and the like) or manufacturers of products sold by local retailers. In these cases it is only necessary for the cable operator to add a local ID.

Other advertisers will look to the cable operator to produce the spot. The cost of the production is separate from the cost of the time. It is usually figured at an hourly rate which may differ for on-location production and studio production (see Appendix G). The advertiser receives an estimate of the cost, which is usually firm whether or not the actual production time exceeds the original expectation.

Scheduling and Insertion

Once the contract is received, the spots are scheduled in the master programming log by the programming or traffic department. The videotape of the commercial is placed on masters for each channel on the schedule. The spots may be in sequence or randomly accessed.

Many cable systems have *automatic insertion equipment.* (Figure 16–1.) The computer programming calls for the proper commercial on the proper channel at the scheduled time. Cable networks cue the computer with an electronic signal when local availabilities are coming up. If a local spot is programmed at that time, it is picked up. If not, the network goes on with its own filler material. Cable systems that do not have automatic insertion equipment, monitor the networks and local origination, switching by hand videocassette players for local commercials.

Billing

The automated insertion equipment is capable of providing a printout of the schedule of spots for a particular advertiser. This serves to verify the schedule and gives the buyer an opportunity to check the schedule against the contract. The invoice also includes or, is itself, an affidavit swearing that the spots ran according to schedule. The bills are usually discounted for payment within 10 days.

Some cable advertising sales may be in the form of *tradeouts.* The advertiser pays for the advertising time in goods or services—the office furniture advertiser in office furniture, the auto dealer in cars and trucks, the catering services advertiser by catering the cable company office Christmas party, and so on.

A cable operator may contract with a specialist in cable advertising sales to handle the entire process. If this is the case, the operator takes a flat fee for all of the advertising availabilities or a percentage of the income after costs.

FIGURE 16-1. A six-channel automated insertion system in Southgate, California.

ASSESSING ADVERTISING

Nielsen provides national audience data for the cable networks with the largest subscriber counts and audience data for cable interconnects, and a few big systems. These networks, interconnects, and systems can supply advertisers and agencies with credible audience data. The cable networks *not* measured by Nielsen must periodically survey a sample of affiliates to gather audience data for programming and advertising purposes. This is very expensive and therefore infrequent. The great majority of cable systems that are not measured by Nielsen only occasionally, if ever, do audience research. Because cable systems have so many channels, it takes very large samples to provide reliable data on any one of them. The large number of channels and frequent channel switching also work against accurate recording or recall of cable use by viewers. For these reasons, conventional television audience methods may underestimate the cable audience.

Cable systems without audience data for their own service areas may make projections from the Nielsen national cable network ratings. A Nielsen rating for a cable network is the average percentage of all households with access to that cable network that are tuned to the network during a particular time period. Usually the time period reported is a total day (24 hours), 7 A.M. to 1 A.M., or prime time. For example, if the USA Network rating is 1.3 percent for prime time and that network has a universe of 31,000,000 subscribing households, it has an average audience of 403,000 households during prime time (.013 × 31,000,000). A cable system with 10,000 subscribers, carrying the USA Network, could be expected to have a prime time audience of 130 households (.013 × 10,000). (Note that the audiences by these ratings are *households*, not individuals. This will change for some audience reports as "people meters" take over.) The estimate for the *local* audience based on the national audience assumes that (1) local audience interests are similar to the national interests, and (2) that the number of channels competing with the cable network in the local system are equal to the average number of competing channels nationally. On the latter point, if the average size of the cable systems carrying the network is 15 channels, and the local system for which we are estimating carries 35, then there is presumably more competition for any one of the channels than in 15-channel systems. It could be anticipated that the USA Network would have a lower audience in a 35-channel system than in a 15-channel system.

Commercial audience ratings services are developing various versions of "people meters." The meters require all household members to log in and out when they are viewing. This results in *individual* viewing data rather than *household* data that register the household as viewing a program whether or not any people are actually in the room where the TV is on. The "people meter" raises problems for cable in comparison to broadcast television because of the more frequent channel switching of cable households and the greater use of cable as a background medium during other activities.

Two-way cable systems have the capability to monitor viewing. The monitoring reports consitute a census of viewing. Every subscriber is included in the aggregate data. These data are far better than any available to estimate broadcast audiences, although they do not, of course, tell how many members of the cable households are viewing, which people are viewing, and indeed whether anyone is actually watching a TV set that is turned on. A system that is not entirely two-way may create a *sample* of the system that *is* two-way for marketing, programming, and advertising research.

Ordinarily a broadcast advertiser is satisfied if the commercials are exposed to a certain number of people at a reasonable cost per thousand. This assessment can be made in purchases of national network time and in the few local cable systems where audience data are available. In most local systems audience information is not available and a qualitative judgment must be affirmed by other information. It is difficult to attribute sales directly to advertising except in direct-response advertising where each sale comes from a commercial. But it is possible to relate sales to cable advertising by *isolating* the cable advertising. Only cable advertising is used. If sales or store traffic increases, the cable advertising is probably responsible.

Or the advertiser may wish to feature a *cable special* that is advertised only on cable. Sales of the special are then attributed to the cable advertising. In this way large advertisers, who cannot afford to isolate advertising in a single medium, can test the effectiveness of cable.

Cable advertising must eventually produce sales, store trafic, or inquiries for advertisers at a level commensurate with the costs. Although it may take an advertiser a while to assess the significance of cable in the marketing program, eventually that determination is made. It is in the best long-term interests of the cable operator and advertisers to make sound qualitative advertising placements, in the absence of appropriate audience measurements, and to systematically evaluate the effect.

Cabletelevision Advertising Bureau

The Cabletelevision Advertising Bureau (CAB) is an industry-supported trade association designed to promote use of cable advertising among advertisers and to help cable operators develop that business. It supplies members and advertisers with research and audience data and members with news, management reports, selling tips, co-op opportunities, success stories, and cable advertising system profiles.

MARKETING / PROGRAMMING / ADVERTISING RESEARCH

We have made reference to cable audience research in Chapter 10, "The Cable Subscriber," Chapter 15, "Marketing," and now in "Advertising Sales." Research is basic to maximizing the potential of cable but at the same time the medium poses some difficult challenges to traditional research methods. The remainder of this chapter considers the applications of research in cable marketing, programming, and advertising as well as some of the special problems.

In-House, Basic Research

Perhaps the most productive and practical market research is that which is conducted in-house as a matter of routine within marketing and customer service departments. This is not much more than specialized record keeping. Ideally, sales profiles are kept for every marketing territory. Demographic characteristics of the territory are determined and related to basic, pay-to-basic, and pay-to-homes-passed figures. These can be compared with system-wide standards and, in the case of an MSO, multiple system standards, and also to Cable Television Advertising and Marketing (CTAM) standards. For each phase of an original launch or remarketing effort, sales figures should be kept so that the productivity of each phase (such as direct mail, telephone, door-to-door) is known. These data should be broken out by meaningful demographic categories, such as assumed neighborhood income level, so that it is known which sales techniques are most productive in particular categories of households.

The same records can be kept for disconnects, upgrades, downgrades, and spins from one pay service to another. This information is very valuable, but difficult to obtain and update because of limitations in billing system software.

In the case of requests for disconnects and downgrades, subscribers are asked "Why?" by the CSR, as a part of the routine procedure for attempting to retain the subscriber.

Survey Research

Survey research can determine audience size and demographics for use in advertising sales. The research may also get product purchase and service usage profiles on cable subscribers for advertisers. The same survey methods may be employed for marketing and program planning, such as subscriber and nonsubscriber attitudes toward services and potential services, packaging schemes, program, channels, and pricing. Survey research may also generate important information for cable company use prior to rate increases. And a survey of community needs may be required prior to renewal.

Telephone and face-to-face personal interviews are most effective. Telephone is cheapest and effective if there is a high proportion of telephone households and a few unlisted numbers. Random-digit dialing is a solution to the latter problem. Nontelephone homes may not be of serious concern in marketing research if it is assumed that most subscribers and the best prospective subscribers have telephone service. Rigorous mail survey techniques, with persistent follow-up mailings can also be productive, but data from low-return, bill stuffer-type questionnaires often used by cable companies cannot be generalized.[5]

It is valuable to have a basic knowledge of survey research methods used in cable marketing, programming, and audience reserach, even if the research is done by outside consultants. Appendix H is a simple, step-by-step presentation and discussion of procedures.

Difficult Research Areas

It should be noted that certain types of survey research are not very successful in cable. It has been long established that program preference research does not produce very useful data. The results of questions about programs that viewers might like to watch, or would want more of, produce a social desirability response that may not predict behavior. People always seem to opt for educational and cultural programs that audience research indicates they do not watch even when available. Respondents know what's "good for them," but other motives often govern behavior.

It is also very difficult to do market research on the likely penetration of cable services *in advance* of any respondent experience with cable. If a community does not have cable, or if a cable service being introduced is quite novel, such as an alarm service or videotex, people are unable to judge their own interest. They may accept the idea readily, when not faced with a decision that requires them to put cash on the line, or they may reject the service simply because they don't understand it. Most people are unable to put themselves realistically in the hypothetical situations that this kind of research asks of them.

One way in which to address this problem is to offer the service or program free, on a test basis, to a limited number of subscribers and then to follow up with a questionnaire to get reactions that help in planning sales and pricing strategies. The test group may later be offered a continuation of the service for a monthly charge to estimate penetration for the whole system.

NOTES

[1]There are at least three books specifically about cable advertising and two on television advertising that are useful to the reader with a desire for more detailed information on this topic:

Kensigner Jones, Thomas F. Baldwin, and Martin P. Block, *Cable Advertising: New Ways to New Business*, Prentice-Hall, 1986.

David Samuel Barr, *Advertising on Cable: A Practical Guide for Advertisers*, Prentice-Hall, 1985.

Ronald B. Kaatz, *Cable Advertiser's Handbook*, Crain Books, 1985.

David F. Poltrack, *Television Marketing: Network/Local Cable*, McGraw-Hill, 1983.

Charles Warner, *Broadcast and Cable Selling*, Wadsworth, 1986.

[2]_____, "Cable TV Facts," 1986, Cabletelevision Advertising Bureau, p. 33.

[3]The Association of National Advertisers and the Cabletelevision Advertising Bureau (ANA/CAB) have develped a standardized claim documentation form that may be acceptable by some manufacturers. According to the manufacturer's policy, some claims may have to be checked by an independent organization such as the Advertising Checking Bureau.

[4]Sometimes broadcast audiences are reported in thousands (the three zeroes are dropped off). In this case the numbers result of dividing cost by audience need not be multiplied by 1,000. However, in cable, the numbers are usually not large enough for reporting in thousands.

[5]*Plan 1 (10 spots):*

$53 (discounted prime rate) × 4 spots	$212
$34 (discounted 12p–6p rate) × 3 spots	102
$25 (discounted ROS rate) × 3 spots	75
Total	$389

Plan 1 (30 spots):

$46 × 10	$460
$27 × 10	270
$20 × 10	200
Total	$930

Plan 1 (60 spots):

$37 × 20	$740
$18 × 20	360
$16 × 20	320
Total	$1,420

Plan 2 (10 spots):

$53 × 5	$265
$25 × 5	125
Total	$390

Plan 2 (30 spots):

$46 × 15	$600
$20 × 15	300
Total	$990

Plan 2 (60 spots):

$37 × 30	$1,110
$16 × 30	480
Total	$1,590

[6]Don A. Dillman, *Mail and Telephone Surveys: The Total Design Method*, John Wiley & Sons, 1978, p. 325.

CHAPTER **17**

International Comparison
of Cable Television Systems

Joseph D. Straubhaar*

Flyer for Vidéotron ltée, Montréal, Canada.

*Joseph D. Straubhaar is an Associate Professor of Telecommunication at Michigan State University.

This chapter will briefly examine cable television as it is developing outside the United States. Since cable has developed most extensively in Western Europe, Japan, and Canada, we will concentrate on those countries although recent developments in several Eastern European and Third World nations will also be reviewed.

In several of these countries, cable is taking a quite different form than in the United States, so this chapter starts with an overview of some of the approaches that other countries have taken in cable television. We review the generations of cable technologies that have developed in various countries and the impacts of other new technologies. We will also look at policy issues such as government versus private ownership of cable, regulation versus deregulation, monopoly and integration with other industries, financing, and government content controls. Some international issues will also be considered: use of foreign satellite channels, restrictions on imported or cross-border programming, cross-border copyright and advertising regulation. We conclude with a look at the development of cable TV in specific countries: the United Kingdom, Canada, Japan, France, West Germany, Netherlands, Belgium, other West European countries, Eastern Europe, and the Third World.

INSTITUTIONAL CONTEXT

In most countries outside the United States and Latin America, broadcasting and now cable television have developed in a context of extensive government involvement. Some countries have made broadcasting and cable a government monopoly, although as we will see, technological and political changes in several countries are leading them toward privatization and deregulation. In most others, government broadcasters are engaged in a restricted oligopolistic competition with a few closely regulated private broadcasters.[1] Broadcast monopolies tend to see cable television as a *threat*. Resources to support broadcast program production may be reduced or fractionalized by cable competition. Broadcast monopolies have also tended to focus on public service—education, high culture, and news—and to avoid advertising. Cable frequently is seen as introducing advertising and an emphasis on entertainment.

Furthermore, telephone systems in other countries have until recently almost universally been started or taken over by government telecommunications monopolies. In many cases, these telecommunications monopolies, known as PTTs or *post, telephone, and telegraph* administrations, have owned and operated the actual physical facilities for broadcasting as well as the various wired telecommunication networks; so they have tended to dominate cable system construction and operation as well.[2]

Because wired networks have most often been PTT monopolies, countries with PTT structures have been hesitant to let cable TV emerge as a separate market-driven industry. Instead, many PTTs, being responsible for telephony, telex, and the like, have thought from the beginning about how to integrate cable television with other wired services. In particular, many of the PTTs have looked to cable's potential for interactive information services rather than concentrating on cable's potential as an expanded but one-way television entertainment medium. In the hopes of several governments, notably Canada, France, Japan, West Germany, and the United Kingdom, cable might serve as the basic infrastructure for the wired city of the new information age.[3]

While the U.S. is currently moving toward less regulation of cable, broadcast, and telecommunications, in many other countries these services are seen as public services that should be carefully controlled. Even when broadcasting and/or cable systems are not government-owned, government authorities or quasigovernment boards have frequently been seen as the proper arbiter of the public interest. In virtually all countries outside the U.S., this has led to the regulation of both broadcasting and cable at the federal level. Some smaller Third World countries have not extended broadcast or other regulatory bodies to cover cable systems begun by private companies, particularly in the Caribbean and Latin America. Some industrialized countries, particularly Japan and the United Kingdom, are considering deregulation of cable or decentralizing control to the local level. However, most seem likely to continue with national control.

History and Technology

In most countries that have had widespread cable development before the 1980s, the history of cable technological development is at least somewhat similar to that in the U.S. Cable systems tended to start with community antenna systems (CATV) or master antenna systems (MATV) for the local retransmissions of broadcast signals. Next was the importation of "distant" broadcast channels. Then efforts were made to provide additional programs. This includes experiments with cable origination, community access channels, pay-TV, and satellite-to-cable channels, to expand cable offerings and segment the audience into more specialized groups. Finally, some experiments have been conducted with interactive systems and some countries now plan to combine cable systems with other telecommunication technologies, usually based in optical fiber and digital switching technology.[4]

Cable for retransmissions, or CATV, is where most countries' cable systems began and where most remain, in fact. In some countries, such as the USSR, People's Republic of China, and several African countries, even radio reception began as a cabled system with a central receiver and a distribution system to loudspeakers. Such systems, known as *wired radio* or *rediffusion*, were widespread not only because of low cost but also because they permitted central control over what people were listening to. Most of these systems broadcast only one channel, usually a government one. In Asia and the Far East (principally the People's Republic of China), 33 percent of all radio received is wired radio; in Eastern Europe and the USSR, the figure is 39 percent.[5]

Cable television systems for retransmission only (CATV) are widespread in some countries, notably in France (36 percent), West Germany (50 percent), and Sweden (45 percent). Smaller penetration of CATV has taken place in Austria (99 percent), Japan (10 percent), and the United Kingdom (14 percent).[6] These systems enable apartment dwellers, isolated villagers, and others to avoid the need for forests of individual tall antennas. A study by the Council of Europe terms these first generation or "passive" cable systems.[7]

In some countries, such as the United Kingdom, cable systems have been forbidden to become more than passive repeaters. Regulators have been reluctant to permit cable operations to originate programming for fear of disturbing a government monopoly or a regulated balance of broadcast channels. Passive cable systems are vulnerable to competition. In Japan and Great Britain, for example, some cable systems have been closed due to competition from expanded VHF or new UHF or DBS stations.[8]

For Canada and the smaller countries on the European continent, there was another transitional phase after these passive systems. Introducing a contentious issue, many of these community systems began to import and distribute signals from several other countries. In Belgium (83 percent cabled), Canada (60 percent), Finland (60 percent) Netherlands (73 percent), Norway (40 percent), and Switzerland (63 percent), most systems repeat television channels from at least one or two neighboring countries.[9] Some of these systems are also negotiating to retransmit satellite channels from other countries. Austria, Belgium, Canada, Finland, France, the Netherlands, Norway, Sweden, Switzerland, and West Germany use satellite signals to get foreign programming delivered to cable systems. (See use by various countries of the U.K.'s Sky Channel, in Table 17–1.)

This tendency to use satellite or imported channels, begins to merge into the second generation of technology. Later than typically occurred in the U.S., there has been a growth of "active" cable systems that originate material beyond that retransmitted from television stations. There have been limited experiments with local access channels, particularly in Belgium and the Netherlands. Cable-originated news programs have existed since 1970 in Canada and can be found on a small fraction of Japanese systems. Pay TV channels for cable are being developed now in some European countries and Canada, but they are a fairly new development. Pay TV is linked to slow changes not only in technology but also in government attitudes toward television service competition and in the use of advertising or premium subscription financing. Other satellite services, paid by advertising or partial government subsidy, are growing faster.

An alternative branch in this pattern of cable technology development for an increasing number of countries is *satellite master antenna television*

(SMATV). Especially when the satellite signals are pirated free, SMATV is an attractive, cheap means of adding diverse cable service to areas without conventional cable or even diverse broadcast fare.[10]

The third generation of cable is the development of interactive systems. Such systems are still largely experimental although some countries, notably the U.K., France, West Germany, and Japan, have moved to encourage interactive development. In more advanced systems, such as Hi-Ovis in Japan, individual households can select special videotaped programs, exchange electronic messages, and input audio or video as well as keypad information. Some Canadian, French, Japanese, and German experiments are using cable to provide videotex-type services: electronic phone books, shopping services, classified advertisements, banking services, travel reservations, electronic mail, and other services using a computerized home terminal.[11]

Such services essentially would integrate cable with other new information technologies. Cable would provide not only video services but also networks for use by computer, telex, and telephone. In Japan, for example, this integration is called the *Integrated Services Network*, a variation on the idea of the integrated services digital network (ISDN). Several countries envision cable providing the basic transmission and switching system for all the services foreseen in the information society of the future.

To summarize the development stages reached by various systems, Table 17–1 compares 1982 and 1985 CATV and MATV penetration in Western Europe with a projection to 1989 for both basic and interactive services, based on a 1982 study and a 1985 study that projects for 1989.

Generations of Cable Technology—Where to Start?

With at least three generations of cable television technology to choose from, a number of countries are having serious debates about which, if any, to adopt. Cable systems, particularly passive or retransmission systems, have grown in response to market demand in a number of countries. Currently, however, a large number of countries are viewing or reviewing cable policy as a major national decision.

Many countries are torn between the lower costs of one-way or limited interactive technologies, the economic benefits of creating a system quickly, and the value of the services that fuller interactivity would provide—a decision process the U.S. has already concluded in favor of the limited one-way systems. Despite the initial rush of enthusiasm that greeted the opening of several countries to cable system construction, repeated experience shows that demand for cable services exists only if costs to the consumer are kept at a very low level. The major demand is for diverse video entertainment, news, and cultural material,[12] except in France where telephone-delivered videotex telephone directories, message services, and other interactive services have be-

TABLE 17-1. Cable, MATV, and Interactive Cable Penetration in Western Europe.

COUNTRY	1982 Cable[a]	1982 MATV[a]	1985 Cable[b]	1985 MATV[b]	1989 Cable[c]	1989 Inter-Active	1986 % Homes Reached by Skychannel[d]
W. Europe	11.1%	29.1%	n.a.	n.a.	11.5%	6.0%	3.8%
Austria	4.0	—	11.0	18.0	9.0	#	7.3
Belgium	85.0	4.0	83.0	n.a.	84.0	21.0	14.5
Denmark	5.0	35.0	—50—		8.0	*	0.0
Finland	7.0	3.5	10.0	50.0	10.0	*	8.8
France	0.4	55.0	2.0	n.a.	4.5	*	0.0
Netherlands	44.0	16.0	51.0	22.0	84.0	21.0	39.5
Norway	11.0	21.0	12.0	19.0	16.0	*	8.9
Spain	n.a.	n.a.	0.0	n.a.	50.0	n.a.	0.0
Sweden	—	50.0	2.0	41.0	6.5	5.0	2.3
Switzerland	35.0	20.0	34.0	29.0	52.0	32.0	28.6
U.K.	14.0	0.5	14.0	1.0	6.5	*	0.6
West Germany	0.7	44.0	n.a.	n.a.	12.0	9.0	3.1

[a]1982 study cited in *Transnational Data Report*, VI:1, Jan. 1983, p. 20.

[b]1985 data from *Euro-Factbook-Basic Hardware and Audience Data*, Dr. Peter Diem, 1986 Edition, ORF Audience Research, London.

[c]1989 estimates are from *Cable TV Communications in Western Europe*, CIT Research, U.K. May 1985, cited in *International Broadcasting*, July 1985, pp. 29–30.

[d]*Cablevision*, April 21, 1986, p. 28.

Note: In 1989 projections, * is a "small fraction" and # is a "large fraction."

come popular, based in part on free terminals supplied by the PTT and relatively low user charges.

While governments would like to create state-of-the-art interactive networks with cable, current demand for more sophisticated and expensive services (beyond video and audio programs) seems unlikely to pay for the cost of such networks, at least in the markets that have been studied, even in technology-oriented Japan.[13] The exception to date has been France, as noted. French plans to integrate videotex into broadband cable have proceeded more slowly. Some countries, such as Italy and Venezuela, seem to be waiting for interactive technologies to become more affordable before encouraging new cable systems. Concern about achieving the economic benefits of creating new industries as soon as possible has led the British to reconsider requiring interactive technology in new cable systems. Other countries, such as France, want to choose a long-term solution now before first- or second-generation cable networks are expanded, requiring parallel construction of networks for integrated telecommunications services.[14]

Cable Competition with Other Technologies

As in the U.S., cable TV faces competition from other video and information technologies. Principal competitors in providing a greater diversity of video entertainment are VCRs and direct broadcast satellites (DBS). VCRs are already widespread in many countries, not only Western Europe but also the Third World. In more affluent countries, VCRs may be seen as complementary to cable, enabling timeshift viewing. In less affluent countries, VCRs plus tape rental shops or exchange clubs may substitute for more costly video distribution mechanisms.[15] In a number of countries in Western Europe, Eastern Europe, and the Third World, policy makers are worried that VCRs will enable audiences to substitute immoral, subversive, or simply distracting material for carefully planned, and often government-controlled, broadcasts. Such policy makers may well prefer cable television to VCRs as a more *controlled* means of providing a greater diversity of television material. This seems to have been the case in French attempts to restrict VCR imports, for example.[16]

In several countries, cable also is preferred by policy makers to DBS because it allows for local or national control. Cross-border DBS will be impossible to control, particularly in Western Europe, where satellite footprints will almost inevitably lap over into other countries, despite international agreements aimed at minimizing such spillover.[17]

More than in the U.S., where satellite-to-cable systems already have extensive penetration, DBS technological developments in Europe may aid SMATV or spur true DBS direct household dish reception at the expense of cable development. In Canada and the Caribbean, DBS via backyard dishes is already widespread, using technology similar to current U.S. TVRO dishes.

Satellite and Other Program Services

Several countries are beginning to distinguish, as the U.S. systems do, between cable system operators and service/program providers. The Netherlands has long had an innovative broadcast system in which the government owns facilities and sets regulations but private groups compete for access or air time. Cable, with multiple channels, facilitates variations on this model and governments that have resisted private broadcasting are considering allowing private cable channels on government or public systems. For example, the French aim for private competition in service provision, although, as of 1986, private ownership of cable physical plants and networks was still limited. In information services, the same trend can be seen in the addition of private service providers to the British Telecom-owned Prestel videotex system, even before the privatization of British Telecom ownership was contemplated.

As in the United States, the major development in several other countries in private cable program services is in satellite services. In Canada, some U.S. services have been allowed in while some areas, such as music, sports, children's programs, and films, have been reserved for Canadian-owned services. In Western Europe, several British ventures, notably Sky Channel (general entertainment), and Music Box (an MTV lookalike) are now sold to other countries. Cable News Network, MTV, and Financial News Network have entered the European market. A multicountry French channel, TV5, is available and a channel is planned by Radio-TV Luxembourg. Several of these are planned as both cable and DBS services, but to date are being marketed more as satellite-to-cable services.[18]

In some countries, notably Canada (the Anik satellites) and India (Insat), DBS, satellite-to-cable and satellite broadcast repeated systems have been employed to reach remote areas with "television" service. The People's Republic of China and others are initiating such programs. These programs tend to be public service government efforts to provide information and entertainment to isolated populations.

Cable Content and Audiences

For a number of the reasons mentioned, a great deal of current cable television investment is being initiated by planned government policy, not pure audience/market demand. Public demand or pull for cable tends to be oriented around known types of content and much of the policy push is oriented toward technology itself or developing new services or content.

Public demand, particularly in Western Europe, has basically sought an expanded diversity of video entertainment, news, and culture. Pent-up demand for entertainment has been particularly prominent, since most European broadcasting has emphasized news, public affairs, and culture much more than U.S. broadcasting has. The dominant public service ethic, although varying from country to country, has emphasized a much less prominent role for pure entertainment. Some observers have noticed a gradual shift toward entertainment in Western European, Socialist, and Third World broadcasting, particularly as audiences have acquired other entertainment options via video and audio cassette recorders, cross-border broadcasting, DBS, pirate broadcast stations, and others with which traditional broadcasters must compete.[19]

In the countries with more activist cable policies, such as the U.K., France, and West Germany, government planners have tended to see increased diversity in television fare as a marketing mechanism to help sell a larger package of cable information and data services. While program diversity is frequently a major goal, the economic value of infrastructure and data services projects often receives more stress. Many governments, as a matter of industrial policy, now stress development of the information sector, including cable, as a means of becoming the anticipated "information societies" of the future. Information and data services also fit better with many countries' long-standing government goals of using media primarily for information, not entertainment.[20]

A similar attitude is often taken towards advertising as a support for the new media, particularly cable. Public broadcasters fear that commercial, advertising-based services will erode their audience and financial bases. Existing commercial media, such as newspapers, magazines, and some broadcasters, also fear that advertising on new cable channels will diminish their share of the overall advertising budgets. Nevertheless, in several countries, such as West Germany and Norway, private ownership or program service involvement in cable is promoted primarily by publishing interests, who regret not having been able to participate in a commercial broadcasting system. In fact, some critics feel that the new technologies are being used as stalking horses for private interests that have been shut out of broadcasting but see a new opportunity in cable television.[21]

The debate over commercialism is fueled by the fact that some governments, such as West Germany, which have avoided commercial broadcasting are considering increased advertising or privately owned cable channels as a means for paying for cable system infrastructure. The wisdom of this compromise is hotly debated in West Germany and Great Britain, where commercialism in broadcasting has been very tightly restrained in favor of the ethic of public service, which emphasizes news, education, and culture over entertainment.[22]

Another concern by public broadcasters and some officials is over the effects of cable narrowcasting. They fear that the segmentation of the mass audience into more specialized groups served by a variety of cable channels will fragment public discourse.[23]

Cable has already shaken up public broadcasting in several countries. In Italy, for example, independent, often illegal cable systems were a challenge to the broadcast monopoly—bringing in distant signals from other countries as a supplement to the government channels. In some cases, like the Netherlands and Taiwan, pirate cable operations not only imported signals but expanded into uncovered hours and offered pornography or foreign newscasts with coverage of proscribed issues (such as Japanese newscasts discussing recognition of Mainland China).[24]

Problem of Imported Cable Programming

Several of the smaller European and Third World countries have used cable from its inception primarily to bring in and retransmit signals from other countries. Some of these countries, such as Belgium, the Netherlands, Switzerland, and Finland have largely accepted the idea of considerable cross-border program flow via cable.[25] For many others, however, the prospect of filling some cable channels primarily with imported material or retransmitted foreign television signals challenges long-standing limits placed on the amount of imported material that could be broadcast.[26]

Great Britain, for example, has limited imported television programs to 14 percent of broadcast time, but plans to relax that rule for cable channels, occasioning a furious debate over the cultural impact that might be made by more programs from the U.S. or elsewhere. France has traditionally had strong similar concerns about protecting its culture. Canada has placed a series of limits on imported U.S. programs.[27] A number of Third World countries are also concerned about "cultural imperialism" from television imports. Concern from Third World, Socialist, and even some Western European countries about the cross-cultural impact of television imports has grown to the degree that the issue is a major aspect of the New World Information Order debate in UNESCO and became a major bone of contention between the U.S. and UNESCO. Of most concern are basic values portrayed in television that may conflict with national cultures: use of violence, sex roles, sexual behavior, consumer habits and ambitions, health practices, racial images, political ideals, and so on. Both programs and advertisements produced in other countries are at issue.[28]

Beside content and cultural issues, there are major economic issues related to cross-border program flow, as well. Questions arise about copyright issues—the allocation of royalties for use of programs originally intended for another country—and how to avoid double exposure by both cable and broadcasters of the same program in one area. When advertisements are included in cross-border broadcasts, there is a question of payment for advertising coverage of a second country beyond the one for which the advertisement was originally intended. These questions already come up in Western Europe due to broadcast signal spillover but will become more intense with cable retransmission of satellite channels and foreign broadcast signals. The Netherlands, Sweden, and Belgium have wrestled with these issues already but have tended to ignore advertisements not originally intended for their citizens.[29]

The Council of Europe has issued a policy "Green Paper" on European "Television Without Frontiers," which stresses the need for liberalized and shared rules on free movement of programs and advertising to cope with cross-border broadcasting and cable reception.[30] Most countries have reiterated instead a determination to apply their existing cross-border broadcast rules to cable and satellite television systems. There are indications that in some cases such rules may be different for PTT or privately owned systems; common carrier rules, rather than broadcast rules, might be applied to PTT operations.[31] Rapid growth of VCRs in Europe has already weakened government power to control cross-border programming. Furthermore, DBS also threatens such control. Faced with these competing technologies, some governments are inclined to loosen imported programming rules somewhat for cable in hopes of maintaining the audience (and their control) on a cabled system.

Both new and existing cable systems foresee an increasing international flow via satellite-to-cable channels. Canada has already had to confront extensive use of U.S. satellite channels, with authorities approving some and trying to restrict others. Sky Channel and Music Box have pressed to obtain permission to operate in a number of countries in Europe. As of 1986 Sky Channel has been approved in countries, including Belgium, the Netherlands, Norway, Finland, Switzerland, and West Germany[32] (Table 17-1). Music Box is less widespread but growing.

Following now are examinations of several countries as case studies of how cable TV is developing.

AREA CASE STUDIES

United Kingdom

Cable initially developed in the United Kingdom after 1946 in small CATV systems. In fact, cable systems were legally forbidden to originate programming until 1983. Retransmission cable systems have been declining slightly in penetration, due to competition from improved broadcast signals, while new increased channel systems have been growing.[33] Under the Thatcher government, however, cable has been targeted as the technology that will wire the country to provide a base for information society development, with expanded entertainment services helping to pay for the infrastructure for more interactive systems. Along with encouraging pay television on cable and private development of cable, the Thatcher government has privatized the PTT (British Telecom) in an overall effort to increase telecommunication growth by incorporating private capital and initiative.

The major policy document of U.K. cable is the April 1983 White Paper issued by the Home Department and the Department of Industry. With a slight change to extend franchise periods, the White Paper recommendations were enacted into law as the Cable and Broadcasting Act of 1984. Following a tradition of concern about private competition with the public British Broadcasting Corporation, the White Paper proposed a Cable Authority modelled on the Independent Broadcasting Authority to regulate private cable initiatives. Although allowing flexibility, the White Paper urged use of switched-star networks and optical fiber to permit gradual development of interactive services. At minimum, it recommended that all ducts be laid in switched-star configuration so switches could be added later. Such networks would allow for interactive services and also encourage British hardware production, since the U.K. is a leader in switched star technology. Older systems were also encouraged to expand by offering four new channels.[34]

In 1983–84, eleven new cable franchises were given out where two-way systems, information services beyond broadcast-type content, local origination, and maximum use of British/European programming were encouraged. These were slow to develop, however, and in 1985 considerable doubt existed as to the viability of some of the smaller and/or more sophisticated proposed systems. At least two of the major initial cable companies, Visionhire and Rediffusion, sold out their interests in several systems. The British government in October 1985 injected $7,050,000 into cable projects to support interactive systems. [35]

One likely problem was that the privatization of the PTT, British Telecom, may have drawn investment funds away from cable. Serious questions also arose about the profitability of either basic or pay services, given fairly high-quality signal reception in most areas and competition from an already diverse set of broadcasting services, including several free from commercials, and a high proliferation of VCRs (around 34 percent penetration of TV households). There is also grave doubt among many in the U.K. that interactive services will become appealing to customers. [36] The experience with interactive videotex, Prestel, has shown very limited household demand, although somewhat greater specialized business demand. The experience of British cable systems by 1985 with demand for interactive services was not encouraging in new systems such as Swindon, where only 12 percent of homes passed had subscribed. Cable companies were also protesting financial burdens in capital allowances, underground cable construction and complex licensing procedures. [37]

The most popular cable TV program services (primarily delivered by satellite) are the Music Box, Sky Channel, Screen Sport, and Children's Channel. A pay movie channel, TEN, collapsed in 1985 and a more modest Home Video Channel was started. Plans to start new channels were also announced by an arts group, British Cable Programmes, for the Arts Channel, a daytime Lifestyle Channel and by the television news service VISNEWS for a British/pan-European news channel to compete with similar plans for European distribution by the U.S.-owned CNN. BBC and the Independent Television companies plan to compete with Superchannel, a projected "best of British" programs from the four BBC and ITV channels. [38]

Ten more franchise areas were offered in 1985. Furthermore, the government in 1985 legalized home earth stations and SMATV, which could compete with cable. However, although SMATV only systems will be authorized in areas without cable franchises, such licenses might be revoked in favor of a cable franchise application. The main goal is to allow franchised cable companies to install SMATV systems to supplement more conventional cable technology, which may revive business by enabling cable systems to add cheaper systems for apartments and small groups of houses to their current coverage areas. [39]

The 1983 White Paper represented a compromise between those who wished to deregulate traditional broadcast program controls for cable and those concerned about weakening the public service tradition of British broadcasting. Cable operators must carry the four existing BBC and ITV television channels, the two public radio channels, and any DBS service financed by public broadcasting license fees. It must also maintain space for the five DBS channels allocated to the U.K. by the International Telecommunication Union. Current broadcast standards for advertising, sexual content, and other areas will be enforced by the Cable Authority. Cable companies will not be required to maintain the strict 14-percent limit on imported programs applied to broadcasters, but will be required to specify the amounts of British, European Community, and other material they plan to use and to steadily increase the amount of British material.[40]

Canada

Cable penetration in Canada has been extensive, preceding U.S. cable development in some areas, with an initial focus on retransmission. However, this not only provides Canadian television services in remote areas but also gives increased access to U.S. material by cable systems carrying American stations. Cable developed early in Canada but has grown slower recently (penetration was 52 percent in 1978, 60 percent in 1985).

Canadian cable development has been like the U.S. system in that government regulates private operators rather than operating systems itself through the PTT. As noted earlier, several Canadian firms, such as Rogers and Maclean Hunter have significant ownership holdings in U.S. cable operations. The Canadian Radio-Television and Telecommunications Commission (CRTC) regulates cable (like all other telecommunication media). Cable regulation is therefore federal, not local. The CRTC applied general broadcast standards, notably Canadian content requirements, to cable, although some of the specific rules vary. Local origination channels are required for systems over 10,000 homes. By 1986, some limited deregulation of rate controls, advertising on local services, and greater initiative in text services was proposed by the CRTC in response to industry demand but most cable operations remained under federal regulation.[41]

In order to stimulate growth in the mature Canadian industry, pay cable was authorized in 1982 by the CRTC. There were to be six channels: four regional, two national (entertainment-oriented First Choice and culture-oriented C-channel). By 1983, C-channel and two regional channels were cancelled. Pay TV penetration has been low (8–12 percent of cable households in 1984).[42] After an appeal to the CRTC in 1984–85 to limit competition, two national channels, First Choice and Allarom, were granted monopolies for Eastern and Western Canada, respectively, to provide premium movie and general entertainment service. Only First Choice was still in business in 1986. Canadian specialty channels for music (Much Music), lifestyle/health (Life Channel), and sports (The Sports Network) had been successfully promoted and authorized as well. All include foreign material, such as Sports Network purchases from the U.S. service ESPN. In 1986, 18 percent of Much Music videos were Canadian, the rest primarily U.S. or British. Subscriber response to these new channels has disappointed cable operators, however. The Life Channel was cancelled in 1986 and the success of others is not yet assured.[43]

Competition for cable systems has grown from satellite TVRO, VCRs, and SMATV. The need to let cable be more competitive forced the CRTC to further liberalize pay TV rules in 1984. The CRTC granted requests to broaden cable appeal by allowing in U.S. cable services, although it banned U.S. movie, music, sports, and childrens' channels to promote national development of such services. The Disney Channel sought entry but the CRTC initiated competition for a national youth channel, which could include Disney material. The CRTC set a two-tier cable structure, allowing first the Canadian services, then up to five of seventeen approved U.S. channels. In 1986, the Canadian Cable Television Association petitioned the CRTC to allow pay-per-view to stimulate more viewer interest. Concerns were raised both about Canadian content rules being implicitly violated, as well as problems for Canadian advertisers arising with U.S.-oriented advertisements. Of all these new channels on the first tier, the most popular were the national music and sports channels, plus the U.S. CNN, CNN Headline News, Nashville Network, and the Arts and Entertainment Network on the second tier. In 1986, however, both CBC and City TV of Toronto have petitioned the CRTC to substitute new all-news channels for CNN, following a government task force report that concluded that CNN was not serving Canadian interests.[44]

Another program service which may become significant for some cable systems is Cancom, a DBS system for television and radio to remote Northern areas started in 1983 to compete with satellite reception of U.S. channels. Cancom relays existing Canadian channels, along with ABC, CBS, NBC, and PBS. By being available by satellite distribution, Cancom's French channel has been picked up by some cable services. However, by 1984, very few cable systems or low-power television operations had picked up Cancom's service and it remained TVRO-oriented.[45]

As in other countries, Canadian government planners would like to see cable systems integrated with other interactive information and data services. Federal and local governments have promoted videotex and other new information systems, particularly the national Telidon system and the Toronto public videotex system, but, unlike some other countries, they have had to work with private companies, not a PTT. Some of the existing carriers, both telephone and cable, such as Rogers and LeGroup Videotron, have shown interest in enhanced services like videotex to add revenue. Rogers and Torstart will test Telidon (videotex) on a cable channel in Toronto, while Videotron will test Videoway (videotex, transactional services, pay-per-view) in Montreal.[46] In general, as 1986 developments indicate, the trend in Canada has been to deregulation and diverse services in the cable industry.

France

France currently has a number of community antenna systems, covering about 35 percent of television households, but plans to move into much more ambitious interactive cable systems. The French government seems to have two major reasons for its very activist policy of promoting cable. First, France, like other countries, hopes to use cable as a major component of infrastructure for a wired information society. Second, the French government has traditionally exercised a great deal of control over French broadcasting and sees cable as a medium that will meet public demand for more diversity yet be more controllable than DBS, VCRs, or other alternative video distribution vehicles.[47]

The French government in 1980 launched the *Telematique* or Telematics program, intending to combine and promote fiber-optic-based cable/telephone lines, computer, and videotex technologies as mass consumer technologies. In particular, the main network is expected to be switched star cable systems using optical fiber and addressable hubs. Built-in capacity is to be 30–60 broadband channels, including video and stereo sound channels; independent connection of a second set; telephony, usually for a second line; videotex access through packet-switched networks to various nationwide host computer centers; home security monitoring; pay TV, including multiple tiers and local two-way services.[48]

In related technologies, the French government has been a very active backer of teletext and videotex, using broadcast and interactive (via telephone lines) technologies. The Antiope teletext system uses broadcast vertical blanking intervals but may be changed for cable delivery. The Teletel or Minitel videotex system of access to various data banks, beginning in several major cities with electronic telephone directories, had 600,000 users in 1984 and 3,000,000 terminals are to be delivered (free to telephone users) by 1986. To date, the system largely operates via telephone lines but is carried on optical fiber cable channels as well.[49]

The French government is committed to operating and, if necessary, subsidizing videotex. In contrast, the Canadian government has provided development funds but now expects videotex to succeed in the market.

Like several other countries, France has begun its step into third generation cable with several pilot projects. The most ambitious is a system in Biarritz. This has distributed services: up to 15 channels of television (including movie pay-TV, text and video pay-per-view, several satellite services-Sky Channel, Music Box, and TV5; broadcast channels from the U.K., Belgium, Switzerland, and Spain); 12 audio channels; and point-to-point interactive services; an electronic phone directory, teleshopping, national Minitel videotex services, advanced telephony services (such as audio-conferencing), two-way videophone services, and exchange of VCR content via optical fiber telephone lines. Services were phased in from 1983 to 1985. The project is being carried out by the PTT to test technologies, services, and subscriber interests. More than many other experiments, Biarritz has built on videotex concepts, expanding two-way videotex data base services with two-way video and sound, using a videotex modem, switching, and videodisc players. A private optical fiber, interactive system will be built in Montpelier in 1987.[50]

To facilitate cable television growth, among other reasons, the French government formally abolished its longstanding broadcasting monopoly in 1982 to allow independent local television stations and abolished its cable monopoly in 1986. In 1984, the government decided to authorize national private stations, such as the Fifth Channel. In 1986, the new conservative government went further by privatizing one of the three national television networks, TF1. Foreign ownership was permitted in the Fifth Channel but remained controversial. The goverment also authorized pay TV services, such as Canal Plus, for cable. The government wants to stimulate new services by private enterprise and let markets test them. It also very much wants to avoid construction of parallel CATV and ISDN telecommunication networks by building a single national broadband network, so even private systems, such as Montpelier's, follow government guidelines for building interactive systems.[51]

Considerable latitude seems to have been given for cable service providers to compete. There will be private and pay networks: PTT, private, and possibly local government switched services. A number of private videotex services using the Minitel and cable systems are emerging. TV5 is a satellite-delivered international French language channel produced in cooperation with the Three French networks, Belgium, Switzerland, and possibly Quebec. A pay television and cable service, Canal Plus, shows movies and is also developing a children's channel. It is owned by several private groups and the news agency HAVAS. It reached 1,200,000 subscribers in 1986 and was then already profitable, unlike Sky Channel or some other private services. Two music channels, STV and Music Channel, have been approved but not implemented. Both would use French and foreign videos.[52]

Besides the Biarritz experiment, advanced optical fiber cable systems are planned for Paris in 1986, Montpelier in 1987, and later for Cannes, Lille, and Lyon. In Paris, the first phase has 14 television channels, including a local channel, broadcast channels from Italy, U.K., and Luxemburg, as well as TV5, CNN, Sky Channel, and a new French children's channel. Audience penetration (of TV households) is projected in reach 1,300,000 homes, or as much as 50 percent, in Paris by 1992. Channels will increase to 30.[53]

Principal problems in France include an anticipated shift of advertising expenditures toward cable from print media and a conflict between motives for cable profitability and cultural protection/promotion, since much of the increased programming will be imported. Nevertheless, there seems to be considerable demand for both entertainment and interactive services, based on videotex. As *The Economist* noted, "France may be the one country where suppressed appetite for video entertainment may actually help foot the bill for cabling the country."[54] Even so, research in 1986-87 indicated that many French cable systems were scaling back their intended degree of interactivity and concentrating on video delivery systems.[54]

West Germany

West Germany has low current penetration of large CATV systems (1 percent) but considerably greater spread of small building or village MATV (44 percent). Cable television was limited by the fact that 98 percent of West German households were reached by three broadcast networks and that until 1982 only retransmission of these networks was permitted on cable, with private channels or local origination forbidden. Expansion of cable systems is currently being promoted by federal, state, and local governments. Cable systems are proposed for reasons similar to those in other European countries: to meet the audience demand for more diversity in television content, to meet that demand with media that still have community and government control—unlike VCRs and DBS, to build an efficient fiber optics network to rationalize communications channels, and to enable the development of new interactive services and new technology to create new industry and employment.[55]

The issue of regulation has been a factor in some delay and confusion in developing cable systems in West Germany. This is due primarily to the extreme decentralization or "federalism" of broadcasting and broadcast regulation. Each state has public broadcasting entities, which form limited national networks, but which are controlled at state level. States vary considerably in approach on some major cable issues: authorization of private channels, private capital participation, pay TV, use of foreign programs, and the need to allow or promote interactive services.[56]

West Germany, like several other nations, has been conducting experiments with various cable, videotex, and integrated optical fiber network technologies and services. Experiments in these areas have been promoted by the recommendations in 1975 of a federal commission on telecommunication development. Although technology and concepts can be promoted by the federal PTT (Bundespost), projects must be implemented by the states, which decided only in 1978 on pilot projects to be supervised by a common media commission of the states, formed in 1983. Pilot cable projects were approved for West Berlin, Ludwigshaven, Dortmund, and Munich. These were originally approved in the 1970s but were delayed by federal-state disputes. These delays have limited their utility as test projects. Nevertheless, some momentum for national cabling is now widely discussed, although major problems in coordination still exist between public and private system owners and installers, state and federal authorities.[57]

A 1980–83 trial tested interactive videotex, known as Bildschirmtext, based on a PTT central computer. Telephone-based videotex will be integrated with cable in a Berlin project to experiment with lowering interactive cable costs. The Berlin project will also test a Broadband Integrated Optical Fiber Local Telecommunications Network (BIGFON). BIGFON, which will also be tested in small trials in other cities, will feature video phones, video conferences, and high-speed text and data services.[58]

The Berlin experiment (now expanding beyond pilot project dimensions) will include a large number of new cable services, some targeted for specific demographic groups (young people, foreigners, senior citizens) and interest groups (film, sports, music), including private and even foreign interests/services, such as the U.K. Music Box and Sky Channel or U.S. services. As in other countries, an effort will be made to stimulate new program production for cable television. Formation of new German program channels, such as Sat 1, is currently being limited by lack of authorization by several states for operation in their areas as either cable or DBS channels. Although pay TV has been successfully launched in Switzerland in 1984, the market was described as still too limited in West Germany in 1985.[59]

Since broadcasting has been conducted by state-level public organizations, the main issue has been whether to allow participation by private firms in cable development. Public broadcasters have resisted private channels or pay TV. However, the attractiveness of including private resources has grown with the threat of outside competition from foreign direct satellite broadcasting and VCRs. Without additional services beyond retransmission, the public has little incentive for cable subscription. Some in West Germany argue that allowing private services may provide such incentive, while others are afraid that private competition will increase entertainment viewing at the expense of education, leading to the "destructuring" and impoverishment of the present public system. In 1986, at least one German pay service was approved, after foreign investors withdrew from the project. The tendency seems to be toward diverse policies by different German states, leading toward a varied pattern of cable development.[60]

The Netherlands and Belgium

The Netherlands has 51-percent cable penetration plus an additional 22 percent with MATV. It has had a system of public broadcasting by various groups (broadcast time proportional to group size represented), which a 1983 White Paper proposes to supplement with private cable narrowcasting.[61] Cable systems are allocated by local governments and usually owned by local public utilities but will use some private channels, local consortiums of electronic and publishing interests, as well as foreign channels such as the French TV5, Sky Channel, or Music Box. Services have been proposed and approved for specialties in films, sports, music, children's programming—several intended for further sale in the rest of the European market—but it is unlikely that all will make it to actual start-up.[62] The private companies, at least the local ones, will probably have rules on local versus imported content to force continued local production. Direct advertising to the Dutch market is still restricted.[63]

The Netherlands has also shown some interest in promoting cable as a base for interactive services. A pilot project has begun in Limburg, one of the few areas not already extensively cabled. The project uses coaxial cable in a hybrid star-tree technology to try to develop services that other current coaxial systems might adopt. Services were planned for off-air retransmission, pay TV, telebanking, teleshopping, and special film, sports, and educational services for a subscription fee. However, by the time the project was begun in 1986, video conferencing, data communication, and video telephony were already offered by private, noncable operations, requiring a scaling down of the planned interactivity.[64]

Belgium has an even higher penetration of cable service, 83 percent, projected for 90–95 percent by 1990. Like the Netherlands, the current system is built around retransmission of local as well as neighboring country television signals. This causes a considerable copyright problem, which is solved by the collection of fees contributed to a central fund for redistribution to foreign broadcasters and other copyright holders.[65]

In 1985 there was reluctant consideration of pay TV channels. Advertising-based foreign channels, like Sky Channel, were either banned or highly controversial. No local advertising exists as yet but it may be allowed to permit local groups to compete with advertising on foreign channels such as Sky Channel and RTL (Luxembourgh). Advertising may be implemented separately in the Flemish and French speaking regions.[66]

Japan

Currently, all but a few experimental cable systems in Japan are small CATV operations: 31,000 systems serving 3,300,000 subscribers. However, an early embrace of the idea of the *information society* has reinforced a perceived need to link or integrate technologies, cultural and information services (and ultimately institutions). This idea has been promoted by the Ministry of Posts and Telephone (MPT) and the Ministry of Trade and Industry (MITI).[67]

There have been two main pilot projects. The Ministry of Posts and Telephone originated the TAMA New Town CCIS, based on CATV but employing limited two-way services (teletext, electronic messaging, still pictures by request, televote/response by keyboard or microphone) and emphasizing visual communication in services (pay channels, facsimile newspapers, and local origination), and experiments. The Ministry of Industry, MITI, created Hi-Ovis (Highly Interactive Optical Visual Information System). Hi-Ovis emphasized optical fiber technology, included channels for requests for videotaped materials from a long menu, and used video as well as audio and keypad feedback from households. As in TAMA, televote feedback was widely popular although the visual component was considered a marginal addition (in fact, it seems to have inhibited some who were shy about their appearance) and the video request service was very popular. "Living information," such as on-line, interactive information, on things such as train schedules or many other types of routine information via cable was not found by subscribers to be more valuable than conventional print or other media.[68]

Also in competition for new interactive services in Japan is the Ministry of Posts and Telephone initiated CAPTAIN network (a videotex-style personal information retrieval system based on telephone networks). After its 1985 introduction, however, a Nippon Telephone and Telegraph (NTT) and the Ministry of Posts and Telephone survey showed audience skepticism about videotex services, such as shopping.[69]

Several cable projects, Ito and Oishi, predict a dim future for cable as an independent medium but predict that it may well become integrated with telephone service, ISDN, interactive services, and broadcasting via optical fiber technology. NTT has stressed the development of the Information Network System, an integrated, digital information processing and transmission system. A 1985 market study in a Tokyo suburb indicates that in a media-rich market many potential customers consider cable too expensive but some expressed interest in interactive health, security, and shopping services.[70]

Beyond pilot projects, the Ministry of Posts and Telephone is providing incentives in ten model cities, where market competition will determine the specific technologies to be used. Based on the relative success of the TAMA and Hi-Ovis experiments, the Ministry of Posts and Telephone will be taking applications for two-way cable systems from private companies in all markets. A number of companies are submitting applications including electronic firms, marketing groups, railways, and some foreign partners, including Viacom. Less attention seems to have been given so far to the formation of any independent video program channels, although NTT has announced a search for information service providers for Information Network Systems. CNN is being carried on some cable systems and some other U.S. firms, such as Viacom, have expressed interest in cable programming as well as participation in system operation.[71]

Unless interactive cable systems prove attractive to potential subscribers, the market for cable in Japan may be quite limited. CATV or retransmission systems are declining in penetration in the face of improved broadcast signals and delivery by NHK in remote areas through DBS. Although DBS may expand its services in Japan, in 1985 it was primarily oriented to remote and rural service, carrying existing NHK signals. Cable may also not be perceived as adding significant diversity in video distribution. With one noncommercial/educational and five commercial/entertainment television networks, there is considerable diversity in conventional broadcasting. With the addition of VCRs (30 percent penetration of TV households), a potential role for cable in adding the element of premium films is also diminished. However, other projections in 1985 show that if costs decrease for satellite delivered services, there might be a considerable audience for U.S.-style cable TV. At present, satellite capacity limits the number of specialized channels that could be offered.[72] Part of this projection of interest in specialized cable channels is based on the growing interest in CNN, now provided to hotels, businesses, and some existing cable systems.

Other Western European Countries

Several Scandinavian countries have proceeded with fairly extensive cable TV development. Denmark resembles the Netherlands and Belgium in being a small country with limited resources, which adds diversity by bringing neighboring countries' signals by cable. Denmark is also building a limited national interactive broadband system. Likewise, Norway and Finland bring in Swedish television for diversity as well as using cable to provide reception for isolated communities. Finland, Norway (particularly the Jevnakker pilot project), and Sweden are expanding the sophistication of their systems, and Finland includes videotex services, satellite services (Sky Channel, Music Box and TV5), and pay channels. Swedish and Norwegian companies are trying to start pan-Scandinavian cable TV channels.[73]

Cable is fairly advanced in Ireland, covering 42 percent of TV homes. Program channels are limited to retransmission of Irish and British broadcasts but may be expanded if regulations are changed to permit pay-TV or advertising-based channels.[74]

The southern European or Mediterranean countries have not yet developed much beyond the MATV stage. Local cable origination flourished in Italy briefly in the early 1970s when a local operation, Tele Biella, challenged the national broadcast monopoly rules. Its successful challenge led in fact to a opening up of local broadcast services, which were less costly than cable, stunting cable growth. Italian authorities anticipate using cable for information services once optical fiber technology becomes more affordable. The city of Barcelona in Spain is planning to cable up to 1,000,000 homes with 16 channels and tentative plans are also being made for Madrid.[75]

Eastern Europe

Despite the extensive use of wired radio in Eastern Europe and the USSR, East Germany is the only country that seems to be developing cable to any extent (2,000,000 homes in 1984).[74] Community antenna TV or MATV is developing in Hungary, East Germany, and some others. Community channels are being added in at least one Hungarian city, Pecs, but remain limited.[77] The USSR, in contrast, had only a few thousand homes cabled although MATV systems seemed to be developing in some areas for retransmission purposes. More extensive plans have been discussed in the USSR, and, with its position in wired radio and satellite transmission technologies, the USSR could move relatively quickly to expand cable. Such expansion seems to have low economic priority and also uncertain political desirability. The USSR is ambivalent about new interactive information technologies and prefers maintaining a system that allows more complete centralization and control, such as centrally produced broadcasting.[78]

Review of the Third World

Cable television is such an expensive technology that it has not been seriously considered in most Third World countries. In the less affluent Third World countries of Africa, television itself is still largely restricted to certain socioeconomic groups and areas near major cities.[79] MATV systems may follow wired radio in some countries. This is already taking place in the People's Republic of China where wired radio has been used since the 1950s and MATV is now being implemented widely. Development from wired radio to cable TV may also take place in Africa, but such beginnings are very tentative.

In countries where resources are highly limited, cable faces even more competition from other video media than in Europe, Japan, or Canada. Some countries, such as India, People's Republic of China, Indonesia, Brazil, Mexico, and the Arab nations are investing heavily in satellite distribution of TV. While MATV distribution on the ground may result, low- or medium-power broadcast repeaters are currently more common, particularly in India and Brazil. Some countries have announced plans for DBS systems but India has the only currently extensive system. Particularly in many Third World countries, the function of expanding the diversity of television offerings has been preempted by a less expensive, more flexible technology, the VCR. The VCR has achieved remarkable penetration in some countries: 90 percent in Kuwait, 20 percent in Hong Kong, 12–14 percent in Malaysia and Venezuela. In some countries such as India, where household penetration of VCRs is still low, many more people have regular access to them in buses, public places, and video screening parlors.[80]

In Third World nations where cable has been adopted, the systems tend to target affluent minority populations with specific language or cultural interests. In this manner, there are small cable systems for English-speaking populations in Mexico City, Costa Rica, South Korea, and others. In some cases, these systems are expanding to include more affluent and educated households that are not primarily in the minority-language group. In the Dominican Republic, for example, small unregulated cable systems pirate U.S. satellite signals and provide them to a relatively restricted audience (estimated 2 percent of Santo Domingo).[81]

There are some small satellite signal-based cable systems operating elsewhere in the Caribbean and Central America. These systems tend to be unregulated, even illegal, using their positions within the footprint of U.S. satellite services to pirate signals for distribution. The problem is sufficiently widespread that U.S. satellite operators, program services, and government officials have made official complaints to several governments, including

Jamaica and the Dominican Republic.[82] In some Caribbean countries multiple multipoint distribution systems (MMDS) are being planned to provide multichannel service at a low capital cost. As other Third World regions fall within the footprint of satellite-based television program services primarily aimed at Western Europe, the Arab World, or India, to name several growing satellite distribution systems, there will almost certainly be a growth in backyard dishes or small SMATV-cable system in these regions as well.

Some of the more affluent "middle income" countries, such as South Korea, Argentina, Hong Kong, and Singapore, are developing cable systems in the major cities and other countries may begin in the next few years.[83] Some of these countries, such as Venezuela, seem likely to wait until technological developments, such as optical fiber, create the possibility for expanded services beyond just extra channels of video fare.[84]

In Argentina, cable systems are most prominent in provincial towns that receive limited broadcast signals. A 1984 study estimated that 37 such systems covered 133,000 homes—an average system size of 3,500 households. In Buenos Aires, two main systems covered 46,000 homes with three nonbroadcast channels. One system emphasized news, service, and movies. The other emphasized culture, sports, and movies.[85]

NOTES

[1]Smith, Anthony, *The Shadow in the Cave*, University of Illinois Press, 1973.

[2]Homet, Roland, *Politics, Cultures and Communication*, New York: Praeger, 1979.

[3]Dutton, Blumler, and Kraemer, "Advanced Wired Cities: Driving Forces and Social Implications," in Dutton, Blumler and Kraemer, eds., *Shaping the Future of Communications: National Visions and Wired City Ventures*, The Washington Program of the Annenberg School of Communications, 1985, Ch. 5. Forthcoming from G.K. Hall, Boston, 1987.

[4]Lhoest, Holde, "The Interdependence of the Media," *Mass Media Files, No. 6*, Council of Europe, 1983.

[5]Head, Sydney, *World Broadcasting Systems*, Wadsworth, 1935, pp. 48–49.

[6]"An EBU survey on cable television in Europe," *EBU Review. Programmes, Administration, Law*, Vol. XXXV, No. 1, January 1984, pp. 31–42.

[7]Lhoest, op. cit.

[8]Ito, Youichi, and Yutaka Oishi, "Lessons from the Tama CCIS and the Ighashi Ikoma Hi-OVIS Experiments," paper prepared for the Forum on "Advanced Wired Cities: Driving Forces and Social Implications," July 1984, Washington Program of the Annenberg Schools. Forthcoming in Dutton, Blumler, and Kraemer, op cit.

[9]EBU Review, op. cit. Muller, Jurgen, "Cable Policy in Europe: The Role of Transborder Broadcasting and Its Effect on CATV," paper for 14th Telecommunication Policy Research Conference, Arlie, VA, April 1986.

[10]*Financial Times' New Media Markets*, Vol. 3:10, May 14, 1985, pp. 1–8.

[11]Ito, Youichi and Oishi, op. cit.

[12]"Study Sees Limited Growth of Cable TV in Europe," *Multichannel News*, November 15, 1982, p. 24. "European Cable: CIT Predict Slow but Steady Progress." *International Broadcasting*, July 1985, pp. 29–30. "Japanese Say Cable-TV Too Expensive." *Electronic Media*, February 7, 1985, p. 18.

[13]Ibid.

[14]Gerin, Francois and Nicolas de Tavernost. "From the Biarritz Project to the French Videocommunications," Advanced Wired Cities Forum, 1984. Forthcoming in Dutton, Blumler and Kraemer, op cit.

[15]Straubhaar, Joseph and Caroline Lin. "A Quantitative Analysis of the Reasons for VCR Penetration Worldwide," for Panel on Growth of Home Video at Telecommunication Policy Research Conference, April 1986.

[16]Boyd, Douglas and Joseph Straubhaar. "The Developmental Impact of the Home Videocassette Recorder in the Third World," *Journal of Broadcasting and Electronic Media*, Winter 1985. Ogan, Christine, "Media Diversity and Communications Policy: Impact of VCRs and Satellite TV," *Telecommunications Policy* 9:1, March 1985, pp. 63–73. Various authors, "Home Video: An *Inter Media* Survey," *Inter Media* 11:4/5, July/September 1983, pp. 17–75. See particularly, "Gunfight at the Poitiers Corrall."

[17]Lhoest, op cit. McQuail, Denis and Karen Siune, *New Media Politics: Comparative Perspectives in Western Europe*, Beverly Hills: Sage Books, 1986.

[18]"Canada—Making Cable Pay," *TV World*, April 1985, pp. 40–44. "Crossing the Border," *CableVision*, August 20, 1984, Plus Section. Lyman, Peter, *Canada's Video Revolution*, Toronto: Lorimer & Co., 1983. "Sky Channel and the Coca-Cola Bird," *Broadcaster*, March 1985. "Sky Channel Eyes Belgium, Will Add Children's Shows," *Multichannel News*, July 8, 1985. "Turner Set to Provide News Service in Europe," *Multichannel News*, July 8, 1985. "Disney Channel Eyes Overseas Expansion," *Multichannel News*, July 8, 1985.

[19]Lhoest, op cit. Powell, David, "Television in the USSR," *Public Opinion Quarterly*, Fall 1975, pp. 287–301. Katz, Elihu and George Wedell, *Broadcasting in the Third World*, Cambridge, MA: Harvard, 1977. Head, op cit.

[20]Homet, op cit. Dizard, Wilson, *The Coming Information Age*, Longman, 1982. Muller, "Cable Policy in Europe," op cit.

[21]Locksley, Gareth, "Cable as Choice," *Telecommunications Policy*, June 1983, pp. 98–99. Collins, Richard, "Will Choice be Extended by Cable?" *Telecommunications Policy*, March 1984, pp. 7–11. Dornan, C.T. "Fear and Longing in the United Kingdom: Cultural Custody and the Expansion of Cable Television," paper at International Communications Association, May 1984, San Francisco.

[22]Ibid. Streeter, Thomas, "The Commercialization of German Broadcasting and the Ideology of the 'New Media,' " paper at the International Communication Association, May 1984, San Francisco.

[23]Dornan, op cit. Homet, op cit.

[24]Head, op cit., pp. 117–123.

[25]"EBU Survey. . ." op cit. "Cable in the Low Countries," *Television/Radio Age International*, February 1984, pp. 38–41. "Scandinavian Countries Begin Studying Cable TV," *Multichannel News*, September 12, 1983, pp 22–23.

[26]Spandler, Richard. "In Europe, Cable and DBS Start Even," *Channels*, March 1985, pp. 27–32. Milne, Alistair, "A View from the Brits: Westward, No!" *Channels*, July 1984, pp. 63–64. Herrmann, Gunter, "Border-crossing radio and television programmes in the Common Market," *EBU Review* XXXVI:1 January 1985, pp. 27–40.

[27]Hill, Arthur, "The American's Are Coming," *Cable Television Business*, November 15, 1982, pp. 50–54. "The Cabling of Europe," *The Economist*, August 13, 1983. Hoskins, Colin, "The Cabling of the UK: Lessons from the Canadian Experience," *Media, Culture and Society*, 1984: 6, pp. 177–189.

[28]Lee, Chin-Chuan, *Media Imperialism Reconsidered*, Beverly Hills: Sage, 1980. Beltran, Luis, "TV Etchings on the Minds of Latin Americans," *Gazette* 24:1, 1978. Tunstall, Jeremy, *The Media are American*, New York: Columbia, 1976.

[29]Herrmann, op cit. Wedell, George, "Television without Frontiers," *EBU Review* XXXVI:1, January 1985, pp. 21–25.

[30]European Economic Community Commission, "Television without Frontiers: The Establishment of the Common Market for Broadcasting, Especially by Satellite and Cable," (COM 84/300 final, 1984).

[31]Herrmann, op cit.

[32]"Sky Channel 1 . . . ," op cit.

[33]"About Churn," *Connections*, September 1985, p. 26. Hollins, Timouthy. *Beyond Broadcasting: Into the Cable Age*. London: Broadcasting Research Unit, British Film Sustitute, pp. 37–40.

[34]The "White Paper" is officially titled "The Development of Cable Systems and Services," Home Department and Department of Industry, Cmd 8866 (April 1983). Ferris, Lloyd, and Casey, "The United Kingdom," *Cable Television Law* Vol. 2, New York: Matthew Bender, 1985, 30:7–18. Salvaggio, Steinfield, and Hood, "An Examination of the UK's Recent Policy Shift Toward a More Competitive Approach: Implications for a Wired Nation," International Communication Association, Honolulu, May 1985. Clement-Jones, Tim, "Cable and satellite TV in the UK and Europe—the emerging legal issues," *Telecommunications Policy*, September 1983, pp. 204–214.

[35]Ibid. Roberge, Paulette, "U.K. Report—A Resounding Vote of No Confidence," *Cable Communications Magazine*, December 1984, pp. 42–43. "UK Cable Mirrors European Woes," *Electronic Media*, January 10, 1985. "U.K. Government Sets Grants to Spur Interactive Services," *Multichannel News*, October 28, 1985, p. 22.

[36]Salvaggio, Steinfield, and Hood, op cit.

[37]Wicklein, John, *Electronic Nightmare: The Home Communications Set and Your Freedom*, Boston: Beacon Press, 1982, Ch. 2. "First British Cable Operations Begin to Gauge Market Potential," *Multichannel News*, July 8, 1985. *Cable Communications*, September 85, p. 43. Dutton, William H. and Blumler, Jay. "The Faltering Development of Cable Television in Britain," Paper presented at the 37th International Communication Association Conference Montreal; May 1987.

[38]Roberge, Paulette, "UK Report—DBS project, pay movie channel fall victim

to viability problems," *Cable Communications*, September 1985, pp. 42–43. "Sky Channel and the Coca-Cola Bird," *Broadcaster*, March 1985. "Sky Channel Eyes Belgium, Will Add Children's Shows," *Multichannel News*, July 8, 1985. "Disney Channel Eyes Overseas Expansion," *Multichannel News*, July 8, 1985. "Trying to Imitate HBO Seen as Mistake for British Pay TV," *Multichannel News*, July 22, 1985, p. 18. "Sky Channel Competitor Planned by BBC, ITC." *Multichannel News*, December 2, 1985, p. 29.

[39]Ibid. "UK Grants Franchises to Five Firms," *Multichannel News*, August 26, 1985.

[40]Ferris, Lloyd, and Casey, op cit.

[41]Lyman, op cit. Ferris, Lloyd, and Casey, op cit. Appelbaum, Simon, "Canadian Cable on the Move," *CableVision*, February 24, 1086, p. 29.

[42]"Crossing the Border," *CableVision*, August 20, 1984, pp. 39–42.

[43]"Canada's Life Channel Suspends Operation," *Multichannel News*, December 8, 1986, p. 21. Ferris, Lloyd, and Casey, op cit. "Canadians protest U.S. cable ad glut," *Electronic Media*, September 27, 1984. "U.S. cable services find gold mine in Canada," *Electronic Media*, August 30, 1984. "Music video service finds favor in Canada," *Electronic Media*, October 11, 1984. "No room for competition in Canadian pay TV," *Cable Communications*, September 1984, pp. 14–23.

[44]"Canadian Firms Seek Approval for 24-Hour Cable News Service," *Multichannel News*, November 10, 1986, p. 3. "Canadian Cable Ops Seek Approval for PPV," *Multichannel News*, October 6, 1986, p. 16.

[45]Easton, Ken, "Clustervision and Villagecable," *Cable Communication*, September 9, 1984, pp. 5–7. "Canadian cable show finds operators feeling aggressive but stymied," *CableVision*, June 9, 1986, p. 24–27.

[46]*Advertising Age*, special section on Canada, December 19, 1983. "Acquisitions, Risky Videotex Venture on Top for Canada's Videotron LTD." *Multichannel News*, May 26, 1986, p. 33. "Canada's Videoway set for April launch," *CableVision*, February 24, 1986, p. 34.

[47]*The Economist*, op cit.

[48]Gerin and Tavernost, op cit.

[49]*Christian Science Monitor*, April 15, 1984, pp. 19–20. "Et Viola! Le Minitel," *New York Times Magazine*, April 19, 1986.

[50]Gerin and Tavernost, op cit. "French Cable System Offers On-Line Library." *Multichannel News*, December 22, 1986, p. 15.

[51]Ibid. Ferris, Lloyd, and Casey, op cit. "French Agency Rejects Most Challenges to Private Channel," *Multichannel News*, April 28, 1986, p. 27. "France to Privatize Largest TV Network," *Multichannel News*, May 26, 1986. "France to End Monopoly on Cable TV Construction," *Multichannel News*, August 11, 1986, p. 15.

[52]*EBU Review, Technical*, 203:29, February 1984. "French Pay TV Network Strikes Deal for U.S. Films," *Multichannel News*, September 29, 1986, p. 25. "New Cable Services Set for Systems in France," *Multichannel News*, January 13, 1986, p. 3. "France Grants Permission for Music Channel Launch," *Multichannel News*, February 10, 1986, p. 20.

[53]"France Set to Accelerate Cable System Construction," *Multichannel News*, September 16, 1985, p. 31. Gerin and Tavernost, op cit. "Paris Cable System Launches to First Subs," *Multichannel News*, December 8, 1986, P. 31.

[54]*The Economist*, op cit., p. 38. Vedel, Thierry. "Cable Policy and Politics in France." Panel on Historical and International Perspectives on Cable Policy, International Communication Association Conference, Montreal May 1987.

[55]Steinfield, Charles, "Introduction: Technology in Search of a Vision (on Germany)." Forthcoming in Dutton, Blumler, and Kraemer, op cit.

[56]"Future of German Private TV Tied to Development of Cable," *Multichannel News*, July 8, 1985, pp. 13–16.

[57]Hege, Hans, "The Berlin Pilot Project—Social Implications of New Technology," Advanced Wired Cities Forum, 1984. Forthcoming in Dutton, Blumler, and Kraemer, op cit.

[58]Kanzow, Jurgen, "The BIGFON Project," Advanced Wired Cities Forum, 1984. Forthcoming in Dutton, Blumler, and Kraemer, op cit.

[59]"Sky Channel . . . ," op cit. Hege, op cit. "Future of German Private TV . . . ," op cit.

[60]Lhoest, op cit. pp. 35–37; Streeter, op cit. "West German Pay Net Wins Government Okay," *Multichannel News*, February 17, 1986, p. 29.

[61]Smith, op cit. *EBU Review*, op cit.

[62]"World's Most Entertaining TV System to Change. Dutch Vie for Cable Access," *Video Age International*, April 1984. "Tug of War in TV Bill," *TV World*, August 1985.

[63]Brent, Kees and Nick Jenkowski, "Cable Television in the Low Countries," XIV International Association for Mass Communication Research Conference, Prague, 1984. "Cable in the Low Countries . . . ," op cit. "The Netherlands—Pay-TV and Satellite: 'Little Threat' to Broadcast," *Tv World*, February 1985, pp. 24–28.

[64]"Netherlands Sets Plans for Pilot Project," *Multichannel News*, September 5, 1983, p. 68. "Netherlands Begins Project Testing Interactive Cable," *Multichannel News*, December 8, 1986, p. 28.

[65]"Cable in the Low Countries . . . ," op cit.

[66]"TV Ads' Slow Road to Acceptance," *TV World*, February 1985, p. 29. *New Media Markets*, op cit.

[67]Ferris, Lloyd, and Casey, op cit. Gonzalez, Ibarra, "Conceptualizing the Wired City in an Information Society: The Japanese Experience," International Communication Association," San Francisco, May 1984.

[68]Masuda, Yoneiji, *Information Society and Post-Industrial Society*, World Future Society, 1982. Gonzalez, op cit. Bowes, John E., "Japan's Approach to an Information Society: a Critical Perspective," in Wilhoit and de Bock, eds., *Mass Communication Yearbook*, Volume 2, Beveraly Hills: Sage, 1981, pp. 67–710. Ito, Youichi, and Oishi, op cit.

[69]Kaneko Hideaki, "New Media Business," *The Oriental Economist*, March 1983, pp. 22–26. Burton, Jack, "Japanese Say Cable TV Too Expensive," *Electronic Media*, February 7, 1985, p. 18.

[70]Ibid. Ito, Youichi, and Oishi, op cit.

[71]Ferris, Lloyd, and Casey, op cit. "NTT Sees Participation of Information Providers for Its INS Model System," NTT press release, April 21, 1983. "CNN finds a following in Japan," *Electronic Media*, September 13, 1984. "Viacom Connects in Japan," *CableVision*, August 22, 1983.

[72]"Cable and Satellite in Japan," *International Broadcasting*, June 1985, p. 60.

[73]"Scandinavian Countries Begin . . . ," op cit. Dahl, Hans, "Reversals in Norway," *Intermedia*, November 1982, pp. 46–47. "The Jevnakker Project," *Nordicom Review*, 1983. "Innovating Cable-TV System in Finland to Expand with Scrambled Pay-per-view," *Videoage*, April 1985. McQuail and Siune, op cit, pp. 57–58. "Cable TV in Sweden to Blossom This Year," *Electronic Media*, February 17, 1986, pp. 16, 30.

[74]"Cable and Satellite in Ireland," *International Broadcasting*, April 1985, p. 15.

[75]Head, op cit. Lhoest, op cit. *Cable Television Spreads Throughout Europe*, London: International Institute of Communication, 1983. "Barcelona Seen Likely Site for Spain's First Cable System," *Multichannel News*, September 8, 1986, p. 17.

[76]Head, op cit, p. 52. Malik, Rex, "Can the Soviet Union Survive Information Technology?" *Intermedia* 1984:3, pp. 10–3. Shahzad, Badhanaseeb, "TV is the Opium of the People," *Intermedia* 1984:3, pp. 24–25.

[77]Szekfu, Andres, "The First Experiment in Community Cable TV in a Socialist Country," Mass Communication Research Center. Budapest, presented at XIV International Association for Mass Communication Research Conference, Prague 1984.

[78]Op cit., Head, Malik, Badhanaseeb.

[79]Katz and Wedell, op cit., Chapters 2–3.

[80]Head, op cit. Boyd and Straubhaar, op cit.

[81]Interview with Bella Mody by author, Michigan State University, October 2, 1985. Interviews by author in Santo Domingo, Dominican Republic, December 3–12, 1984.

[82]Ibid. Silver, Rosalind, "Pirating the Caribbean," *Media and Values*, Summer 1985, pp. 3–8.

[83]Noguer, Jorge, *Radiodifusion in Argentine*, Buenos Aires: Bien Comun, 1985, pp. 525–527. Demuth, Nick, "Green Light for Cable," *Asia-Pacific Broadcasting and Telecommunications*, August 1986, p. 28. Mody, interview, op cit. Kim, Young Y., "Diffusion of Cable-TV in Korea," unpublished paper, March 1984, Michigan State University.

[84]Interviews by author in Caracas, Venzuela, June 14–23, 1985.

[85]Noguer, op cit.

CHAPTER **18**

Impact of New Communication Technologies

Backyard dish capable of bringing in more than 100 television channels.

Cable and other communication technologies will have an impact on the media environment and human behavior. It is tempting to become overexcited by the possibilities of unlimited, broadband communication channels with two-way capability. One can envision a communications environment perfectly suited to each individual's interests and needs, interactive communication with the rest of the world, and the home of the future where work, school, play, and home management resources are all efficiently available from within the household at a minimum expenditure of time and energy. Seeing the advantages and horrors of such a communication environment is barely a challenge to the imagination.

But fast as these new communication technologies seem to be coming, it will be argued here that the most immediate changes will be peripheral. Entertainment options have multiplied quickly. The quality of television for both entertainment and information has been enhanced. Nevertheless, these are somewhat marginal increments in the usefulness of television. Major changes will evolve much more slowly. The benefits of two-way communication in broadband context will be tested by trial and error over a long time as the technology, services, and marketing programs develop. The social change resulting from greater, more convenient access to information will certainly be slow as people accept, adapt to, and absorb information in new forms. Home management, working, and schooling, aided by broadband communication, may replace some earlier methods of approaching the same functions, but the force of habit and existing institutions will not submit easily and quickly. Human behavior, dependent on well established family and institutional values, is not to be suddenly corrupted, or enhanced, by new communication technologies.

The most important changes will be slow enough so that the impacts may be assessed and the communication systems adapted to desired goals. This is not to say that we will not institutionalize some undesirable characteristics of the new communication systems before we recognize more appropriate policies. But there will be time to contemplate, understand, and control the more profound aspects of our new communication environment as it develops.

In this chapter we look at cable and other communication technologies as they affect each other, the quality of telecommunications services, and the media environment. The results of the introduction of new communication technologies and services are not universal across the society. Rural, small town, urban, and suburban communities will realize different effects, both positive and negative. Some communities in the United States have no local mass media at all; others may have a weekly newspaper and faint signals from distant radio and television stations.[1] From this point, media availability ranges over a variety of combinations of radio, television, and newspapers up to the media-saturated metropolitan communities. Different media environments may be distinguished at one level by enumerating media—television stations, cable channels, radio stations, daily newspapers, and weekly newspapers. Of the mass media, only magazines are equally accessible, at least by mail, to all media environments. At another level, the character of mass media content and the relationship of media to individuals in the target audience also differs from community to community. It is essential to account for these differences in media environment in assessing mass-media service and impact. Communication technologies can be developed and structural changes made to remedy deficiencies in media service, where they exist, and approach a parity in access to media for all citizens. This process is well underway.[2]

DIRECT BROADCAST SATELLITES

Definition

Direct broadcast satellite (DBS) service has been defined as a "radiocommunication service in which signals transmitted or retransmitted by space stations are intended for direct reception by the general public.[3] In the broadcasting-satellite service, two kinds of direct reception are possible: *individual reception* refers generally to simple domestic installations with small antennas; *community reception* involves a more complex system with larger antennas intended for use by a group at one location or through a distribution system such as cable covering a limited area. Such systems are referred to as *Satellite Master Antenna Systems* (SMATV) and are described later in this chapter. DBS can be divided into two categories; C-band, which uses the satellites presently programming to cable systems, and Ku-band, a higher frequency band reserved for direct broadcast.

Technology, C-band

In 1987 there were more than one-and-a-half million satellite earth stations owned by individuals.[4] These earth stations, sometimes called *backyard dishes*, receive the same satellites that serve the cable industry. Complete systems, including the antenna, preamplifier, and receiver, can cost under $2,000.

In the U.S., the home dish must have a clear line-of-sight in a southerly direction to the satellites in equatorial orbit. The very weak signals are amplified up to a million times, downconverted in the house to a TV frequency that can be seen on any TV receiver. The dish can be rotated, from inside the home, to aim at several satellites serving cable systems or broadcasters. Over 100 channels are available.

The rapid growth of home satellite dishes can be attributed mainly to the fact that programming costs nothing to the subscriber. By purchasing a dish for $2,000, a homeowner could get a large selection of programming, more channels than available on the local cable system. Though most dishes were sold to people living in rural areas where cable was not available, a growing number were sold to homes within cable systems. Prior to the introduction of scrambling (encryption) of their signals in 1985, programmers had no effective way of preventing individual dish owners from receiving their channels. Individual dish owners, of course, were receiving their programming at no cost, while cable operators were paying the programmers for their product.

Scrambling

Under the 1984 Cable Act the right of home owners to satellite signals was established. As long as the signal is not scrambled and not marketed to dish owners, it is available for private, noncommercial use in the home. As soon as the signal is scrambled and a marketing system is in place to sell it, the home dish owner is obligated to pay.

Under pressure from cable operators, who were beginning to see home earth stations appear within their systems, and also to take advantage of the income potential from the dish owners, the pay networks began to scramble their signals. By 1988 most of the national pay networks and the major basic satellite networks were scrambled. The suppliers of scrambled programming have set up ways in which individual dish owners may purchase programming.

With scrambling, dish owners have to pay for programming at a rate comparable to what they would pay for the same services over cable. It is unlikely that many people living within a cable area will purchase a dish and pay for programming after scrambling is fully implemented. Therefore, most C-band dish sales in the future are likely to be in rural areas.

When scrambling was first tested by HBO in 1985, there was much concern about compatibility. Three different scrambling technologies had been introduced, each requiring a different descrambler. The decision of Showtime and The Movie Channel to adopt the same system as HBO (the General Instrument Videocipher II) assured that only one standard would be used for C-band programs.

In order to receive scrambled programs, a home earth station owner first purchases or leases a descrambler from a retail dealer (or, in some cases, from the cable system in the area). Descramblers cost around $400. The customer then places the order for programming by calling a toll-free number operated by the programmer, or calls the cable system ("one-stop shopping" for all channels desired) if the customer lives within a cable franchise area. The programmer relays the service request to a computer center operated by the descrambler manufacturer, which sends the authorization command over the satellite transponder to the customer's descrambler. The customer is then billed either by the cable company or by the program supplier. Only one descrambler is required, regardless of the number of program sources requested.

The FCC has attempted to protect the home TVRO owner from predatory zoning regulations surreptitiously sponsored by competing multichannel technologies. The FCC will preempt zoning ordinances that do not have a clearly defined health, safety, or aesthetic objective. The ordinance cannot impose unreasonable limitations or costs on the C-band dish user.

Technology, Ku-band

Future DBS systems may utilize the Ku-band (12–14 GHz). At these frequencies, smaller dish sizes can be used, reducing the price of the home receiving system.

C-band satellites serving the United States are low-powered, using about 5 watts per transponder. To transmit acceptable signals to small home antennas, Ku-band DBS will have to use 100–200 watts per transponder. Because of this great increase in the power requirement, only six channels of TV programming will be available instead of the potential of 24 channels offered by present satellites. The Ku-band transmission is sensitive to atmospheric disturbances, and the use of high-power, high-frequency transponders with new and less tested technology increases the risk of system failure. By the time Ku-band DBS comes into existence, however, the technology will be thoroughly perfected.

Ku-band satellites have a *footprint* that covers only a small portion of the U.S. (C-band satellites cover the entire country). To cover the entire U.S. in Ku-band, several satellites are needed. Each satellite would broadcast to one time zone in the contiguous United States. Depending on its position relative to the satellite's footprint, a few locations in the United States may not be able to receive all of the channels programmed for that area. To receive the signals, one must be in the satellite's footprint and have a clear line-of-sight orientation between the satellite and the receiving terminal. In mountainous and urban areas, this can be a problem as natural or man-made obstructions interfere. Once the signal has been received, the high-frequency microwaves are converted to the lower, conventional TV channel frequencies. The home terminal includes a decoding system that unscrambles the programming. These decoders will be addressable, under the control of a central office, which can communicate to turn the decoding system on or off on a per-program or per-channel basis. This will allow tiered services. To serve the entire country with DBS on a local time basis, with four or more channels, would require 10 satellites; each satellite would serve about half a time zone.[5]

Programming

Ku-band DBS programming would be from the same sources that supply cable—the movie industry, sports, and independent producers. It is likely, in fact, that the same services which program for the cable industry and are marketing their channels to the C-band dish market will be programming Ku-band DBS. This would mean that the rural, low-density areas would have access to the best of the national programs on cable. Localized interests would not be served, of course, although some special programs could be created to suit the interests of the rural audiences, for example, a television version of the rural audience magazine *Grit*.

Economic Future

In markets where cable is established, it is not likely that DBS would be competitive. Cable can provide more channels for about the same monthly cost and lower installation charges. Because cable already serves most urban and suburban areas, DBS will be limited to sparsely populated countryside outside metropolitan areas and the rural areas. It has been estimated that there are about 6 million such households. The 6 million homes could shrink considerably, however. Cable systems are at present wiring areas with under 30 homes per mile of density. In the future, cable may serve areas with even lower homes per mile of density, if rates equivalent to those proposed by DBS could be charged. Telephone companies may also provide video service to very low-density areas. MMDS and low-power television could reach some of these homes, but with fewer channels.

One company, United Satellite Communications Inc. (USCI), attempted a Ku-band DBS service in 1983. USCI used an existing Ku-band satellite for a four-channel service, charging about $30 per month. After a year, USCI had attracted only about 20,000 subscribers in the northeast and midwest United States. The service was discontinued in 1985.

It has been proposed that the FCC establish DBS as a high-definition television distribution medium. (See discussion of HDTV in Chapter 5.) This would delay the initiation of service while the technology is developed. Consumers would need a receiver designed to the new standard. HDTV via DBS could mean a different market for DBS among high-income people anxious to improve television quality. Even here, there would be competitors, in cable systems, videocassettes, and videodiscs.

SUBSCRIPTION TELEVISION

Definition

Subscription television (STV) is a "system whereby subscription television broadcast programs . . . are transmitted and intended to be received in intelligible form by members of the public only for a fee or charge."[6] Each STV station is a commercially licensed broadcast facility that has received authorization from the FCC to operate a subscription service. The television signals are scrambled when transmitted so that only paying households with leased or purchased decoders may receive them.

Technology

STV uses standard UHF television channels for broadcast. The signal is scrambled at the transmitter, and the subscriber uses a standard receiving antenna with a special decoder to view the channel.

The service area of an STV station is the same as that for any other UHF station. The quality of the signal received varies depending on the transmitting power, antenna height above average terrain, operating frequency, obstructions (natural or man-made), receiving antenna, TV set quality, and decoder. As with any conventional television broadcasting station, STV stations must obtain licenses for three-year periods. Spectrum availabilities are predetermined according to the table of assignments for TV channels. If an applicant is granted a standard broadcast license, the licensee may operate the STV service.

Only areas of the country that have available UHF or VHF channels can have the standard STV service. Any commercial television broadcast station licensee or holder of a construction permit for a new station may recieve authorization to provide STV service. An STV license applicant need not ascertain community needs as does an applicant of a conventional broadcast television license.

Low-power television stations may also provide subscription TV service, especially in rural areas. (See "Low-Power Television," later in this chapter.)

Programming

There are no restrictions on the kind of programming that STV stations can offer other than those imposed on all broadcasters. The FCC removed all pay TV programming restrictions on STV (designed to protect broadcast TV) after a similar set of rules pertaining to pay cable was struck down by the D.C. Circuit of Appeals.[7]

Typically, STV offerings include movies, specials, sports, or other entertainment.

Economic Future

The first formally licensed STV station under the new rules began broadcasting in the Newark, N.J. area on March 1, 1977. By 1982, almost 30 STV stations were on the air. However, as cable systems were built in the suburbs of the cities where these stations were located, subscriber counts began to drop. After peaking in 1983 at about 1.5 million subscribers, STV began a rapid decline both in the number of subscribers and in the number of stations. By 1987, there were less than 100,000 STV subscribers, and only 2 stations remaining on the air.[8]

One way in which STV may survive is through low-power TV in small communities. It is possible that networks of low-power STV stations could be interconnected by satellite, perhaps with a full-power STV station, to share the administrative and operating costs of programming a subscription service.

MULTIPOINT DISTRIBUTION SERVICE

Definition

Multipoint Distribution Service (MDS) is a common carrier closed-circuit microwave system transmitting a signal addressed to multiple, fixed receiving points.[9] Until recently, only two MDS channels in the 2,150- to 2,160-MHz portion of the spectrum were authorized. In 1984, the FCC reallocated the spectrum in the 2,500–2,690-MHz band among the MDS, instructional television fixed service (ITFS), and operational fixed services (OFS). Thirty-one underutilized channels were reallocated between the other two services, which would allow eight more channels to MDS for a total of ten in each location. Licenses for the new channels were awarded on a lottery basis in two groups of four channels. This expanded MDS service is called *Multichannel Multipoint Distribution Service* (MMDS) or *Multichannel Television* (MCTV). In addition, the FCC authorized ITFS frequencies to be leased by commercial users. By making arrangements with several licensees, a MMDS operator could offer ten or more channels of programming. After a slow start, MMDS services were available by 1987 in Cleveland, San Francisco, Washington, D.C., Milwaukee, and New York.

Technology

MDS stations are intended to provide one-way transmission (usually in an omnidirectional pattern) from a stationary transmitter to multiple receiving facilities located at fixed points designated by the subscriber. Provision is made, however, for licensing of return microwave channels to provide two-way interactive service.

The receiving stations use directive antennas to pick up the line-of-sight microwave transmissions. The signal is converted by a downconverter from the microwave frequency to a selected lower frequency, then passed through a decoder (in an addressable or scrambled system), conducted by coaxial cable to the TV receiver and displayed on an unused channel on the TV set.

Receiving equipment for MMDS is similar to MDS, except that the downconverter is tunable to allow reception of multiple channels. The downconverter also tunes over-the-air broadcast channels, so that only one tuning knob is needed to receive all television programming.

The service area extends from 15 to 25 miles or more, depending on the terrain. The picture quality is similar to cable service, though heavy rain can attenuate the signals. The transmission range is a function of transmitting power, the character and size of the receiving antennas, line-of-sight orientation, and obstructions such as terrain shielding, foliage, or intervening buildings.[10] MDS service is quite reliable up to the 25-mile primary service area. Not all homes can be served by MDS, since line-of-sight reception is required. In hilly areas, less than 20 percent of the homes may be capable of MDS reception, and even with flat terrain, tall trees and buildings often prevent 20 to 40 percent of the homes from receiving an adequate signal.

Programming

Single-channel MDS generally offered subscription television programming on a 24-hour basis. HBO or Showtime were the most common services. MMDS systems offer one channel of subscription programming and several satellite-delivered channels, such as CNN, MTV, and the USA Network.

Economic Future

The important cost variables in MDS are the downconverter package, antenna, transmitter, installation and maintenance, marketing, and transaction costs. The costs for the downconverter can range from under $50 to $350 or more. Special equipment features, such as low noise capability, automatic fine tuning (prevents drifting), or filtering, add to the cost. Antennas cost from about $20.[11] Households farther from the transmitter need larger, more costly antennas.

Generally, power for MDS stations is limited to 10 watts.[12] Under some circumstances a waiver can be obtained to transmit at 100 watts. This requires an additional amplifier to upgrade the service. A 10-watt transmitter costs about $17,000 whereas a 100-watt amplifier costs $13,500.[13] An FCC staff study noted that, if initial construction grants were issued permitting carriers to use a single piece of equipment for 100-watt transmitters, instead of the add-on amplifier, there would be a cost savings of several thousands of dollars while increasing the technical reliability of the equipment.[14] To this cost must be added the costs, which vary considerably, of transmitting antennas, towers, and monitoring, switching, and originating equipment. The total would be $150,000 or less.

In the past, costs for MDS have declined fairly steadily. This trend is expected to continue into the future but with a slower rate of decline. There is competition in the equipment manufacturing industry; thus hardware options are available at competitive prices. The big breakthrough was the dramatic cost reduction in installation packages (receiving antenna/downconverter). Since 1976 costs have declined from about $1,500 per installation to an average price range of under $100 to about $300. At this price, it has become possible to market to single-family dwellings.[15]

Subscription prices for pay television delivered via MDS average about $15 per month and range from $7 to $30 in different communities. Deposits of $10–$100 are sometimes required. Consumer costs vary as a function of the type of dwelling (single-family or multiple units), quality of equipment, and service tiers or packages purchased. Usually it is significantly less expensive to provide service to apartments and other multiple-unit dwellings with a single antenna and downconverter for the whole group of dwellings.

MDS as a common carrier receives statutory protection against unauthorized reception.[16] Security for MDS systems adds to the total cost. Typical methods of guarding against piracy are the use of scrambled signals or addressable downconverters. Presently most MDS signals are not scrambled, but as the equipment becomes more available to the general public, piracy is more likely, and MDS systems will have to begin encoding the signals and installing decoders. All of the proposed MMDS systems will use addressable downconverters and scrambling.

MDS systems can operate in a smaller market than can conventional high-powered STV. Under 10,000 subscribers may sustain MDS at a breakeven rate or better. This relatively small number of subscribers makes MDS viable in a larger number of cities than STV, which typically requires a base of 35,000–40,000 subscribers for breakeven costs.[17] However, cable is usually well established in these smaller communities.

The MDS industry has no restrictions on cross-ownership or multiple ownership or any other restrictions normally associated with broadcasting because of its common carrier status. MMDS operators are now barred from programming the channels themselves.[18] The FCC has proposed elimination of this rule.[19]

There is no restriction on the number of MDS licenses that can be awarded to any individual or group. Thus, wholly owned networks can be developed.

The MDS business has been fairly active. By 1980, some 131 MDS stations were authorized by the FCC, with 86 MDS stations completely constructed and offering service and another 467 applications pending. The growth of cable in major markets, however, brought the expansion of MDS to a halt in the early 1980s. Most MDS stations have been losing subscribers in recent years, and their survival depends upon their success in converting to multichannel operation.

MMDS, cable, and perhaps DBS could compete for the same multichannel television market. MMDS might have a slight price advantage over some of the competitors. In comparison with the two competitors, MMDS requires far less capital investment and therefore may serve smaller numbers of people. When compared to cable, MMDS has the advantage of a low initial capital investment in the transmitter, and investment in receiving equipment only as subscribers are added. This allows MMDS to be profitable with a relatively small number of subscribers spread over a large area. In addition, MMDS systems do not require a long time to construct, as do cable systems. A MMDS system with ten channels, plus the local broadcast stations, would provide a 15- to 20-channel system, which might be quite competitive with cable; particularly at the lower rates that would be permitted by the lower capital costs of the wireless MMDS.

However, MMDS faces many technical and operational problems. First, the line-of-sight transmission limits the potential number of homes a MMDS system can serve. Predicting which home can receive service is impossible, so on-site testing must be done before a MMDS installation is made, greatly increasing the installation cost.

In addition, most installations require a fifty-foot-high antenna, which many single-family residents find unsightly. These antennas are easily damaged by high winds. Finally, though MMDS operators can offer ten or so channels, they cannot offer the same variety of pay and basic programming as cable and the cable operators are likely to be aggressive competitors. MMDS operators may be successful in providing service to apartment units and condominiums through SMATV systems.

SATELLITE MASTER ANTENNA TELEVISION (SMATV)

Definition

SMATV systems, also referred to as *private cable*, are small cable systems serving residents of a single apartment or condominium complex, trailer park, or other residential area that is entirely on privately owned property. Because no city or county streets are crossed, and since FCC regulations exempt systems serving dwelling units under common ownership or management from being classified as cable systems, SMATV operators do not require franchises and are not regulated by the FCC.

Legal Status

SMATV systems operate through leases made with the owners of apartment buildings or with condominium associations. These leases are often exclusive, prohibiting cable systems from offering service on the property. In return, the landlord receives a payment, usually 3 to 6 percent of the gross revenues of the SMATV system. There has been much litigation over the legality of these leases and the rights of the landlords, SMATV operators, and cable systems. Some state laws resolve most of these issues.

Technology

SMATV systems are small cable systems and utilize the same technology, though most are constructed with lower-quality equipment. Because of the small number of units served, headend costs are a significant consideration, and few SMATV systems offer over 12 channels. Converters are rarely used, and signal security is usually provided by traps.

Programming

SMATV systems utilize C-band earth stations to receive the same programming that cable systems carry. Until recently, program suppliers, under pressure from cable operators, were reluctant to provide access to their channels. Now, however, virtually all programmers make their channels available to the SMATV industry, directly or through the local cable franchise.

Economic Future

As the cost of earth station equipment dropped, SMATV became practical in relatively small complexes, and today, SMATV is viable in complexes with as few as 150 units. These systems generally provide five or more pay and advertiser-supported channels (such as HBO, ESPN, CNN, WTBS, and USA) for about $20 per individual subscriber or $10 per unit on a *bulk rate* (where the apartment management adds the service into the rent).

For many years the SMATV industry was made up of very small operators who generally lacked the financial, technical, and management resources to succeed. Recently, however, consolidation of systems has taken place, and larger players are emerging. Now that programming is widely available, and firms with expertise and funds are taking over the business, SMATV has become a serious competitor to cable in many markets. In 1987 there were about 500,000 SMATV subscribers.[20]

Cable operators, who have traditionally wired apartments with no payment to the owners, are now having to negotiate for the right to provide service. The success of SMATV in a given market often depends upon the inability of the cable operator to successfully deal with apartment owners.

A significant limitation of SMATV is the fact that no city or county streets may be crossed. This makes 35- to 50-channel systems impractical, since an expensive headend would be required at each location. SMATV operators look to AML microwave (see Chapter 2) as a way of serving many apartment complexes with 50 to 100 channels in direct competition with cable. Two-way services may also be provided in this type of system by using a reverse microwave path. AML licenses are granted only to cable operators at present. However, with the FCC's deregulatory attitude, it is likely that SMATV will, at some time in the future, be able to link apartments together using microwave.

A second method of interconnection uses infrared light links. It is possible to modulate light to carry several television channels for short distances. Since light wave communications are not regulated by the FCC, SMATV systems may be tied together without any governmental regulation. Such systems are limited to 1,000–2,000 feet at present, and the technology is experimental.

A final way of providing interconnection of SMATV systems is MMDS. With MMDS, ten or more channels could be broadcast to apartment buildings, where over-the-air channels could be added. In this way, a 15- or 20-channel service could be provided with a very low headend cost at each building.

TELEVISION BROADCAST TRANSLATOR AND LOW-POWER STATIONS

Definitions

A television broadcast *translator station* is "a station in the broadcast service operated for the purpose of retransmitting the programs and signals of a television broadcast station, without significantly altering any characteristic of the original signal other than its frequency and amplitude for the purpose of providing television reception to the general public."[21] A few translators are allowed limited origination of programming.

A *primary* television station provides programs and signals that are then retransmitted on a different TV channel by a translator station to extend a service area. Translator stations operate in both the UHF and VHF bands. These stations have been in existence for years, and there are thousands now serving rural areas at a distance from television markets.

Low-power television stations (LPTV) rely on virtually the same technology but are different in that they may originate programming or receive and retransmit satellite programs without any limitation. They may be commercial or noncommercial and may be STV stations. The FCC licenses LPTV stations by monthly lottery from among many thousands of applications. A construction permit is issued to the lottery winners. The station must be on the air one year later. If the station goes on the air and meets its technical requirements, it is licensed. Most never make it and some have failed after a few months. There were 409 LPTV stations on the air in 1987.[22]

Technology

Translators and LPTV stations are reduced-scale broadcast television systems, limited in power to fit into the present allocation scheme and table of assignments for higher-powered VHF and UHF broadcast stations. There is room for about 4,000 such stations. The quality of the equipment for low-power transmission and low-cost production has improved over the years to permit origination. The reliability of the transmitting equipment has also improved, minimizing the risk of interference. Since translators and LPTV stations are assigned frequencies already in use by VHF and UHF television broadcasters, interference must be minimized. It is possible for a low-power station to interfere with the operation of an adjacent or co-channel television station for many miles beyond its effective service area. It should be noted that translators and LPTV stations can interfere with a cable system, too. If a cable systems picks up a distant TV station over the air and a translator is put in the community on the same or an adjacent channel, it can interfere with the cable system's reception of that channel.

The service area and potential for interference are limited by power, antenna height, the type of transmitter used, and the quality of the receiving antenna installed by users of the service. Limits on power—10 watts for VHF and 1,000 watts for UHF—are placed on LPTV transmitters. Coverage, however, is determined by effective radiated power, which is multiplication of transmitter power (called *gain*) as a result of the transmitting antenna concentrating the signal. The coverage of VHF and UHF low-power stations at varying power and antenna heights is illustrated in Table 17–1, prepared by the CPB.[23]

The low-power station provides a *secondary* service on a frequency that is in use by a full-powered station. It must regulate its signal, if necessary, to protect the primary station. If interference cannot be prevented, the LPTV station must go off the air.

To retransmit another station's signal, low-power or translator stations may be used: a television translator relay station, a television intercity relay station, a television studio-transmitter link (STL) station, CARS (cable antenna relay service), common carrier microwave station, or satellite service. It appears that most of the applicants for LPTV licenses have in mind a satellite-interconnected network of many stations. LPTV used for STV will have the same security problems discussed in the STV section.

Programming

At this point it is not clear what type of programming will be most feasible for LPTV. Everything is possible. In the mid-1980s there were STV, music video, nonprofit educational, religious, and municipal stations. A few were commercial broadcasters with some local origination.[24]

Presumably, in urban areas with many broadcast television stations in service and cable bringing in more programming, LPTV would be narrowcast to special interests. In small town and rural areas, LPTV could supplement broadcast television station availability.

The FCC has deliberately kept low-power television stations free to try any kind of programming, with only a few restrictions. LPTV stations must comply with the reasonable access and the equal time requirements of the Communications Act of 1934. The same laws that prohibit obscenities and lotteries for broadcasters apply to LPTV. The copyright laws also apply to LPTV.

Economic Future

A 10-watt VHF transmitter systems costs about $15,000, while a 1,000-watt UHF systems is around $75,000. Usually, space is leased to install the transmitting antenna on an existing tower or building. For satellite-delivered programming, an earth station is required (approximately $5,000).

These capital investments would suffice for LPTV stations providing a satellite-delivered network, such as ESPN or CNN. It would cost a few thousand dollars more to automate the insertion of local commercials.

If the station were an LPTV-STV station, it would make another investment in decoders of about $150 per subscribing household (see discussion of STV). This would be the major capital expense. Total cost for transmitting equipment, site lease, and antenna is $100,000 to $150,000.[25]

Operating costs would depend on programming. Cost to the station of a pay television satellite network would be about $4 per subscriber per month; an advertiser-supported network, perhaps a few cents per household covered. Operating expenses for the transmitting equipment would be about $5,000 annually.[26] Lease of the tower space could also be a significant cost. Administration and marketing of a passive station not originating any programming could be as low as $50,000. Production costs would be entirely dependent on the scale and whether the production crew were full-time or contracted for on an ad hoc basis.

As translators have already done, LPTV stations can add programming options to small communities and rural areas underserved by television. The cost of stations designed to broadcast satellite services may be sufficiently low to cover these media-poor areas with one or more additional signals, providing the most popular programs, available on cable, to noncabled homes on a subscription basis.

In urban areas, LPTV may have difficulty in competition with conventional broadcast television and cable. But LPTV might find a niche that could be sustained by subscription, advertising, or a combination.

A major problem with low-power television is the circumscribed service area. In hilly areas, reception is limited to homes with a line-of-sight path to the transmitter. Similarly, buildings block reception in downtown areas. Reception with indoor antennas is usually limited to about 3–5 miles from the transmitter. Few residents, especially in cable areas, will invest in an outdoor antenna to receive LPTV broadcasts.

A few cable systems carry LPTV stations, although carriage is not required by the FCC. If the station has local origination of any consequence, it is difficult for the cable operator to ignore. If LPTV programming duplicates cable programming and/or channel capacity is limited, it is difficult for the cable operator not to ignore, hoping to drive the LPTV competition out of business. Some LPTV stations pay cable systems for carriage in cash and in local advertising availabilities that may be used for cable service promotion.

VIDEODISCS/CASSETTES

Definition

Videodiscs and videocassette machines come under the broad label of home video equipment—"electronic equipment that enables the consumer to record and/or play back sound and images on a standard TV set."[27] These machines provide a service that enables the home viewer to "play" programs on the television screen in the manner of records played on a phonograph.

The videocassette recorders (VCRs) and disc systems differ both in the nature of their software and in their technical capabilities. The videocassette machines can record television programming as well as play back existing cassettes. The VCR software medium is magnetic cassette tape of varying widths. Because the disc system cannot record, it must be used exclusively to play prerecorded material. The discs are manufactured with a protective coating that makes them almost indestructible.

The videodisc unit cannot now record or edit, capacities available on the videocasette recorder. The disc systems can support excellent stereo sound, good frame indexing, and fast random access, which makes the system desirable for industrial and educational uses.

The cassette recorders can record one program even while another program is being viewed, and can duplicate other tapes as well as edit existing program tapes. Although cassette tapes are reusable, they have a limited life.

Some playback-only machines are available for the consumer with a desire to use only rental or purchased tapes without ever recording.

Technology

The videocassette recorders read a magnetic tape. The original recorders were quadruplex (four-head) units, which were large, cumbersome, and suitable for commercial use only. Helical scanning, developed in Japan, enabled the machines to move from four-head readers to one head, and from 2-inch-wide tape to 3/4- to 1/2-inch tape. The cassette was marketed in 1972 with Sony's U-Matic, a 3/4-inch tape system wound in a cassette. In 1975 Sony produced the Betamax recorder and began the home video industry. The Betamax used 1/2-inch tape and two heads, which were oriented separately.

Competing with Beta is VHS which, although using the same technical components, has a slower speed. This increases the length of time available for recording. At first, Beta had significant advantages over VHS, including higher video quality and stero sound. Later, VHS technology improved, and today the market is dominated by the VHS format.

A new format, 8 millimeter, has emerged to be used for light-weight combination camera-recorders (*camcorders*). All manufacturers of 8-mm systems have agreed on a standard, thus assuring compatibility. The 8-mm format includes a very high-quality audio track that will allow excellent stereo reproduction.

The videodisc machine scans and plays encoded material from a disc by laser beam, which involves no physical contact with the disc and thus no scratching or wear in the grooves. Once encoded, the disc is protected in a plastic sleeve. The encoded signal is converted to electrical impulses, which are then modulated to a VHF frequency tunable on a television set. Playing time depends among other things on the rate at which the disc spins. Currently discs spin as fast as 1,800 revolutions per minute. The disc machine is the size and shape of an audio disc turntable. Discs are available in different sizes. A 12-inch disc can hold up to 60 minutes of programming on each side. Discs are mass produced by molding from metal dies that have been etched by a master disc. The frames are numbered and addressable. The discs have dual audio potential (for example, stereo sound).

Predictions of new capabilities for disc machines include automatic changers, smaller equipment with comparable capacity for playing time, and the capability to serve as a mass storage medium for home computers.

Programming

The range of potential programming is limitless, but at present the movie industry dominates the software market.[28] The average videocassette rental shop in 1986 had about 2,400 different titles and each "A" title is rented an average of 80 times.[29] By 1985, virtually all movies released by Hollywood were also released on videocassettes. Tape rental, through a large and growing number of retail outlets, has become a big business. In 1985, the movie industry's revenue from sale of prerecorded tapes was double its revenue from pay cable.[30] Videocassette rental revenues have now exceeded the domestic movie theater revenues.[31] Pornographic movies make up a significant percentage of tapes sold and rented. Information and "how to" cassettes are growing in importance.

In using cassette rentals, the television viewer has more personal control. A choice can be made from a huge catalog of titles. Playing time is the viewer's choice and intermissions can be taken as desired without missing anything.

Recording of programs off the air, called *timeshifting*, is a major use of VCRs. Because many recorders have remote controls, viewers can skip over, *zap*, commercials when playing back their tapes. As the penetration of VCRs increases, this is becoming of great concern to the advertising industry.

Economic Future

In 1986, a typical VCR cost under $400, with some units selling for $250 or less. Blank tapes sold for $4 or so, and prerecorded tapes of major films were priced at $20 to $80 with rental around $1 or $2 per day.

Videocassette recorders have become a standard item in many homes. By mid-1986, 38 percent of the homes in the U.S. had at least one VCR,[32] and industry leaders predict that eventually the penetration will reach 70 percent.[33] The rapid growth of VCRs caught both the cable industry and the movie industry by surprise, and the long-term implications are not fully understood.

Clearly, VCR tape rental and purchase will continue to be a major force in television and movies. VCR use takes time away from other media. Home VCR users are a different market from theatrical moviegoers, demographically. Usage may be stimulated by new product and by more sophisticated marketing techniques. Mass market and vending machine outlets may make access more convenient for the most popular titles. The home video industry will have increasing competition from pay-per-view cable. Despite the current movie industry revenues greatly favoring home video the industry is anxious to encourage PPV because of a general dislike for the *first sale doctrine*. This doctrine permits the film industry revenues *only* in the original sale of a video cassette to the video store; it does not share in the rental revenues.

The advertising industry has two major concerns as VCRs multiply. The viewers of commercial-free cassettes are likely to be displaced from the audience for advertiser-supported broadcast or cable television, and, as noted, commercials can be skipped in playback of recorded advertiser-supported programs.

VIDEOTEX AND HOME COMPUTERS

Definition

As discussed in Chapter 5, videotex is a service that interconnects a terminal located in a subscriber's home or business with a central computer to allow retrieval of data from or interaction with the central computer.

Presently, the telephone network is used to provide videotex service. Cable has the potential to transmit videotex data more economically than the phone network, and to allow higher speed of transmission than telephone lines, making possible services that intermix text and video.

Either a videotex user may have a terminal specially designed for the purpose, or a home computer may be used as a terminal. Relatively inexpensive software packages for the newer home computers can convert the computers to a videotex terminal.

Teletext is essentially a simulation of videotex in a one-way transmission system. Although more limited than videotex, it appears to function in the same way in accessing limited databases. The interactive functions, such as home banking or reservation services, are not possible in teletext.

Teletext decoders are much simpler than videotex. A complete decoder is now available on a set of integrated circuits (chips) for installation inside TV sets. These chips add about $25 to the cost of a television set, and the price is likely to drop.

Technology

A videotex terminal consists of a modem to interface it with the telephone line or cable system, a processing unit (which in reality is a small computer), and built-in software (which controls the operation of the terminal). A television set is used to display the information, and a small keyboard is used to input information. Videotex terminals cost around $600.

A home computer consists of a central processing unit (CPU, usually on a single chip on home computers), a keyboard, and a CRT (cathode ray tube, or monitor) or television screen. In addition, computers have two types of memory in which data may be stored. Temporary storage is in random access memory (RAM), which is made up of a group of chips. When the computer is turned off, the data in these chips is lost. A typical home computer has 64 to 256 *kilobytes* of RAM storage. (A byte is the equivalent of a single letter or number, and the prefix "kilo" means 1,000; 64 kilobytes is therefore 64,000 bytes.)

For permanent storage, a magnetic disk is used. The most common is the *floppy disk*, a 3- to 5-inch plastic disk that stores up to 1 million bytes of data. *Hard disks*, which store 10 million or more bytes, are becoming more common in personal computers as their prices drop.

A home computer system with a single floppy disk drive costs about $700, while a system with a hard disk and a large amount of RAM memory costs over $2,000. Prices for computer hardware have dropped dramatically in the last ten years, and the trend is expected to continue.

Programming

One of the difficulties in launching the home computer as a household utility has been the lack of software. It may cost as much as $50,000 to create a particular program. This cost must be spread over a great many users.

After a flurry of sales of very low-cost home computers in the early 1980s, sales dropped to a trickle. Very little practical programming had been developed, and the uses for computers in the home were limited.

At present, personal computers are used almost exclusively for business purposes. Those in use in homes are used to perform functions that relate to work rather than household activities. Word processing, spreadsheet creation (financial planning), and database creation and management are the most common uses of personal computers.

Videotex information "kiosks" in public places are expected to become quite common before 1990. The information accessed will be pertinent to the location of the kiosk, such as transportation, hotels, and sightseeing in airports, or shopping information in shopping malls.

Through telephone lines the home computer can access data banks of infinite variety. Many such databases are already available (as discussed in the section on videotex in Chapter 9). Several experiments are being conducted to test the market for such information.

Economic Future

Computer literacy and *user-friendly machines* are necessary to reach the market beyond business users. In addition, programs will have to be developed to provide real benefit within the home. Few people are adept at using computers, and some have a fear of the computer technology. Home computer hardware and software must be designed to be more compatible with inexperienced users. A major problem in accessing information databases is the cumbersome menu selection process.

Videotex system operators have had a great deal of trouble finding services for which people are willing to pay. For a service to have economic utility, it must either be more convenient to use, more timely, or more specialized than other information sources. Newspapers and television news provide a large amount of information in a very convenient form.

Before videotex services become widespread, several problems must be solved. First, the cost of the hardware must drop (the use of already existing home computers will help). Secondly, a more effective means of locating the desired information must be found. The present method, using a series of menus, is tedious and cumbersome. Finally, operators must find services that appeal to a large segment of the population.

The use of cable to provide videotex service, or to interconnect home computers, is some time in the future. Cable systems must have operational two-way capability, and must be maintained to a high degree of reliability. Modems must be developed that are as reliable and inexpensive as those available for telephone lines. Finally, cable systems will have to serve a higher percentage of the homes in the country, and systems will have to be interconnected in order to allow a market-wide videotex service. Once videotex is successful over phone lines, it is likely that cable videotex will follow.[34]

SOCIAL IMPACT

Home Communication Environment

All of the technologies described so far will expand the availability of television programming. They will have the effect of equalizing the availability of choice in television across geographical areas, bringing a large diversity to urban, suburban, and rural sections of the country.

SMATV will provide an alternative to cable in urban markets, perhaps creating parallel competition and thereby keeping rates down and improving quality. SMATV will have little impact outside of major television markets, however.

MMDS and DBS will expand multichannel video programming to areas not served by cable; downtown areas where cable is too expensive to install, or areas on the fringe of cable franchises where homes are too remote for cable service.

The videocassette or videodisc machines have the potential to equalize television programming availability, except for current affairs programming. Presumably, everything that is now available on television, over the air, and via cable or satellites (that is not dependent for its value on immediacy) could be available on cassettes. Thus, places without multichannel technologies could have almost everything that is available in the television-rich environments, except the most perishable of the content.

Great stores of information are available to the home computer or videotex terminal owner through telecommunication linkages. These devices working with other computers and databases will supplement printed reference resources with more up-to-date information.

For communities without extensive reference resources, the home computer or videotex/teletext makes these resources more accessible. For communities where the resources are available, access is much more convenient.

As more video technologies become available, consumers become indifferent to the distribution medium. The programming material or software is paramount. The distribution systems may vary across media environments (from small towns to large cities) but demand for some types of entertainment and information may be constant. The mix of existing technologies may also vary from household to household to accommodate individual needs and interests.

Beyond the effects of new communication technologies on the numbers of communication channels and the amount of entertainment in the media environment lies the larger question of social impact. We have suggested at the outset that these technologies are not likely to be revolutionary in their influence on the human experience. None of the "revolutionary" communication advances since the Guttenberg press has had the predicted impact. But each technological change (high-speed presses, telegraph, telephone, color printing, radio, television, and the like) has made differences in the way in which people interact with one another and their world environment. It is important to contemplate the differences that may result from the newest wave of technology in the continuing effort to understand our society.

Impact on Children

Children are perhaps most vulnerable to communication impact, because they are relatively unformed, but also because they are more accepting of new technologies. We know a little about how children react to some of the new technologies. We can make some educated speculation about other effects based on a knowledge of what new programs and services will be available.

The especially heavy television viewing of cable households may be a result of the slightly larger size of the cable household as suggested in Chapter 10. The fact that cable subscribers have to be especially interested in television and its content to begin with must also contribute. The new television menu available to the cable subscriber encourages even more viewing. Children are participating in this additional viewing.

Most people believe that parents have some responsibility to teach children how to use television to maximize its benefits and minimize presumed negative effects of particular programs or too much viewing. Some also believe that the schools have a responsibility here. Fully developed, multiple-channel, two-way cable greatly complicates the task. There is much more to select from and much more to avoid. Assigning priorities to the available material is made more difficult by the lack, in most cable systems, of adequate program guides and descriptions. Where the cable operator, aided by program suppliers, does provide a full program guide, the programs are "promoted" rather than described. Perhaps various types of guides, developed independently of the industry, will supply an alternate, more critical perspective, helping children and adults to make more rational choices.

Psychiatrists and psychologists have long recommended family viewing of television as a means of helping children to adjust to the world presented by television. The presence of adults makes some of television less fearful. Dramatic license (whether in portraying violence or sex and racial and occupational stereotypes) may be discounted by parental comments. All the content can be explained and discussed. But one can speculate what cable television and VCRs will do to family viewing. On separate sets young children may be viewing Nickelodeon, and an older group of children may watch "Ghostbusters" on a pay channel or VCR. Their mother may be viewing a talk show on Lifetime, while the father is engrossed in ESPN, and the grandparents at still another set watching CBN. This is, of course, an extreme, but cable television *fragments* audiences, even within a household. It could have a negative effect on family communication patterns and parenting. As noted in Chapter 10, there is less parental mediation of child television viewing in cable households.

Children will be exposed to more sex on television. Where parental discipline is used to prevent viewing of these programs, it is likely to break down occasionally with even more excitement over the special permission or clandestine viewing. Broadcast television, lest its programming seems bland by comparison to cable and videocassette, is likely to be forced into further liberalities in programming.

Little is known about the effects of mediated sexual intimacy on children. Exposure to such content no doubt predates formal methods of teaching children about sex. There is no question that one of the uses of sexually explicit media and pornography is to stimulate sexual arousal. It should certainly have this effect on postpubescent teens.

On cable there is wider availability of programming attractive to children in all day parts. In the broadcast television world, the child must accommodate, most of the time, to programming aimed primarily at adults. With their greater buying power, the adults are more valuable to advertisers. Cable provides programs produced for children throughout the day and adds greatly to the number of choices of off-network, syndicated programming carried by the broadcast stations available on cable. Among these programs are heavy child favorites (such as "Happy Days"). Movies are also appealing to children. Some of the pay channels run a number of G movies that are essentially child fare.

Commercial broadcast television was never able to offer much *age-specific* programming to children. Cable networks and local origination can be addressed to narrow audiences, making the programming more appropriate to those audiences, be they preschoolers or adolescent rock music lovers.

Two-way participatory television could be very good for kids, making better use of their energy than more passive viewing. Children may take special delight in manipulating images, as in TV games, in testing themselves in quiz programs, in voting along with others, or in ordering their own specialized information through videotex. Instructional programs such as "Sesame Street" could be more engrossing in the two-way mode, and the learning enhanced through reinforcement. On the other hand, children could become more prone to interact with television than with peers or adults, delaying some kinds of social learning.

Traditionalists may view the personal computer, linked to reference resources through phone or cable lines, as a crutch for children, which may prevent their familiarity with conventional library research methods. Children may be less motivated to learn traditional methods in the knowledge that many of the resources are available at the home keyboard and screen. Nevertheless, the child must learn both the traditional and new methods to take advantage of the breadth of research resources available.

Children seem to thrive on an identification with school and community. To the extent that cable television can provide coverage of schools and neighborhoods within metropolitan areas, urban children should have the advantage of smaller town children where the media routinely cover events, sports, and people from their own environment.

General Impact

The emergence and acceptance of cable and VCRs may be quite disturbing to some people. We know that older people are resistant to cable and sometimes excessively negative, suggesting suspiciousness, a vague fear of the new medium, and a clear unwillingness to make the change. Learning to use cable to the fullest advantage may be more difficult for adults than for children. Adults are less comfortable with electronic gadgetry of MTS receivers, VCRs, and cable converters than are children of the present generation. Adults are somewhat set in their habits, having practiced them longer. They may experience some "future shock" in adapting.

Most of the 40-year-old social concerns about television are magnified by the prospect of *more* television via cable and VCRs. Most viewing will be passive. Viewers may give up even more of their leisure time to the enticements of popular culture. On the other hand, more channel space, the opportunity for the expression of intensity of demand through subscription fees or pay-per-view options, and the wider field for talent development may raise the general level of television and make a place for the highest levels of entertainment and information for those who demand it. At this level of quality, it may be argued that television can be as intellectually engaging as any of the arts and, therefore, *not passive*.

If the viewer chooses, commercials on television may be entirely avoided by zapping or using public broadcasting stations, a few basic cable networks, pay networks, local cable channels, and videocassettes.

In the worst of the "1984" scenarios, the home with interactive broadband communication capability is self-contained. Its members need not leave the premises for most activities. Artists have drawn shocking pictures of the human being evolving from such an environment—enlarged head, popping eyes, tiny body. At the other extreme, one might take another view of this development as essentially *liberating*. The household may be freed from some of the demeaning commutes to a distant workplace, some routine shopping trips, and noisy classrooms. Working couples might be together for more time. Time saved may be used to develop more satisfactory social contact.

Television may come to serve a broader purpose with its variety of formats. It is probable that video programming will be used as "background" as well as "foreground" in audience attentiveness. In this regard television may take over some of the function of radio. Certain programs—"Today," "Good Morning America," "Tonight"—are designed with an otherwise occupied audience in mind. Cable adds significantly to the kind of programming that might be on for long periods of time while engaging the attention only occasionally—

CNN, the video music channels, weather, adult entertainment, public access programming, minor sports, gavel-to-gavel coverage of meetings, second and third viewings of movies, and religious networks may all be examples. This is perhaps a valuable function of television.

The new technologies, computer and cable, along with the larger channel capacity and capability for narrowcasting, vastly expand the information available to the home. The home can have the capacity of a public library, and more. Information and news can be very local, giving even urban residents the neighborhood news on demand or on a regular schedule. This could bring local government, institutions, and people closer together. A better identification with the immediate community may enhance community pride and reduce alienation.

While new technologies can localize information more practically than can the old, they may also divert attention from the local news. We reported earlier the tendency of local audiences to be seduced from their local outlets by slicker, more dramatic programming from larger metropolitan areas. Attendance to news material, electronic or print, may be a ritual act built on an habituated appetite for political intrigue, crime, scandal, and human interest features, all of which repeat endlessly in the news. Time and place may not be very important.[35] The best play on these interests may capture the audience. Local programming may have the least of the resources, in terms of the actual events and in gathering and producing the news, to attract these interests. If diversion from local news to a more exciting and fulfilling compendium of news from a wider area takes place, some of the potential for a more informed, less alienated local citizen is lost.

Information may be tailored better to individual interests. Through videotex, a unique budget of information may be ordered by each household. This may greatly improve the efficiency of information seeking. At the same time it reduces the opportunity for random exposure to information, as we now have in broadcast and print news, that may broaden viewer interests. It can be hoped that individuals will plan exposure to general information as well as to highly specific news to meet specialized interests.

Whatever the case, new access to communication channels of much greater content diversity permits a much more individualized exposure. The media user is no longer *forced* into the entertainment and information formulas that have been shaped by mass tastes, limited channels, and the commercial system. The size of the "mass" audience may be greatly reduced; the concept of "mass" media may be less meaningful.

With the diverse media opportunities and the resultant highly individualized media exposure, we lose the common media experience that was once dictated by the system and may have had significant values related to community and national unity. Despite diversity, there will always be media events that will draw community and nation together both for serious and for the less serious entertainment purposes. We will still have the opportunity to develop national symbols and a common ground for communication—through crisis news, blockbuster movies that both theater and home audiences view, and popular entertainers capable of finding a common denominator.

The advantages of the new communication technologies will not be shared equally across all members of the society. Almost all the new services bear a consumer cost, much beyond the cost of advertiser-supported broadcasting and print media. The full range of services—for example, the top tier of cable, VCRs and tape libraries, videotex and interactive games, depending on usage—could cost $50 and more monthly per household. Low-income persons will be excluded from the full benefit, thus widening the information gap between low-income and high-income people. At the lowest cost level of cable services (basic), much of the news and information services will be provided, principally advertiser-supported. It is in the higher levels of entertainment and in interactive information retrieval where the poor are most clearly disadvantaged. It is possible that some sort of "information subsidy" may be provided by government if the balance shifts too far in the direction of the high-income households.

Entertainment television for low-income people may be all second run. People who can afford full-service cable and VCRs will have first access to movies and sports. Even popular series programming may get to broadcasting only as syndicated shows after cable runs.

The new costs of media and communication services will result in a major income redistribution for everyone. In just a few years, communication costs will rise from an inconsequential to a significant proportion of the household budget.

In the optimistic view, the new communication technologies project a net social gain—in quality of entertainment, in the cultural level, in convenience, in education, in knowledge of public affairs, and in freedom of choice. There are, however, economic and perhaps social costs that must be monitored to maintain that gain.

NOTES

[1]Some of the material in this section was developed under Grant No. DAR 7910614 from the National Science Foundation. Thomas F. Baldwin, John D. Abel, and Richard V. Ducey, "The Media Environment: Consumption and Function of Media Under Different Conditions of Access to Media in Isolated Communities, Small Town, Medium Cities and Metropolitan Areas," Michigan State University. A portion of this chapter was written in collaboration with Richard V. Ducey.

[2]For a basic background in the technical characteristics of the media systems discussed in this chapter, see Geroge F. Whitehouse, *Understanding the New Technologies of the Mass Media*, Englewood Cliffs, N.J.: Prentice-Hall, 1986, pp. 190.

[3]International Telecommunication Union, *Final Protocol: Space Telecommunication* (197), 23 N.S.T. 1573–1574, July 17, 1971.

[4]David Bollier, "There's Life After Scrambling," *Channels 1987 Field Guide*, p. 66.

[5]David M. Rice, "Direct Broadcast Satellites: Legal and Policy Options," Communication Media Law Center of New York University Law School, prepared for the FCC Network Inquiry Staff, Washington, D.C. 1979.

[6]47 C.F.R. 73.641 (a) and (b).

[7]The case regarding pay cable was *Home Box Office v. FCC*, 567 F. 2d 9 (D.C. Cir., 1977), cert. denied 434 U.S. 829 (1977).

[8]Paul Kagan, *The Pay TV Newsletter*, October 31, 1985, p. 5.

[9]Kristin Booth Glen, *Report on Multipoint Distribution Service*, prepared for the Network Inquiry of the FCC, Washington, D.C., November 1979, p. 1.

[10]Glen, *Report on Multipoint Distribution Service*, p. 64.

[11]Glen, *Report on Multipoint Distribution Service*, p. 71.

[12]47 C.F.R. 21.904.

[13]Glen, *Report on Multipoint Distribution Service*, p. 64.

[14]Glen, *Report on Multipoint Distribution Service*, p. 65.

[15]Glen, *Report on Multipoint Distribution Service*, p. 70.

[16]47 C.F.R. 21.903 (b) (2).

[17]Glen, *Report on Multipoint Distribution Service*, p. 104.

[18]47 C.F.R. 21.903 (b) (2).

[19]J.L. Freeman, "FCC Proposes Easing Regulations Over MDS," *Multichannel News*, June 2, 1986, p. 6.

[20]_____, "SMATV," *Channels 1987 Field Guide*, p. 70.

[21]47 C.F.R. 74.701 (a).

[22]"Summary of Broadcasting as of April 30, 1987," *Broadcasting*, July 6, 1987.

[23]"Low-Power Television Guidebook," Corporation for Public Broadcasting.

[24]Jill Marks, "What Happened to LPTV," *Cable Television Business*, December 1, 1987, p. 44.

[25]"FCC's Grass Roots Ploy Sprouts Confusion Over TV's Future," *CableVision*, April 20, 1981, p. 65.

[26]"Comments of the Association of Maximum Service Telecasters," in FCC Docket No. 78–253, January 10, 1979.

[27]Shiela Máhony, Nick Demartino, and Robert Stengle, "The Home Video Market," in *Keeping Pace with the New Television: Public Television and Changing Technology*, New York: The Carnegie Corporation of New York, VNY Books International, 1980. This section was originally written with Janet Bridges.

[28]Agostino et al., *Home Video*, p. 50–51.

[29]_____, "1986 VSDA Annual Survey," Video Software Dealers Association.

[30]Paul Kagan, *The Pay TV Newsletter*, November 30, 1985, p. 2.

[31]_____, "Tape Me, Tape Me," *Orbit*, January 1987, p. 28.

[32]_____, "VCR Penetration Hits 38 Percent," *Multichannel News*, (reported from Arbitron, July 1986), October 13, 1986, p. 46.

[33]David Lachenbruch, "The Makers' Lament: Not-So-Fast Forward," *Channels 1987 Field Guide*, p. 88.

[34]The new communication technologies are updated in *Channels of Communications*, annual "Field Guides." Subscription Service Dept., Box 2001, Mahopac, NY 10541.

[33]This concept, called the "ludenic theory of newsreading," is developed in William Stephenson, *The Play Theory of Mass Communications*, University of Chicago Press, 1963, pp. 147–159.

Sample
Public Access Channel Rules

UNITED CABLE TELEVISION OF
MID-MICHIGAN PUBLIC ACCESS CHANNEL RULES
East Lansing and Meridian Township

I. PREAMBLE

1.1 The purpose of these Public Access Rules is to clearly define the rights and responsibilities of the United Cable Television of Mid-Michigan (UNITED CABLE) and the applicant in the use of Public Access facilities provided by United Cable. The primary purpose of these facilities is to encourage East Lansing and Meridian Twp. residents to take the opportunity provided by Public Access in the production of localized television programming.

II. APPLICATIONS

2.1 A user will be defined as any East Lansing or Meridian Twp. resident applying for use of the Public Access production facilities.

2.2 Any individual or group may use the access channel first-come, first-serve, and non-discriminatory.

2.3 If the applicant is under 18 years of age s/he must have an adult co-sign the application form and agreement. The co-signer is then responsible, along with the applicant for any financial responsibility connected with the use of the Company's facilities other than normal wear and tear.

2.4 Any user charged with one period of studio use during one week shall not be prevented from assisting in other studio productions during the same week.

III. USER RESPONSIBILITY

3.1 The applicant assumes full responsibility for use of United Cable production facilities, other than normal wear and tear.

3.2 Under the terms of the Public Access Channels Usage Agreement, the public access producer assumes all liability for program content and agrees to idemnify United Cable, the City of East Lansing, the Charter Township of Meridian and its representatives for all liability or other injury due to program content.

3.3 Persons utilizing the Public Access production facilities will, at all times, be under the supervision and authority of United Cable Public Access Coordinators. All handling of the facilities or granting of authority to do so will be done by the Public Access Coordinator in charge at the time of the production. Users of the channel may not present any material designed to promote the sale of commercial products or services. This includes any advertising by, or on the behalf of, candidates for public office.

3.4 The applicant must be sure that, if the use of music and non-music copyrighted material is involved, the appropriate copyright clearances have been obtained. Before presenting a program, s/he must have signed the Public Access Channels Usage Agreement (Appendix A).

3.5 Regarding the use of studio facilities the following stipulations will be observed:

a. Users are requested to arrive at least 15 minutes before their scheduled times of appearance.

b. Users and other participants are asked to keep the control room clear of non-essential personnel during the production.

c. Technical help and users shall not smoke, snack, or bring drinks into the control room.

d. Users and other participants may not interfere with the production and/or studio usage time of another user.

3.6 United Cable reserves the right to temporarily refuse the use of the Access Channel and facilities to any person under the influence of alcohol, drugs or otherwise not under full control of his/her senses.

3.7 Any individual/program is responsible for canceling reserved studio/porta-pac time if that time will not be used. All public access users are also expected to be prompt for their reserved activities, a grace period of 15 minutes from the scheduled time will be allowed afterwhich the studio/porta-pak time will be given to other users. Any individual/program that does not cancel scheduled time will be kept record of, after three such violations the program/individual will be put on a 60-day probation during which if another violation occurs a 30-day suspension from public access will be authorized.

3.8 Removal of United Cable videotapes off the premises is strictly prohibited (except with use of porta-paks in the field). If absolutely necessary to sign them out full cost of tape is required as a deposit.

IV. UNITED CABLE TELEVISION OF MID-MICHIGAN's RESPONSIBILITY

4.1 Basically, there are two ways to provide programming in order to utilize public access time. A program may be produced by using United Cable's portable facilities, or studios; or, a prerecorded tape (or film) may be supplied to United Cable. In either case, the tape (or film) is then scheduled for showing on the designated Public Access Channel as outlined in the "Public Access Schedule Procedure."

4.2 United Cable will provide a qualified person to offer technical and programming assistance to channel users in order to assume optimum technical quality.

4.3 Applications to use the Public Access Channel for the showing of a prerecorded program is outlined in the section designated "Public Access Schedule Procedure." The Company will not edit, or alter in any way, the content of Public Access material without permission of the user. The necessity of duplication or any other type of alteration will be discussed with Public Access Coordinator when application is made.

4.4 United Cable will keep for public inspection all applications for use of the Public Access Channel and a complete record of the names and addresses of all persons or groups who request access time. The company will retain all records for a period of two years.

V. FACILITIES MADE AVAILABLE BY UNITED CABLE

5.1 All facilities will be offered on a first-come, first-serve basis. Public Access facilities are to be used only in the production of Public Access programming.

5.2 For remote productions, "Porta-Pak" type equipment and the necessary supplemental production gear will be supplied free of charge to any applicant qualifying under the terms of these Public Access Rules on 24 hours notice for a period not to exceed 24 hours. Facilities will also be made available for electronic editing of the taped footage produced with the Porta-Pak.

5.3 In cases where studio production facilities are required, application should be made at least one week in advance, for the convenience of the user. This time limitation may be excepted in cases of public interest.

5.4 United Cable Public Access Production Studio may be reserved one week in advance, by applicants for one continuous 90-minute period of time on any given work day free of charge.

5.5 Each finished program produced/recorded using the public access facilities will be allowed one free dubbing (copying) per episode, afterwards a charge of $10 per hour (one-hour minimum required) will be levied. The dubbing of an episode must be approved by the program producer and time scheduled through a public access coordinator.

VI. PUBLIC ACCESS SCHEDULE PROCEDURE

6.1 First, all users must sign the Public Access Channel Usage Agreement, included as Appendix A of these rules. Producers of live public access programs must complete the Public Access Channel Use Agreement prior to commencement of cablecasting.

6.2 Access programming will normally be videotaped; except in order to achieve the timeliness of a program. United Cable may permit live presentations.

6.3 There are no theoretical limitations imposed on the running time of any Public Access program. There are, however, several practical and logistical factors which may dictate the limitations on program length. These limitations may include:

a. Public Access production facility availability

b. Channel time availability

6.4 United Cable will deviate from the established telecast schedule only upon its approval of written request, by a group or individual affected by the deviation. It is not the intent of the Company to herein inhibit or restrict the use of the Access Channel, but to insure in as fair a manner as possible, that all persons and groups wishing to use the Public Access facilities have an equal and fair opportunity to take advantage of communication potentials in Public Access. The utilization in a monopolistic manner by one or few select groups or individuals is not deemed to be in the public interest. Channel use limitations will be applied only in cases where the public interest is not being maintained.

6.5 United Cable resumes responsibility for rescheduling programming which is delayed or interrupted for a duration of 10% of its total time, if such delays or interruptions are beyond control of the user.

VII. REGULATION OF OPERATION

7.1 United Cable reserves the right to waive any self-imposed regulation when such waiver is judged by the Company to be in the public interest.

7.2 Any violation of these rules may, at the Company's discretion, cause United Cable to withhold the use of its facilities from the violator.

7.3 All applicants for use of United Cable facilities should be aware that they may be held accountable for their actions by the same laws that govern any public activity.

7.4 Any violation of an access volunteer's right to use the facilities in accordance with these Public Access Rules will be reported to the East Lansing or Meridian Township Cable Communications Commission for appropriate action.

VIII. MISCELLANEOUS

8.1 As experience shows a need, these Rules shall be subject to periodic revision, upon approval of United Cable, the East Lansing and Meridian Township Cable Communications Commissions.

8.2 The Company will have available, information regarding services offered to Public Access users by the East Lansing and Meridian Township Cable Communications Commissions. This information will be prepared and supplied by the Commission.

PUBLIC ACCESS CHANNEL USE AGREEMENT

Name: _____

Phone: _____

Address: _____

_____

Program Title: _____

Length: _____

Comment: _____

REQUESTED DATE OF PLAYBACK:

Date: _____   Time: _____

Date: _____   Time: _____

May we schedule your program at additional times?

    YES _____   NO _____

FORMAT

Tape: 1/2″ RR _____   3/4″ Cassette _____

Betamax _____   Live Studio _____

"Applicant" herewith applies to United Cable Television of Mid-Michigan (United Cable) for use of the designated public access channel on the following terms and conditions:

1. No charge shall be made for the use of United Cable's public access channel.

2. Applicant will not cablecast any advertising material designed to promote the sale of commercial products or services, including advertising by and on behalf of candidates for public office.

3. Applicant will not cablecast a lottery or any advertisement of or information concerning a lottery.

4. Applicant will not cablecast any obscene or indecent material.

5. Applicant agrees to make all appropriate arrangements with, and to obtain all clearances from broadcast stations, networks, sponsors, without limitation from the foregoing, any and all other persons (natural and otherwise) as may be necessary to transmit its program material over the Company's cable television system.

6. In recognition of the fact that the company has no control over the content of the Applicant's public access cablecast, Applicant agrees to indemnify and hold the Company harmless from any and all liability or other injury (including reasonable costs of the defending claims or litigations) arising from or in connection with claims for failure to comply with any applicable laws, rules regulations, or other requirements of local, state or federal authorities; for claims of libel, slander, invasion of privacy, or infringement of common law or statutory copyright; for unauthorized use of trademark, trade name, or service mark; for breach of contractual or other obligations owing to third parties by company; and for any other injury or damage in law or equity which claims result from the Applicant's use of the United Cable designated public access channel.

7. Applicant recognizes that the Federal Communications Commission requires Company to maintain available for public inspection a record of all persons applying for use of designated public access channel, and agrees that this application may be used for such record.

8. Applicant states that he/she has read Company's "Public Access Rules" governing use of cable public access channel and agrees to abide by the terms and conditions contained therein.

_____

Signature

_____

UNITED CABLE REPRESENTATIVE

_____

Date

REQUEST FOR PORTABLE EQUIPMENT

Name _____

Today's Date _____

Address _____

Phone _____

Driver's License Number _____

For How Long? From _____ To _____
                        (Date)   (Time)   (Date)   (Time)

EQUIPMENT REQUESTED

Porta-Pac Kit # _____   _____ AC Power Adaptor

_____ Video Tape Recorder   _____ 3-to-2 prong AC Adaptor

_____ Carrying Case   _____ Ear Speaker

_____ Internal Battery Pack   _____ AC Extension Cord

_____ Rf Unit & Cable   _____ EV-635A Microphone w/cable

_____ Tripod   _____ Video Camera w/zoom lens

Portable Lighting Kit # _____

_____ Lighting Instruments   _____ Spare Lamps

_____ Stands   _____ Barn Doors

_____ Power Cords

Videotape

_____ 1/2-Hour # _____

_____ 1-Hour #   _____

Other _____

_____

_____

I agree to assume complete financial responsibility for use of United Cable's equipment, normal wear and tear excepted.

Signature _____

Checked out by _____ Date _____

Checked in by _____ Date _____

FCC Local Origination Rules

76.205 ORIGINATION CABLECASTS BY CANDIDATES FOR PUBLIC OFFICE

(a) *General requirements.* If a cable television system operator shall permit any legally qualified candidate for public office to use the system's origination channel(s) and facilities therefore, the system operator shall afford equal opportunities to all other such candidates for that office: *provided, however,* that such cable television system operator shall have no power of censorship over the material cablecast by any such candidate, *and provided, further,* that an appearance by a legally qualified candidate on any

(1)   bona fide newscast,

(2)   bona fide interview,

(3)   bona fide news documentary (if the appearance of the candidate is incidental to the presentation of the subject or subjects covered by the news documentary), or

(4)   on-the-spot coverage of bona fide news events (including but not limited to political conventions and activites incidental thereto) shall not be deemed to be use of the facilities of the system within the meaning of this paragraph.

(b) *Rates and practices.* (1) The rates, if any, charged all such candidates for the same office shall be uniform, shall not be rebated by any means direct or indirect, and shall not exceed the charges made for comparable origination use of such facilities for other purposes.

(2) In making facilites available to candidates for public office, no cable television system operator shall make any discrimination between candidates in charges, practices, regulations, facilities, or services for or in connection with the service rendered or make or give any preference to any candidate for public office or subject any such candidate to any prejudice or disadvantage; nor shall any cable television system operator make any contract or other agreement which shall have the effect of permitting any legally qualified candidate for any public office to cablecast to the exclusion of other legally qualified candidates for the same public office.

(c) *Records, inspections.* Every cable television system operator shall keep and permit public inspection of a complete record of all requests, for origination cablecasting time made by or on behalf of candidates for public office, together with an appropriate notation showing the disposition made by the operator of such requests, the charges made, if any, and the length and time of cablecast, if the request is granted. Such records shall be retained for a period of two years.

(d) *Time of request.* A request for equal opportunities for use of the origination channel(s) must be submitted to the cable television system operator within one (1) week of the day on which the first prior use, giving rise to the right of equal opportunities, occurred, provided, however, that, where a person was not a candidate at the time of such first prior use, he shall submit his request within one (1) week of the first subsequent use after he has become a legally qualified candidate for the office in question.

(e) *Burden of proof.* A candidate requesting such equal opportunities of the cable television system shall have the burden of proving that he and his opponent are legally qualified candidates for the same public office.

76.209 PERSONAL ATTACKS; POLITICAL EDITORIALS

Section (a), Fairness Doctrine, has now been deleted, (b), (c) and (d) still apply.

(b) When, during origination cablecasting, an attack is made upon the honesty, character, integrity, or like personal qualities of an identified person or group, the cable television system operator shall, within a reasonable time and in no event later than one (1) week after the attack, transmit to the person or group attacked (1) notification of the date, time, and identification of the cablecast; (2) a script or tape (or an accurate summary if a script or tape is not available) of the attack; and (3) an offer of a reasonable opportunity to respond over the system's facilities.

(c) The provisions of paragraph (b) of this section shall not be applicable (1) to attacks on foreign groups or foreign public figures; (2) to personal attacks which are made by legally qualified candidates, their authorized spokesmen, or those associated with them in the campaign, on other such candidates, their authorized spokesmen, or persons associated with the candidates in the campaign; and (3) to bona fide newscasts, bona fide news interviews, and on-the-spot coverage of a bona fide news event (including commentary or analysis contained in the foregoing programs, but the provisions of paragraph (b) of this section shall be applicable to editorials of the cable television system operator).

(d) Where a cable television system operator, in an editorial, (1) endorses or (2) opposes a legally qualified candidate or candidates, the system operator shall, within 24 hours of the editorial, transmit to respectively (i) the other qualified candidate or candidates for the same office, or (ii) the candidate opposed in the editorial, (1) notification of the date, time, and channel of the editorial; (2) a script or tape of the editorial; and (3) an offer of a reasonable opportunity for a candidate or spokesman of the candidate to respond over the system's facilities, *provided, however,* that where such editorials are cablecast within 72 hours prior to the day of the election, the system operator shall comply with the provisions of this paragraph sufficiently far in advance of the broadcast to enable the candidate or candidates to have a reasonable opportunity to prepare a response and to present it in a timely fashion.

76.213 LOTTERIES

(a) No cable television system operator, except as in paragraph (c), when engaged in origination cablecasting shall transmit or permit to be transmitted on the origination cablecasting channel or channels any advertisement of or information concerning any lottery, gift enterprise, or similar scheme, offering prizes dependent in whole or in part upon lot or chance, or any list of the prizes drawn or awarded by means of any such lottery, gift enterprise, or scheme, whether said list contains any part or all of such prizes.

(b) The determination whether a particular program comes within the provisions of paragraph (a) of this section depends on the facts of each case. However, the commission will in any event consider that a program comes within the provisions of paragraph (a) of this section if in connection with such program a prize consisting of money or thing of value is awarded to any person whose selection is dependent in whole or in part upon lot or chance, if as a condition of winning or competing for such prize, such winner or winners are required to furnish any money or thing of value or are required to have in their possession any product sold, manufactured, furnished, or distributed by a sponsor of a program cablecast on the system in question.

(c) The provisions of paragraphs (a) and (b) of this section shall not apply to advertisements or lists of prizes or information concerning a lottery conducted by a state acting under the authority of state law when such information is transmitted (1) by a cable system located in that state, (2) by a cable system located in an adjacent state which also conducts such a lottery, or (3) by a cable system located in another state which is integrated with a cable system described in (1) or (2) herein, if termination of the receipt of such transmission by the cable system in such other state would be technically infeasible.

(d) For the purposes of paragraph (c) "lottery" means the pooling of proceeds derived from the sale of tickets or chances and allotting those proceeds or parts therof by chance to one or more chance takers or ticket purchasers. It does not include the placing or accepting of bets or wagers on sporting events or contests.

76.221 SPONSORSHIP IDENTIFICATION, LIST RETENTION, RELATED REQUIREMENTS

(a) When a cable television system operator engaged in origination cablecasting presents any matter for which money, service, or other valuable consideration is either directly or indirectly paid or promised to, or charged or accepted by such operator, the operator, at the time of the cablecast, shall announce (1) that such matter is sponsored, paid for, or furnished, either in whole or in part, and (2) by whom or on whose behalf such consideration was supplied, provided, however, that "service or other valuable consideration" shall not include any service or property furnished either without or at a nominal charge for use on, or in connection with, a cablecast unless it is so furnished in consideration for an identification of any person, product, service, trademark, or brand name beyond an identification reasonably related to the use of such service or property on the cablecast.

For the purposes of this section, the term "sponsored" shall be deemed to have the same meaning as "paid for."

(b) Each cable television operator engaged in origination cablecasting shall exercise reasonable diligence to obtain from employees, and from other persons with whom the system operator deals directly in connection with any matter for cablecasting, information to enable such system operator to make the announcement required by this section.

(c) In the case of any political origination cablecast matter or any origination cablecast matter involving the discussion of public controversial issues for which any film, record, transcription, talent, script, or other material or service of any kind is furnished, either directly or indirectly, to a cable television system operator as an inducement for cablecasting such matter, an announcement shall be made both at the beginning and conclusion of such cablecast on which such material or service is used that such film, record, transcription, talent script, or other material or service has been furnished to such operator in connection with the transmission of such cablecast matter, provided, however, that in the case of any cablecast of 5 minutes' duration or less, only one such announcement need be made either at the beginning or conclusion of the cablecast.

(d) The announcement required by this section shall, in addition to stating the fact that the origination cablecasting matter was sponsored, paid for or furnished, fully and fairly disclose the true identity of the person or persons or corporation, committee, association or other unincorporated group, or other entity by whom or on whose behalf such payment is made or promised, or from whom or on whose behalf such services or other valuable consideration is received, or by whom the material or services referred to in paragraph (c) of this section are furnished. Where an agent or other person or entity contracts or otherwise makes arrangements with a cable television system operator on behalf of another, and such fact is known or by the exercise of reasonable diligence, as specified in paragraph (b) of this section, could be known to the system operator, the announcement shall disclose the identity of the person or persons or entity on whose behalf such agent is acting instead of the name of such agent. Where the origination cablecasting material is political matter or matter involving the discussion of a controversial issue of public importance and a corporation, committee, association or other unincorporated group, or other entity is paying for or furnishing the matter, the system operator shall, in addition to making the announcement required by this section, require that a list of the chief executive officers or members of the executive committee or of the board of directors of the corporation, committee, association or other unincorporated group, or other entity shall be made available for public inspection at the local office of the system. Such lists shall be kept and made available for a period of two years.

(e) In the case of origination cablecast matter advertising commercial products or services, an announcement stating the sponsor's corporate or trade name, or the name of the sponsor's product, when it is clear that the mention of the name of the product constitutes a sponsorship identification, shall be deemed sufficient for the purposes of this section and only one such announcement need be made at any time during the course of the cablecast.

(f) The announcement otherwise required by this section is waived with respect to the origination cablecast of "want ad" or classified advertisements sponsored by an individual. The waiver granted in this paragraph shall not extend to a classified advertisement or want ad sponsorship by any form of business enterprise, corporate or otherwise. Whenever sponsorship announcemnts are omitted pursuant to this paragraph, the cable televison system operator shall observe the following conditions: (1) Maintain a list showing the name, address, and (where available) the telephone number of each advertiser (2) Make this list available to members of the public who have a legitimate interest in obtaining the information contained in the list.

(g) The announcements required by this section are waived with respect to feature motion picture film produced initially and primarily for theatre exhibition.

NOTE—The waiver heretofore granted by the commission in its Report and Order, adopted November 16, 1960 (FCC 60-1369, 40 FCC 95), continues to apply to programs filmed or recorded on or before June 20, 1963, when §73.654(e), the predecessor television rule, went into effect.

(h) Commission interpretations in connection with the provisions of the sponsorship identification rules for the broadcasting services are contained in the commission's Public Notice, entitled "Applicability of Sponsorship Identification Rules," dated May 6, 1963 (40 FCC 141), as modified by Public Notice, dated April 21, 1975 (FCC 74-418). Further interpretations are printed in full in various volumes of the Federal Communications Commission reports. The interpretations made for the broadcasting services are equally applicable to origination cablecasting.

APPENDIX **C**

Basic Satellite Networks (1986)

This appendix lists the basic satellite networks with national distribution. (Excluded are other networks that are mainly regional, that do not account for many hours of telecasting within a week, or that do not have many subscribers.) The programming descriptions attempt to characterize the programming, without being exhaustive, and to indicate unique programs or formats. Program titles and topics in parentheses are illustrative of content. "Off-network syndication" means reruns of commerical broadcast network series. "Local availabilities" means that, in addition to national advertising sales, the network makes available commercial time to be sold locally by the affiliated cable systems.

The audiences reported are based on the Nielsen Home Index Special Ratings in 1986. The figures are for *households*, not individuals. In most cases the number of individuals viewing would be greater than the number of households. The first number is the average households viewing over all of the hours the network is cablecasting. The second is the average number of households viewing during prime time (defined for most of the networks as 8 to 11 P.M., Monday–Saturday and 7 to 11 P.M. Sunday). Only a few of the cable networks qualify for Nielsen measurement.

Fees to affiliates—the amount networks charge to affiliates for the program services—vary depending on affiliation contracts, total number of subscribing households and other factors. A big MSO may pay only half the announced rate.

Subscriber and affiliate counts are drawn from *Cablevision*, November 17, 1986, p. 36 and *Broadcasting*, December 1, 1986, p. 66, *Multichannel News*, May 11, 1987 and *Cable Strategies*, March 15, 1987. The Nielsen audience figures were reported in *Multichannel News*, January 19, 1987, p. 38

or *Multichannel News,* September 8, 1986, p. 39. Other sources include network promotional literature, program guides, press releases, and trade publications.

SUPERSTATIONS

WTBS

Atlanta, Georgia. Movies, off-network syndication, news, Atlanta Braves baseball, NBA basketball, college football, college basketball, original programming (first run on the station) such as "Portrait of America" with programs exploring each of the United States and the U.S. territories, Jacques Cousteau specials featuring expeditions to distant ocean locations, the "World of Audubon" series produced by WTBS and the National Audubon Society on environmental issues, deliberately prosocial made-for-cable sitcoms "Safe at Home" and "Down to Earth," and "Good News" a weekly half hour of positive news. National Geographic Explorer series for children and adults about natural history, science, exploration, and archeology (programs on tornados, the Everglades swamp, Peru's Lake Titicaca, sharks, Skycam video cameras). Affiliates 8,373, subscribing households 37,582,000. Average households viewing during all operating hours 615,000, prime time 923,000. Fee to affiliates 20 cents per subscriber per month. Commercial (*no* local availabilities for affiliates). 24-hour.

WOR

New York. Sports, movies, and off-network syndication. Affiliates 1,999; subscribing households 8,669,134. Fee to affiliates 10 cents per subscriber per month. Commercial (*no* local availabilities for affiliates). 24-hour.

WPIX

New York. Movies, off-network syndicated series, news, New York Yankees baseball, New York Giants preseason football, Eastern college football, Big East basketball, original New York events such as Macy's 4th of July fireworks. Affiliates 239, subscribing households 2,412,000. Fee to affiliates, 10 cents per subscriber per month. Commercial (*no* local availabilities for affiliates). 24-hour.

NEWS AND INFORMATION

CNN (Cable News Network)

Atlanta, Georgia. Regularly scheduled general newscasts throughout the day. News bureaus in many U.S. cities and world capitals. Extended live

coverage of major events. Special topic newscasts on sports, business, entertainment, weather, medicine, science, style, international. Panel interview
programs, investigative reports, phone-in. Exchanges news with Soviet Union
and other Eastern Bloc nations. Affiliates 9,955. Subscribing households,
35,834,000. Average households viewing during all operating hours (CNN and
CNN Headline) 338,000, prime time 427,000. Fee to affiliates 15 to 22 cents
per month per subscriber for CNN and CNN Headline. Commerical (local
availabilities for affiliates). 24-hour.

CNN Headline News

News recycled and updated every 30 minutes. In four segments separated
by commercials. The fourth segment, weather and features, may be preempted
for local news headlines. Affiliates 3,538. Subscribing households, 20,975,000.
Fee to affiliates, free to 22 cents per subscriber month depending on whether
system takes CNN. Commercial (local availabilities for affiliates). 24-hour.

C-SPAN I (Cable Satellite Public Affairs Network)

24-hour complete live coverage of political events. Live proceedings from
the floor of the U.S. House of Representatives, House Committee hearings,
viewer participation (call-in) programs on current issues with members of Congress, other political leaders, journalists. High school students questioning
government leaders. National Press Club speeches. "A Day in the Life of . . . "
series featuring daily activities of media institutions, federal departments,
members of Congress. Extensive election coverage. Affiliates, 2,300. Subscribing households, 25,250,000. Fee to affiliates, negotiated (about 4 cents per
months per subscriber). Underwriting. Nonprofit cable industry cooperative.
24-hour. C-SPAN II, separate channel for the U.S. Senate.

FNN (Financial News Network)

NYSE, AMEX, and NASDAQ tickers. Business and consumer news,
market analyses. "Score," a night-time/weekend sports ticker with video
highlights of games. Affiliates 1,360, subscribing households 22,300,000. Fee
to affiliates $195 to $345 per month. Score, 3.5 cents per household per month.
Commercial (local availabilities for affiliates). 24-hour.

The Weather Channel

Atlanta, Georgia. National and international weather reports based on
the entire meterological database of the world. Eight times hourly, 75 seconds
of character-generated local weather including current conditions, seasonal
averages, immediate and extended forecasts (simultaneously in 700 zones).

Special reports for various sports, travelers, gardeners, aviators, farmers. Regional, international, city-by-city reports. Affiliates 2,200, subscribing households 22,000,000. Average households viewing during all operating hours 56,000, prime time 62,000. Fee to affiliates 6–10 cents per subscriber per month. Commercial (local availabilities for affiliates). 24-hour.

RELIGIOUS

ACTS (American Christian Television System)

Ft. Worth, Texas. Family Christian entertainment; country and gospel music, sports interviews, "classic" movies. Affiliates 236, subscribing households 4,600,000. Free to affiliates. Commercial (local availabilities). 24-hour.

EWTN (Eternal World Television Network)— Catholic Cable Network

Birmingham, Alabama. Religious magazine, interview, and call-in formats anchored by network founder "Mother Angelica Live" (Catholic charismatic renewal). Some syndicated family entertainment. Affiliates 314, subscribing households 4,800,000. Free to affiliates. Noncommercial. 8 P.M. to 12 M.

The Inspirational Network

Charlotte, North Carolina. Talk and variety. Christian music. Children's programming. Evangelists. Religious news. "Sound Effects" (contemporary religious music videos). Foreign language programs. Christian drama such as "Pattern for Living." Women's interest. Affiliates 1,300, subscribing households 13,000,000. Pay affiliates 3 cents per month per subscriber. Noncommerical. 24-hour.

TBN (Trinity Broadcasting Network)

Santa Ana, California. Religious music and variety, bible studies, talk shows, including "Praise the Lord" hosted by Paul and Jan Crouch. Evangelists. Affiliates 444, subscribing households 6,304,000. Compensates affiliates 10 cents per year per subscriber up to a maximum of $25,000. Noncommerical. 24-hour.

ETHNIC

BET (Black Entertainment Television)

Washington, D.C. Black-oriented music videos and gospel, black college football and basketball, news, entertainment news. Family entertainment such as Nipsey Russell's "Juvenile Jury." Affiliates 590, subscribing households 14,000,000. Fee to affiliates 3 cents per subscriber per month. Commercial (local availabilities for affiliates). 24-hour.

SIN (Spanish International Network)

New York. Spanish language programming including news, movies, music, live sports, childrens' programs, and novellas. World Cup Soccer. Affiliates 383, subscribing households 33,673,300 (3,860,500 Spanish-speaking households). Free to affiliates. Commercial. 24-hour.

CHILDREN

Nickelodeon

New York. Child-oriented cartoons, interview shows, music videos, animal programs, off-network syndication. Family programming (Nick at Nite) in the evening. Affiliates 4,501, subscribing households 29,000,000. Average households viewing during 7 A.M. to 1 P.M., 225,000. Fees to affiliates 15 cents per subscriber per month. Commercial (local availabilities for affiliates). 7 A.M. to 8 P.M.

SPORTS

ESPN—The Total Sports Network

Bristol, Connecticut. Major and minor amateur and professional sports. NFL football. Live and replays. Sports news. Dramatized sports biographies. Affiliates 13,878, subscribing households 36,900,000. Average households viewing during all operating hours 266,000 (1985). Prime time 553,000 (1986). Fees to affiliates 23 cents per subscriber per month. Commercial (local availabilities for affiliates). 24-hour.

EDUCATIONAL

The Discovery Channel

Landover, Maryland. Films on geography, science and technology, nature, adventure, and history. Launched June 1985. Affiliates 932, subscribing households 9,500,000. Fee to affiliates 5 cents with a rebate based on national advertising sales. Commercial (local availabilities). 3 P.M. to 3 A.M.

The Learning Channel

Washington, D.C. Information and instruction for adults. Learning ("Spanish for Survival," "American Government I," "Introducing Biology"). Income ("Personal Time Management," "Investment Talk," "Planning a New Business"). Pastimes ("Inside Your Schools," "Parents as Partners," "Search for Solutions"). Some for college credit. Affiliates 775, subscribing households 7,000,000. Fee to affiliates 5 cents per subscriber per month. Noncommercial. 20 hours.

MUSIC

Country Music Television

Nashville, Tennessee. Music videos, live performance specials, brief interviews—all country music. Affiliates 438, subscribing households 6,200,000. Free to affiliates. Commercial (local availabilities for affiliates). 24-hour.

MTV (Music Television)

New York. Rock music videos (some exclusive) with VJs (video jockeys). "Top 20 Video Countdown," music news, humor, interviews, concert tour information. Comedy series (BBC's *The Young Ones*). Affiliates 4,269, subscribing households 30,700,000. Average households viewing during 7 A.M. to 1 A.M. 268,000 (1985), prime time 309,000 (1985). Fee to affiliates 10–15 cents per subscriber per month. Commercial (local availabilities for affiliates). 24-hour.

VH-1 (Video Hits One)

New York. Music videos featuring "contemporary" music (Elton John, Billy Joel), "soft rock" (Neil Diamond, Melissa Manchester), softer R&B (Diana Ross, Smokey Robinson), cross-over country (Willie Nelson, Kenny Rogers). VJs. Affiliates 718, subscribing households 8,900,000. Free to affiliates. Commercial (local availabilities). 24-hour.

GENERAL

A&E (Arts and Entertainment Network)

New York. Stage drama, biography, series, music, dance. Right of first refusal on U.S. television distribution of BBC programs. Usually first-run for

U.S. television. Locally produced documentaries on "The Most Livable Cities." Intended for a relatively discriminating audience. Affiliates 2,200, subscribing households 19,700,000. Average households viewing, prime time, 134,000. Fee to affiliates 7 cents (charter affiliates) and 10 cents (new affiliates) per subscriber per month. Commercial (Local availabilities to affiliates). 8 A.M. to 4 A.M.

CBN Cable Network

Virginia Beach, Virginia. Family-oriented programming. Off-network series (such as "Gentle Ben," "Wagon Train," "The Carol Burnett Show," "Patty Duke Show," "The Farmer's Daughter"). "The 700 Club" and other inspirational programming. Movies, game shows. "Doris Day's Best Friends" (pets). Strong on the western genre in television series and movies. Many childrens' programs. Some original production of family entertainment series ("The Campbells," "Butterfly Island"). News. Affiliates 7,260, subscribing households 32,338,000. Average households viewing during all operating hours 332,000. Prime time 356,000. Fee to affiliates 6 cents per subscriber per month. Commercial (local availabilities for affiliates). 24-hour.

Lifetime

New York. Talk television built around anchor personalities. Some syndicated series. Much oriented to women. Exercise, weight control, lifestyle. "Good Sex! with Dr. Ruth Westheimer." Medical news and instruction for doctors on Sundays. Affiliates 2,987, subscribing households 25,800,000. Average households viewing during all operating hours 115,000, prime time 196,000. Fee to affiliates 6 cents per subscriber per month. Commercial (local availabilities for affiliates). 24-hour.

Tempo Television

Tulsa, Oklahoma. Oriented toward people over 45 years. Movies, religion, news, business, telephone auctions, real estate, outdoor recreation, foreign language programs. Mainly time sales to outside producers. Affiliates 640, subscribing households 12,500,000. Fee to affiliates 5 cents per subscriber per month. Commercial (local availabilities for affiliates). 24-hour.

Silent Network

Beverly Hills, California. Entertainment, information and education for the deaf. Affiliates 206, subscribing households 6,800,000. Free to affiliates. Commercial. 10 A.M. to 12 P.M. Saturday.

TNN (The Nashville Network)

Nashville, Tennessee. Country entertainment. Music ("Grand Ole Opry Live," country music entertainers in concert, country music videos, gospel). Sports (NASCAR races, rodeos, tractor pulls, swamp buggy races, hot rodding, fishing). Variety, talk, games, comedy. Affiliates 3,770, subscribing households 26,600,000. Average households viewing during all operating hours 167,000, prime time 306,000. Fee to affiliates 5 to 10 cents per subscriber per month. Commercial (local availabilities for affiliates). 24-hour.

USA Network

New York. Off-network and off-pay movies and specials, off-network series, NHL hockey, college football and basketball, wrestling. Women-oriented daytime mix of movies, music, off-network soaps, and lifestyle. Children's programming, news, business information, and music videos. Affiliates 6,500, subscribing households 33,000,000. Average households viewing during all operating hours 275,000, prime time 447,000. Fee to affiliates 10 cents per subscriber per month. Commercial (local availabilities for affiliates). 24-hour.

SHOPPING NETWORKS

Home Shopping Network (HSN)

Clearwater, Florida. General merchandise. Affiliates 324, subscribing households 8,000,000. Pay affiliates about 5 percent of gross sales from within franchise. Local availabilities to affiliates. 24-hour.

Cable Value Network (CVN)

Minneapolis, Minnesota. Merchandise presented in program format by category of product. Affiliates 500, subscribing households 13,000,000. Pays affiliates 5 percent of sales in franchise territory. 24-hour.

Innovations in Living (HSN2)

Clearwater, Florida. High-tech products. Affiliates 340, subscribing households 6,200,000. Pay affiliates negotiated percentage of sales in franchise, amount depends on whether system also carries HSN1. 24-hour.

The Travel Channel

Travel information and offers of trip packages. Affiliates 150, subscribing households 7,000,000.

Pay Services (1986)

Nationally Distributed Pay Services	Affiliates	Subscribing Households
American Movie Classics	270	1,000,000
Bravo	251	350,000
Cinemax	3,300	3,700,000
The Disney Channel	2,958	2,550,000
Home Box Office (HBO)	6,900	14,600,000
Home Theater Network (HTN)	350	300,000
Showtime	3,200	5,400,000
The Movie Channel	3,250	3,250,000
The Nostalgia Channel	175	700,000
The Playboy Channel	580	732,000
Regional Pay Services		
Dodgervision	50	615,000
Home Sports Entertainment—Houston/Dallas	114	225,000
Home Team Sports	76	350,000
New England Sports Network	160	170,276
PRISM-Philadelphia	90	370,000
Pro-Am Sports	115	115,000
SportsChannel (New England)	110	228,900
SportsChannel (New York)	66	562,000
SportsVision (Chicago)	39	532,000

Source: "Cable Stats," *CableVision,* November 17, 1986, p. 36.

"Cable Specialty Services Subscriber Counts," *Multichannel News,* December 29, 1986, p. 32.

Cable Communications Policy Act of 1984

Public Law 98–549
98th Congress

An Act

To amend the Communications Act of 1934 to provide a national policy
regarding cable television.

*Be it enacted by the Senate and House of Representatives of the United
States of America in Congress assembled,*

SHORT TITLE; TABLE OF CONTENTS

SECTION 1. (a) This Act may be cited as the "Cable Communications
Policy Act of 1984".

(b) The table of contents for this Act is as follows:

AMENDMENT OF COMMUNICATIONS ACT OF 1934

SEC. 2. The Communications Act of 1934 is amended by inserting after title V the following new title:

"TITLE VI—CABLE COMMUNICATIONS

"Part I—General Provisions

"PURPOSES

"SEC. 601. The purposes of this title are to—
    "(1) establish a national policy concerning cable communications;

"(2) establish franchise procedures and standards which encourage the growth and development of cable systems and which assure that cable systems are responsive to the needs and interests of the local community;

"(3) establish guidelines for the exercise of Federal, State, and local authority with respect to the regulation of cable systems;

"(4) assure that cable communications provide and are encouraged to provide the widest possible diversity of information sources and services to the public;

"(5) establish an orderly process for franchise renewal which protects cable operators against unfair denials of renewal where the operator's past performance and proposal for future performance meet the standards established by this title; and

"(6) promote competition in cable communications and minimize unnecessary regulation that would impose an undue economic burden on cable systems.

"DEFINITIONS

"SEC. 602. For purposes of this title—

"(1) the term 'affiliate', when used in relation to any person, means another person who owns or controls, is owned or controlled by, or is under common ownership or control with, such person;

"(2) the term 'basic cable service' means any service tier which includes the retransmission of local television broadcast signals;

"(3) the term 'cable channel' or 'channel' means a portion of the electromagnetic frequency spectrum which is used in a cable system and which is capable of delivering a television channel (as television channel is defined by the Commission by regulation);

"(4) the term 'cable operator' means any person or group of persons (A) who provides cable service over a cable system and directly or through one or more affiliates owns a significant interest in such cable system, or (B) who otherwise controls or is responsible for, through any arangement, the management and operation of such a cable system;

"(5) the term 'cable service' means—

"(A) the one-way transmission to subscribers of (i) video programming, or (ii) other programming service, and

"(B) subscriber interaction, if any, which is required for the selection of such video programming or other programming service;

"(6) the term 'cable system' means a facility, consisting of a set of closed transmission paths and associated signal generation, reception, and control equipment that is designed to provide cable service which includes video programming and which is provided to multiple subscribers within a community, but such term does not include (A) a facility that serves only to retransmit the television signals of 1 or more television broadcast stations; (B) a facility that serves only subscribers in 1 or more multiple unit dwellings under common ownership, control, or management, unless such facility or facilities uses any public right-of-way; (C) a facility of a common carrier which is subject, in whole or in part, to the provisions of title II of this Act, except that such facility shall be considered a cable system (other than for purposes of section 621(c)) to the extent such facility is used in the transmission of video programming directly to subscribers; or (D) any facilities of any electric utility used solely for operating its electric utility systems;

"(7) the term 'Federal agency' means any agency of the United States, including the Commission;

"(8) the term 'franchise' means an initial authorization, or renewal thereof (including a renewal of an authorization which has been granted subject to section 626), issued by a franchising authority, whether such authorization is designated as a franchise, permit, license, resolution, contract, certificate, agreement, or otherwise, which authorizes the construction or operation of a cable system;

"(9) the term 'franchising authority' means any governmental entity empowered by Federal, State, or local law to grant a franchise;

"(10) the term 'grade B contour' means the field strength of a television broadcast station computed in accordance with regulations promulgated by the Commission;

"(11) the term 'other programming service' means information that a cable operator makes available to all subscribers generally;

"(12) the term 'person' means an individual, partnership, association, joint stock company, trust, corporation, or governmental entity;

"(13) the term 'public, educational, or governmental access facilities' means—

> "(A) channel capacity designated for public, educational, or governmental use; and
>
> "(B) facilities and equipment for the use of such channel capacity;

"(14) the term 'service tier' means a category of cable service or other services provided by a cable operator and for which a separate rate is charged by the cable operator;

"(15) the term 'State' means any State, or political subdivision, or agency thereof; and

"(16) the term 'video programming' means programming provided by, or generally considered comparable to programming provided by, a television broadcast station.

"PART II—USE OF CABLE CHANNELS AND CABLE OWNERSHIP RESTRICTIONS

"CABLE CHANNELS FOR PUBLIC, EDUCATIONAL, OR GOVERNMENTAL USE

"SEC. 611. (a) A franchising authority may establish requirements in a franchise with respect to the designation or use of channel capacity for public educational, or governmental use only to the extent provided in this section.

"(b) A franchising authority may in its request for proposals require as part of a franchise, and may require as part of a cable operator's proposal for a franchise renewal, subject to section 626, that channel capacity be disignated for public, educational, or governmental use, and channel capacity on institutional networds be designated for educational or governmental use, and may require rules and procedures for the use of the channel capacity designated pursuant to this section.

"(c) A franchising authority may enforce any requirement in any franchise regarding the providing or use of such channel capacity. Such enforcement authority includes the authority to enforce any provisions of the franchise for services, facilities, or equipment proposed by the cable operator which relate to public, educational, or governmental use of channel capacity, whether or not required by the franchising authority pursuant to subsection (b).

"(d) In the case of any franchise under which channel capacity is designated under subsection (b), the franchising authority shall prescribe—

"(1) rules and procedures under which the cable operator is permitted to use such channel capacity for the provision of other services if such channel capacity is not being used for the purposes designated, and

"(2) rules and procedures under which such permitted use shall cease.

"(e) Subject to section 624(d), a cable operator shall not exercise any editorial control over any public, educational, or governmental use of channel capacity provided pursuant to this section.

"(f) For purposes of this section, the term 'institutional network' means a communication network which is constructed or operated by the cable operator and which is generally available only to subscribers who are not residential subscribers.

"CABLE CHANNELS FOR COMMERCIAL USE

"SEC. 612. (a) The purpose of this section is to assure that the widest possible diversity of information sources are made available to the public from cable systems in a manner consistent with growth and development of cable systems.

"(b)(1) A cable operator shall designate channel capacity for commercial use by persons unaffiliated with the operator in accordance with the following requirements:

"(A) An operator of any cable system with 36 or more (but not more than 54) activated channels shall designate 10 percent of such channels which are not otherwise required for use (or the use of which is not prohibited) by Federal law or regulation.

"(B) An operator of any cable system with 55 or more (but not more than 100) activated channels shall designate 15 percent of such channels which are not otherwise requried for use (or the use of which is not prohibited) by Federal law or regulation.

"(C) An operator of any cable system with more than 100 activated channels shall designate 15 percent of all such channels.

"(D) An operator of any cable system with fewer than 36 activated channels shall not be required to designate channel capacity for commercial use by persons unaffiliated with the operator, unless the cable system is required to provide such channel capacity under the terms of a franchise in effect on the date of the enactment of this title.

"(E) An operator of any cable system in operation on the date of the enactment of this title shall not be required to remove any service actually being provided on July 1, 1984, in order to comply with this section, but shall make channel capacity available for commercial use as such capacity becomes available until such time as the cable operator is in full compliance with this section.

"(2) Any Federal agency, State, or franchising authority may not require any cable system to designate channel capacity for commercial use by unaffiliated persons in excess of the capacity specified in paragraph (1), except as otherwise provided in this section.

"(3) A cable operator may not be required, as part of a request for proposals or as part of a proposal for renewal, subject to section 626, to designate channel capacity for any use (other than commercial use by unaffiliated persons under this section) except as provided in sections 611 and 637, but a cable operator may offer in a franchise, or proposal for renewal thereof, to provide, consistent with applicable law, such capacity for other than commercial use by such persons.

"(4) A cable operator may use any unused channel capacity designated pursuant to this section until the use of such channel capacity is obtained, pursuant to a written agreement, by a person unaffiliated with the operator.

"(5) For the purposes of this section—

"(A) the term 'activated channels' means those channels engineered at the headend of the cable system for the provision of services generally available to residential subscribers of the cable system, regardless of whether such services actually are provided, including any channel designated for public, educational, or governmental use; and

"(B) the term 'commercial use' means the provision of video programming, whether or not for profit.

"(6) Any channel capacity which has been designated for public, educational, or governmental use may not be considered as designated under this section for commercial use for purpose of this section.

"(c)(1) If a person unaffiliated with the cable operator seeks to use channel capacity designated pursuant to subsection (b) for commercial use, the cable operator shall establish, consistent with the purpose of this section, the price, terms, and conditions of such use which are at least sufficient to assure that such use will not adversely affect the operation, financial condition, or market development of the cable system.

"(2) A cable operator shall not exercise any editorial control over any video programming provided pursuant to this section, or in any other way consider the content of such programming, except that an operator may consider such content to the minimum extent necessary to establish a reasonable price for the commercial use of designated channel capacity by an unaffiliated person.

"(3) Any cable system channel designated in accordance with this section shall not be used to provide a cable service that is being provided over such system on the date of the enactment of this title, if the provision of such programming is intended to avoid the purpose of this section.

"(d) Any person aggrieved by the failure or refusal of a cable operator to make channel capacity available for use pursuant to this section may bring an action in the district court of the United States for the judicial district in which the cable system is located to compel that such capacity be made available. If the court finds that the channel capacity sought by such person has not been made available in accordance with this section, or finds that the price, terms, or conditions established by the cable operator are unreasonable, the court may order such system to make available to such person the channel capacity sought, and further determine the appropriate price, terms, or conditions for such use consistent with subsection (c), and may award actual damages if it deems such relief appropriate. In any such action, the court shall not consider any price, term, or condition established between an operator and an affiliate for comparable services.

"(e)(1) Any person aggrieved by the failure or refusal of a cable operator to make channel capacity available pursuant to this section may petition the commission for relief under this subsection upon a showing of prior adjudicated violations of this section. Records of previous adjudications resulting in a court determination that the operator has violated this section shall be considered as sufficient for the showing necessary under this subsection. If the commission finds that the channel capacity sought by such person has not been made available in accordance with this section, or that the price, terms, or conditions established by such system are unreasonable under subsection (c), the Commission shall, by rule or order, require such operator to make available such channel capacity under price, terms, and conditions consistent with subsection (c).

"(2) In any case in which the Commission finds that the prior adjudicated violations of this section constitute a pattern or practice of violations by an operator, the Commission may also establish any further rule or order necessary to assure that the operator provides the diversity of information sources required by this section.

"(3) In any case in which the Commission finds that the prior adjudicated violations of this section constitute a pattern or practice of violations by any person who is an operator of more than one cable system, the Commission may also establish any further rule or order necessary to assure that such person provides the diverstiy of information sources required by this section.

"(f) In any action brought under this section in any Federal district court or before the Commission, there shall be a presumption that the price, terms, and conditions for use of channel capacity designated pursuant to subsection (b) are reasonable and in good faith unless shown by clear and convincing evidence to the contrary.

"(g) Notwithstanding sections 621(c) and 623(a), at such time as cable systems with 36 or more activated channels are available to 70 percent of households within the United States and are subscribed to by 70 percent of the households to which such systems are available, the Commission may promulgate any additional rules necessary to provide diversity of information sources. Any rules promulgated by the Commission pursuant to this subsection shall not preempt authority expressly granted to franchising authorities under this title.

"(h) Any cable service offered pursuant to this section shall not be provided, or shall be provided subject to conditions, if such cable service in the judgment of the franchising authority is obscene, or is in conflict with community standards in that is is lewd, lascivious, filthy, or indecent or is otherwise unprotected by the Constitution of the United States.

"OWNERSHIP RESTRICTIONS

"SEC. 613. (a) It shall be unlawful for any person to be a cable operator if such person, directly or through 1 or more affiliates, owns or controls, the licensee of a television broadcast station and the predicted grade B contour of such station covers any portion of the community served by such operator's cable system.

"(b)(1) It shall be unlawful for any common carrier, subject in whole or in part to title II of this Act, to provide video programming directly to subscribers in its telephone service area, either directly or indirectly through an affiliate owned by, operated by, controlled by, or under common control with the common carrier.

"(2) It shall be unlawful for any common carrier, subject in whole or in part to title II of this Act, to provide channels of communications or pole line conduit space, or other rental arrangements, to any entity which is directly or indirectly owned by, operated by, controlled by, or under common control with such common carrier, if such facilities or arrangements are to be used for, or in connection with, the provision of video programming directly to subscribers in the telephone service area of the common carrier.

"(3) This subsection shall not apply to any common carrier to the extent such carrier provides telephone exchange service in any rural area (as defined by the Commission).

"(4) In those areas where the provision of video programming directly to subscribers through a cable system demonstrably could not exist except through a cable system owned by, operated by, controlled by, or affiliated with the common carrier involved, or upon other showing of good cause, the Commission may, on petition for waiver, waive the applicability of paragraphs (1) and (2) of this subsection. Any such waiver shall be made in accordance with section 63.56 of title 47, Code of Federal Regulations (as in effect September 20, 1984) and shall be granted by the Commission upon a finding that the issuance of such waiver is justified by the particular circumstances demonstrated by the petitioner, taking into account the policy of this subsection.

"(c) The Commission may prescribe rules with respect to the ownership or control of cable systems by persons who own or control other media of mass communications which serve the same community served by a cable system.

"(d) Any State or franchising authority may not prohibit the ownership or control of a cable system by any person because of such person's ownership or control of any media of mass communications or other media interests.

"(e)(1) Subject to paragraph (2), a State or franchising authority may hold any ownership interest in any cable system

"(2) Any State or franchising authority shall not exercise any editorial control regarding the content of any cable service on a cable system in which such governmental entity holds ownership interest (other than programming on any channel designated for educational or governmental use), unless such control is exercised through an entity separate from the franchising authority.

"(f) This section shall not apply to prohibit any combination of any interests held by any person on July 1, 1984, to the extent of the interests so held as of such date, if the holding of such interests was not inconsistent with any applicable Federal or State law or regulations in effect on that date.

"(g) For purposes of this section, the term 'media of mass communications' shall have the meaning given such term under section 309(i)(3)(C)(i) of this Act.

"PART III—FRANCHISING AND REGULATION

"GENERAL FRANCHISE REQUIREMENTS

"SEC. 621. (a)(1) A franchising authority may award, in accordance with the provisions of this title, 1 or more franchises within its jurisdiction.

"(2) Any franchise shall be construed to authorize the construction of a cable system over public rights-of-way, and through easements, which is within the area to be served by the cable system and which have been dedicated for compatible uses, except that in using such easements the cable operator shall ensure—

   "(A) that the safety, functioning, and appearance of the property and the convenience and safety of other persons not be adversely affected by the installation or construction of facilities necessary for a cable system;

   "(B) that the cost of the installation, construction, operation, or removal of such facilities be borne by the cable operator or subscriber, or a combination of both; and

   "(C) that the owner of the property be justly compensated by the cable operator for any damages caused by the installation, construction, operation, or removal of such facilities by the cable operator.

  "(3) In awarding a franchise or franchises, a franchising authority shall assure that access to cable service is not denied to any group of potential residential cable subscribers because of the income of the residents of the local area in which such group resides.

  "(b)(1) Except to the extent provided in paragraph (2), a cable operator may not provide cable service without a franchise.

  "(2) Paragraph (1) shall not require any person lawfully providing cable service without a franchise on July 1, 1984, to obtain a franchise unless the franchising authority so requires.

  "(c) Any cable system shall not be subject to regulation as a common carrier or utility by reason of providing any cable service.

  "(d)(1) A State or the Commission may require the filing of informational tariffs for any intrastate communications service provided by a cable system, other than cable service, that would be subject to regulation by the Commission or any State if offered by a common carrier subject, in whole or in part, to title II of this Act. Such informational tariffs shall specify the rates, terms, and conditions for the provision of such service, including whether it is made available to all subscribers generally, and shall take effect on the date specified therein.

  "(2) Nothing in this title shall be construed to affect the authority of any State to regulate any cable operator to the extent that such operator provides any communication service other than cable service, whether offered on a common carrier or private contract basis.

  "(3) For purposes of this subsection, the term 'State' has the meaning given it in section 3(v).

  "(e) Nothing in this title shall be construed to affect the authority of any State to license or otherwise regulate any facility or combination of facilities which serves only subscribers in one or more multiple unit dwellings under common ownership, control, or management and which does not use any public right-of-way.

"FRANCHISE FEES

"SEC. 622. (a) Subject to the limitation of subsection (b), any cable operator may be required under the terms of any franchise to pay a franchise fee.

"(b) For any twelve-month period, the franchise fees paid by a cable operator with respect to any cable system shall not exceed 5 percent of such cable operator's gross revenues derived in such period from the operation of the cable system. For purposes of this section, the 12-month period shall be the 12-month period applicable under the franchise for accounting purposes. Nothing in this subsection shall prohibit a franchising authority and a cable operator from agreeing that franchise fees which lawfully could be collected for any such 12-month period shall be paid on a prepaid or deferred basis; except that the sum of the fees paid during the term of the franchise may not exceed the amount, including the time value of money, which would have lawfully been collected if such fees had been paid per annum.

"(c) A cable operator may pass through to subscribers the amount of any increase in a franchise fee, unless the franchising authority demonstrates that the rate structure specified in the franchise reflects all costs of franchise fees and so notifies the cable operator in writing.

"(d) In any court action under subsection (c), the franchising authority shall demonstrate that the rate structure reflects all costs of the franchise fees.

"(e) Any cable operator shall pass through to subscribers the amount of any decrease in a franchise fee.

"(f) A cable operator may designate that portion of a subscriber's bill attributable to the franchise fee as a separate item on the bill.

"(g) For the purposes of this seciton—

"(1) The term 'franchise fee' includes any tax, fee, or assessment of any kind imposed by a franchising authority or other governmental entity on a cable operator or cable subscriber, or both, solely because of their status as such;

"(2) the term 'franchise fee' does not include—

"(A) any tax, fee, or assessment of general applicability (including any such tax, fee, or assessment imposed on both utilities and cable operators or their services but not including a tax, fee, or assessment which is unduly discriminatory against cable operators or cable subscribers);

"(B) in the case of any franchise in effect on the date of the enactment of this title, payments which are required by the franchise to be made by the cable operator during the term of such franchise for, or in support of the use of, public, educational, or governmental access facilities;

"(C) in the case of any franchise granted after such date of enactment, capital costs which are required by the franchise to be incurred by the cable operator for public, educational, or governmental access facilities;

"(D) requirements or charges incidental to the awarding or enforcing of the franchise, including payments for bonds, security funds, letters of credit, insurance, indemnification, penalties, or liquidated damages; or

"(E) any fee imposed under title 17, United States Code.

"(h)(1) Nothing in this Act shall be construed to limit any authority of a franchising authority to impose a tax, fee, or other assessment of any kind on any person (other than a cable operator) with respect to cable service or other communications service provided by such person over a cable system for which charges are assessed to subscribers but not received by the cable operator.

"(2) For any 12-month period, the fees paid by such person with respect to any such cable service or other communications service shall not exceed 5 percent of such person's gross revenues derived in such period from the provision of such service over the cable system.

"(i) Any Federal agency may not regulate the amount of the franchise fees paid by a cable operator, or regulate the use of funds derived from such fees, except as provided in this section.

"REGULATION OF RATES

"SEC. 623. (a) Any Federal agency or State may not regulate the rates for the provision of cable service except to the extent provided under this section. Any franchising authority may regulate the rates for the provision of cable service, or any other communications service provided over a cable system to cable subscribers, but only to the extent provided under this section.

"(b)(1) Within 180 days after the date of the enactment of this title the Commission shall prescribe and make effective regulations which authorize a franchising authority to regulate rates for the provision of basic cable service in circumstances in which a cable system is not subject to effective competition. Such regulations may apply to any franchise granted after the effective date of such regulations. Such regulations shall not apply to any rate while such rate is subject to the provisions of subsection (c).

"(2) For purposes of rate regulation under this subsection, such regulations shall—

"(A) define the circumstances in which a cable system is not subject to effective competition; and

"(B) establish standards for such rate regulation.

"(3) The Commission shall periodically review such regulations, taking into account developments in technology, and may amend such regulations, consistent with paragraphs (1) and (2), to the extent the Commission determines necessary.

"(c) In the case of any cable system for which a franchise has been granted on or before the effective date of this title, until the end of the 2-year period beginning on such effective date, the franchising authority may, to the extent provided in a franchise—

"(1) regulate the rates for the provision of basic cable service, including multiple tiers of basic cable service;

"(2) require the provision of any service tier provided without charge (disregarding any installation or rental charge for equipment necessary for receipt of such tier); or

"(3) regulate rates for the initial installation or the rental of 1 set of the minimum equipment which is necessary for the subscriber's receipt of basic cable service.

"(d) Any request for an increase in any rate regulated pursuant to subsection (b) or (c) for which final action is not taken within 180 days after receipt of such request by the franchising authority shall be deemed to be granted, unless the 180-day period is extended by mutual agreement of the cable operator and the franchising authority.

"(e)(1) In addition to any other rate increase which is subject to the approval of a franchising authority, any rate subject to regulation pursuant to this section may be increased after the effective date of this title at the discretion of the cable operator by an amount not to exceed 5 percent per year if the franchise (as in effect on the effective date of this title) does not specify a fixed rate or rates for basic cable service for a specified period or periods which would be exceeded if such increase took effect.

"(2) Nothing in this section shall be construed to limit provisions of a franchise which permits a cable operator to increase any rate at the operator's discretion; however, the aggregate increases per year allowed under paragraph (1) shall be reduced by the amount of any increase taken such year under such franchise provisions.

"(f) Nothing in this title shall be construed as prohibiting any Federal agency, State, or a franchising authority, from—

"(1) prohibiting discrimination among customers of basic cable service, or

"(2) requiring and regulating the installation or rental of equipment which facilitates the reception of basic cable service by hearing impaired individuals.

"(g) Any State law in existence on the effective date of this title which provides for any limitation or preemption of regulation by any franchising authority (or the State or any political subdivision or agency thereof) of rates for cable service shall remain in effect during the 2-year period beginning on such effective date, to the extent such law provides for such limitation or preemption. As used in this section, the term 'State' has the meaning given it in section 3(v).

"(h) Not later than 6 years after the date of the enactment of this title, the Commission shall prepare and submit to the Congress a report regarding rate regulation of cable services, including such legislative recommendations as the Commission considers appropriate. Such report and recommendations shall be based on a study of such regulation which the Commission shall conduct regarding the effect of competition in the marketplace.

"REGULATION OF SERVICES, FACILITIES, AND EQUIPMENT

"SEC. 624. (a) Any franchising authority may not regulate the services, facilities, and equipment provided by a cable operator except to the extent consistent with this title.

"(b) In the case of any franchise granted after the effective date of this title, the franchising authority, to the extent related to the establishment or operation of a cable system—

    "(1) in its requests for proposals for a franchise (including requests for renewal proposals, subject to section 626), may establish requirements for facilities and equipment, but may not establish requirements for video programming or other information services; and

    "(2) subject to section 625, may enforce any requirements contained within the franchise—

        "(A) for facilities and equipment; and

        "(B) for broad categories of video programming or other services.

"(c) In the case of any franchise in effect on the effective date of this title, the franchising authority may, subject to section 625, enforce requirements contained within the franchise for the provision of services, facilities, and equipment, whether or not related to the establishment or operation of a cable system.

"(d)(1) Nothing in this title shall be construed as prohibiting a franchising authority and a cable operator from specifying, in a franchise or renewal thereof, that certain cable services shall not be provided or shall be provided subject to conditions, if such cable services are obscene or are otherwise unprotected by the Constitution of the United States.

"(2)(A) In order to restrict the viewing of programming which is obscene or indecent, upon the request of a subscriber, a cable operator shall provide (by sale or lease) a device by which the subscriber can prohibit viewing of a particular cable service during periods selected by that subscriber.

"(B) Subparagraph (A) shall take effect 180 days after the effective date of this title.

"(e) The Commission may establish technical standards relating to the facilities and equipment of cable systems which a franchising authority may require in the franchise.

"(f)(1) any Federal agency, State, or franchising authority may not impose requirements regarding the provision or content of cable services, except as expressly provided in this title.

"(2) Paragraph (1) shall not apply to—

"(A) Any rule, regulation, or order issued under any Federal law, as such rule, regulation, or order (i) was in effect on September 21, 1983, or (ii) may be amended after such date if the rule, regulation, or order as amended is not inconsistent with the express provisions of this title; and

"(B) any rule, regulation, or order under title 17, United States Code.

"MODIFICATION OF FRANCHISE OBLIGATIONS

"SEC. 625. (a)(1) During the period a franchise is in effect, the cable operator may obtain from the franchising authority modifications of the requirements in such franchise—

"(A) in the case of any such requirement for facilities or equipment, including public, educaitonal, or governmental access facilities or equipment, if the cable operator demonstrates that (i) it is commercially impracticable for the operator to comply with such requirement, and (ii) the proposal by the cable operator for modification of such requirement is appropriate because of commercial impracticability; or

"(B) in the case of any such requirement for services, if the cable operator demonstrates that the mix, quality, and level of services required by the franchise at the time it was granted will be maintained after such modification.

"(2) Any final decision by a franchising authority under this subsection shall be made in a public proceeding. Such decision shall be made within 120 days after receipt of such request by the franchising authority, unless such 120 day period is extended by mutual agreement of the cable operator and the franchising authority.

"(b)(1) Any cable operator whose request for modification under subsection (a) has been denied by a final decision of a franchising authority may obtain modification of such franchise requirements pursuant to the provisions of section 635.

"(2) In the case of any proposed modification of a requirement for facilities or equipment, the court shall grant such modification only if the cable operator demonstrates to the court that—

"(A) it is commercially impracticable for the operator to comply with such requirement; and

"(B) the terms of the modification requested are appropriate because of commercial impracticability.

"(3) In the case of any proposed modification of a requirement for services, the court shall grant such modification only if the cable operator demonstrates to the court that the mix, quality, and level of services required by the franchise at the time it was granted will be maintained after such modification.

"(c) Notwithstanding subsections (a) and (b), a cable operator may, upon 30 days' advance notice to the franchising authority, rearrange, replace, or remove a particular cable service required by the franchise if—

"(1) such service is no longer available to the operator; or

"(2) such service is available to the operator only upon the payment of a royalty required under section 801(b)(2) of title 17, United States Code, which the cable operator can document—

"(A) is substantially in excess of the amount of such payment required on the date of the operator's offer to provide such service, and

"(B) has not been specifically compensated for through a rate increase or other adjustment.

"(d) Notwithstanding subsections (a) and (b), a cable operator may take such actions to rearrange a particular service from one service tier to another, or otherwise offer the service, if the rates for all of the service tiers involved in such actions are not subject to regulation under section 623.

"(e) A cable operator may not obtain modification under this section of any requirement for services relating to public, educational, or governmental access.

"(f) For purposes of this section, the term 'commercially impracticable' means, with respect to any requirement applicable to a cable operator, that it is commercially impracticable for the operator to comply with such requirement as a result of a change in conditions which is beyond the control of the operator and the nonoccurrence of which was a basic assumption on which the requirement was based.

"RENEWAL

"SEC. 626. (a) During the 6-month period which begins with the 36th month before the franchise expiration, the franchising authority may on its own initiative, and shall at the request of the cable operator, commence proceedings which afford the public in the franchise area appropriate notice and participation for the purpose of—

"(1) identifying the future cable-related community needs and interests; and

"(2) reviewing the performance of the cable operator under the franchise during the then current franchise term.

"(b)(1) Upon completion of a proceeding under subsection (a), a cable operator seeking renewal of a franchise may, on its own initiative or at the request of a franchising authority, submit a proposal for renewal.

"(2) Subject to section 624, any such proposal shall contain such material as the franchising authority may require, including proposals for an upgrade of the cable system.

"(3) The franchising authority may establish a date by which such proposal shall be submitted.

"(c)(1) Upon submittal by a cable operator of a proposal to the franchising authority for the renewal of a franchise, the franchising authority shall provide prompt public notice of such proposal and, during the 4-month period which begins on the completion of any proceedings under subsection (a), renew the franchise or, issue a preliminary assessment that the franchise should not be renewed and, at the request of the operator or on its own initiative, commence an administrative proceeding, after providing prompt public notice of such proceeding, in accordance with paragraph (2) to consider whether—

"(A) the cable operator has substantially complied with the material terms of the existing franchise and with applicable law;

"(B) the quality of the operator's service, including signal quality, response to consumer complaints, and billing practices, but without regard to the mix, quality, or level of cable services or other services provided over the system, has been reasonable in light of community needs;

"(C) the operator has the financial, legal, and technical ability to provide the services, facilities, and equipment as set forth in the operator's proposal; and

"(D) the operator's proposal is reasonable to meet the future cable-related community needs and interests, taking into account the cost of meeting such needs and interests.

"(2) In any proceeding under paragraph (1), the cable operator shall be afforded adequate notice and the cable operator and the franchise authority, or its designee, shall be afforded fair opportunity for full participation, including the right to introduce evidence (including evidence related to issues raised in the proceeding under subsection (a)), to require the production of evidence, and to question witnesses. A transcript shall be made of any such proceeding.

"(3) At the completion of a proceeding under this subsection, the franchising authority shall issue a written decision granting or denying the proposal for renewal based upon the record of such proceeding, and transmit a copy of such decsion to the cable operator. Such decision shall state the reasons therefor.

"(d) Any denial of a proposal for renewal shall be based on one or more adverse findings made with respect to the factors described in subparagraphs (A) through (D) of subsection (c)(1), pursuant to the record of the proceeding under subsection (c). A franchising authority may not base a denial of renewal on a failure to substantially comply with the material terms of the franchise under subsection (c)(1)(A) or on events considered under subsection (c)(1)(B) in any case in which a violation of the franchise or the events considered under subsection (c)(1)(B) occur after the effective date of this title unless the franchising authority has provided the operator with notice and the opportunity to cure, or in any case in which it is documented that the franchising authority has waived its right to object, or has effectively acquiesced.

"(e)(1) Any cable operator whose proposal for renewal has been denied by a final decision of a franchising authority made pursuant to this section, or has been adversely affected by a failure of the franchising authority to act in accordance with the procedural requirements of this section, may appeal such final decision or failure pursuant to the provisions of section 635.

"(2) The court shall grant appropriate relief if the court finds that—

"(A) any action of the franchising authority is not in compliance with the procedural requirements of this section; or

"(B) in the event of a final decision of the franchising authority denying the renewal proposal, the operator has demonstrated that the adverse finding of the franchising authority with respect to each of the factors described in subparagraphs (A) through (D) of subsection (c)(1) on which the denial is based is not supported by a preponderance of the evidence, based on the record of the proceeding conducted under subsection (c).

"(f) Any decision of a franchising authority on a proposal for renewal shall not be considered final unless all administrative review by the State has occurred or the opportunity therefor has lapsed.

"(g) For purposes of this section, the term 'franchise expiration' means the date of the expiration of the term of the franchise, as provided under the franchise, as it was in effect on the date of the enactment of this title.

"(h) Notwithstanding the provisions of subsections (a) through (g) of this section, a cable operator may submit a proposal for the renewal of a franchise pursuant to this subsection at any time, and a franchising authority may, after affording the public adequate notice and opportunity for comment, grant or deny such proposal at any time (including after proceedings pursuant to this section have commenced). The provisions of subsections (a) through (g) of this section shall not apply to a decision to grant or deny a proposal under this subsection. The denial of a renewal pursuant to this subsection shall not affect action on a renewal proposal that is submitted in accordance with subsections (a) through (g).

"CONDITIONS OF SALE

"SEC. 627. (a) If a renewal of a franchise held by a cable operator is denied and the franchising authority acquires ownership of the cable system or effects a transfer of ownership of the system to another person, any such acquisition or transfer shall be—

"(1) at fair market value, determined on the basis of the cable system valued as a going concern but with no value allocated to the franchise itself, or

"(2) in the case of any franchise existing on the effective date of this title, at a price determined in accordance with the franchise if such franchise contains provisions applicable to such an acquisition or transfer.

"(b) If a franchise held by a cable operator is revoked for cause and the franchising authority acquires ownership of the cable system or effects a transfer of ownership of the system to another person, any such acquisition or transfer shall be—

"(1) at an equitable price, or

"(2) in the case of any franchise existing on the effective date of this title, at a price determined in accordance with the franchise if such franchise contains provisions applicable to such an acquisition or transfer.

"PART IV—MISCELLANEOUS PROVISIONS

"PROTECTION OF SUBSCRIBER PRIVACY

"SEC. 631. (a)(1) At the time of entering into an agreement to provide any cable service or other service to a subscriber and at least once a year thereafter, a cable operator shall provide notice in the form of a separate, written statement to such subscriber which clearly and conspicuously informs the subscriber of—

"(A) the nature of personally indentifiable information collected or to be collected with respect to the subscriber and the nature of the use of such information;

"(B) the nature, frequency, and purpose of any disclosure which may be made of such information, including an identification of the types of persons to whom the disclosure may be made;

"(C) the period during which such information will be maintained by the cable operator;

"(D) the times and place at which the subscriber may have access to such information in accordance with subsection (d); and

"(E) the limitations provided by this section with respect to the collection and disclosure of information by a cable operator and the right of the subscriber under subsections (f) and (h) to enforce such limitations.

In the case of subscribers who have entered into such an agreement before the effective date of this section, such notice shall be provided within 180 days of such date and at least once a year thereafter.

"(2) For the purposes of this section, the term 'personally identifiable information' does not include any record of aggregate data which does not identify particular persons.

"(b)(1) Except as provided in paragraph (2), a cable operator shall not use the cable system to collect personally identifiable information concerning any subscriber without the prior written or electronic consent of the subscriber concerned.

"(2) A cable operator may use the cable system to collect such information in order to—

"(A) obtain information necessary to render a cable service or other service provided by the cable operator to the subscriber; or

"(B) detect unauthorized reception of cable communications.

"(c)(1) Except as provided in paragraph (2), a cable operator shall not disclose personally identifiable information concerning any subscriber without the prior written or electronic consent of the subscriber concerned.

"(2) A cable operator may disclose such information if the disclosure is—

"(A) necessary to render, or conduct a legitimate business activity related to, a cable service or other service provided by the cable operator to the subscriber;

"(B) subject to subsection (h), made pursuant to a court order authorizing such disclosure, if the subscriber is notified of such order by the person to whom the order is directed; or

"(C) a disclosure of the names and addresses of subscribers to any cable service or other service, if—

"(i) the cable operator has provided the subscriber the opportunity to prohibit or limit such disclosure, and

"(ii) the disclosure does not reveal, directly or indirectly, the—

"(I) extent of any viewing or other use by the subscriber of a cable service or other service provided by the cable operator, or

"(II) the nature of any transaction made by the subscriber over the cable system of the cable operator.

"(d) A cable subscriber shall be provided access to all personally identifiable information regarding that subscriber which is collected and maintained by a cable operator. Such information shall be made available to the subscriber at reasonable times and at a convenient place designated by such cable operator. A cable subscriber shall be provided reasonable opportunity to correct any error in such information.

"(e) A cable operator shall destroy personally identifiable information if the information is no longer necessary for the purpose for which it was collected and there are no pending requests or orders for access to such information under subsection (d) or pursuant to a court order.

"(f)(1) Any person aggrieved by any act of a cable operator in violation of this section may bring a civil action in a United States district court.

"(2) The court may award—

"(A) actual damages but not less than liquidated damages computed at the rate of $100 a day for each day of violation or $1,000, whichever is higher;

"(B) punitive damages; and

"(C) reasonable attorneys' fees and other litigation costs reasonably incurred.

"(3) The remedy provided by this section shall be in addition to any other lawful remedy available to a cable subscriber.

"(g) Nothing in this title shall be construed to prohibit any State or any franchising authority from enacting or enforcing laws consistent with this section for the protection of subscriber privacy.

"(h) A governmental entity may obtain personally identifiable information concerning a cable subscriber pursuant to a court order only if, in the court proceeding relevant to such court order—

"(1) such entity offers clear and convincing evidence that the subject of the information is reasonably suspected of engaging in criminal activity and that the information sought would be material evidence in the case; and

"(2) the subject of the information is afforded the opportunity to appear and contest such entity's claim.

"CONSUMER PROTECTION

"SEC. 632. (a) A franchising authority may require, as part of a franchise (including a franchise renewal, subject to section 626), provisions for enforcement of—

"(1) customer service requirements of the cable operator; and

"(2) construction schedules and other construction-related requirements of the cable operator.

"(b) A franchising authority may enforce any provision, contained in any franchise, relating to requirements described in paragraph (1) or (2) of subsection (a), to the extent not inconsistent with this title.

"(c) Nothing in this title shall be construed to prohibit any State or any franchising authority from enacting or enforcing any consumer protection law, to the extent not inconsistent with this title.

"UNAUTHORIZED RECEPTION OF CABLE SERVICE

"SEC. 633. (a)(1) No person shall intercept or receive or assist in intercepting or receiving any communications service offered over a cable system, unless specifically authorized to do so by a cable operator or as may otherwise be specifically authorized by law.

"(2) For the purpose of this section, the term 'assist in intercepting or receiving' shall include the manufacture or distribution of equipment intended by the manufacturer or distributor (as the case may be) for unauthorized reception of any communications service offered over a cable system in violation of subparagraph (1).

"(b)(1) Any person who willfully violates subsection (a)(1) shall be fined not more than $1,000 or imprisoned for not more than 6 months, or both.

"(2) Any person who violates subsection (a)(1) willfully and for purposes of commercial advantage or private financial gain shall be fined not more than $25,000 or imprisoned for not more than 1 year, or both, for the first such offense and shall be fined not more than $50,000 or imprisoned for not more than 2 years, or both, for any subsequent offense.

"(c)(1) Any person aggrieved by any violation of subsection (a)(1) may bring a civil action in a United States district court or in any other court of competent jurisdiction.

"(2) The court may—

"(A) grant temporary and final injunctions on such terms as it may deem reasonable to prevent or restrain violations of subsection (a)(1);

"(B) award damages as described in paragraph (3); and

"(C) direct the recovery of full costs, including awarding reasonable attorneys' fees to an aggrieved party who prevails.

"(3)(A) Damages awarded by any court under this section shall be computed in accordance with either of the following clauses:

"(i) the party aggrieved may recover the actual damages suffered by him as a result of the violation and any profits of the violator that are attributable to the violation which are not taken into account in computing the actual damages; in determining the violator's profits, the party aggrieved shall be required to prove only the violator's gross revenue, and the violator shall be required to prove his deductible expenses and the elements of profit attributable to factors other than the violation; or

"(ii) the party aggrieved may recover an award of statutory damages for all violations involved in the action, in a sum of not less than $250 or more than $10,000 as the court considers just.

"(B) In any case in which the court finds that the violation was committed willfully and for purposes of commercial advantage or private financial gain, the court in its discretion may increase the award of damages, whether actual or statutory under subparagraph (A), by an amount of not more than $50,000.

"(C) In any case where the court finds that the violator was not aware and had no reason to believe that his acts constituted a violation of this section, the court in its discretion may reduce the award of damages to a sum of not less than $100.

"(D) Nothing in this title shall prevent any State or franchising authority from enacting or enforcing laws, consistent with this section, regarding the unauthorized interception or reception of any cable service or other communications service.

"EQUAL EMPLOYMENT OPPORTUNITY

"SEC. 634. (a) This section shall apply to any corporation, partnership, association, joint-stock company, or trust engaged primarily in the management or operation of any cable system.

"(b) Equal opportunity in employment shall be afforded by each entity specified in subsection (a), and no person shall be discriminated against in employment by such entity because of race, color, religion, national origin, age, or sex.

"(c) Any entity specified in subsection (a) shall establish, maintain, and execute a positive continuing program of specific practices designed to ensure equal opportunity in every aspect of its employment policies and practices. Under the terms of its program, each such entity shall—

"(1) define the responsibility of each level of management to ensure a positive application and vigorous enforcement of its policy of equal opportunity, and establish a procedure to review and control managerial and supervisory performance;

"(2) inform its employees and recognized employee organizations of the equal employment opportunity policy and program and enlist their cooperation;

"(3) communicate its equal employment opportunity policy and program and its employment needs to sources of qualified applicants without regard to race, color, religion, national origin, age, or sex, and solicit their recruitment assistance on a continuing basis;

"(4) conduct a continuing program to exclude every form of prejudice or discrimination based on race, color, religion, national origin, age, or sex, from its personnel policies and practices and working conditions; and

"(5) conduct a continuing review of job structure and employment practices and adopt positive recruitment, training, job design, and other measures needed to ensure genuine equality of opportunity to participate fully in all its organizational units, occupations, and levels of responsibility.

"(d)(1) Not later than 270 after the effective date of this section, and after notice and opportunity for hearing, the Commission shall prescribe rules to carry out this section.

"(2) Such rules shall specify the terms under which an entity specified in subsection (a) shall, to the extent possible—

"(A) disseminate its equal opportunity program to job applicants, employees, and those with whom it regularly does business;

"(B) use minority organizations, organizations for women, media, educational institutions, and other potential sources of minority and female applicants, to supply referrals whenever jobs are available in its operation;

"(C) evaluate its employment profile and job turnover against the availability of minorities and women in its franchise area;

"(D) undertake to offer promotions of minorities and women to positions of greater responsibility;

"(E) encourage minority and female entrepreneurs to conduct business with all parts of its operation; and

"(F) analyze the results of its efforts to recruit, hire, promote, and use the services of minorities and women and explain any difficulties encountered in implementing its equal employment opportunity program.

"(3) Such rules also shall require an entity specified in subsection (a) with more than 5 full-time employees to file with the Commission an annual statistical report identifying by race and sex the number of employees in each of the following full-time and part-time job categories:

"(A) officials and managers;

"(B) professionals;

"(C) technicians;

"(D) sales persons;

"(E) office and clerical personnel;

"(F) skilled craft persons;

"(G) semiskilled operatives;

"(H) unskilled laborers; and

"(I) service workers.

The report shall include the number of minorities and women in the relevant labor market for each of the above categories. The statistical report shall be available to the public at the central office and at every location where more than 5 full-time employees are regularly assigned to work.

"(4) The Commission may amend such rules from time to time to the extent necesary to carry out the provisions of this section. Any such amendment shall be made after notice and opportunity for comment.

"(e)(1) On an annual basis, the Commission shall certify each entity described in subsection (a) as in compliance with this section if, on the basis of information in the possession of the Commission, including the report filed pursuant to subsection (d)(3), such entity was in compliance, during the annual period involved, with the requirements of subsections (b), (c), and (d).

"(2) The Commission shall, periodically but not less frequently than every five years, investigate the employment practices of each entity described in subsection (a), in the aggregate, as well as in individual job categories, and determine whether such entity is in compliance with the requirements of subsections (b), (c), and (d), including whether such entity's employment practices deny or abridge women and minorities equal employment opportunities. As part of such investigation, the Commission shall review whether the entity's reports filed pursuant to subsection (d)(3) accurately reflect employee responsibilities in the reported job classifications.

"(f)(1) If the Commission finds after notice and hearing that the entity involved has willfully or repeatedly without good cause failed to comply with the requirements of this section, such failure shall constitute a substantial failure to comply with this title. The failure to obtain certification under subsection

(e) shall not itself constitute the basis for a determination of substantial failure to comply with this title. For purposes of this paragraph, the term 'repeatedly', when used with respect to failures to comply, refers to 3 or more failures during any 7-year period.

"(2) Any person who is determined by the Commission, through an investigation pursuant to subsection (e) or otherwise, to have failed to meet or failed to make best efforts to meet the requirements of this section, or rules under this section, shall be liable to the United States for a forfeiture penalty of $200 for each violation. Each day of a continuing violation shall constitute a separate offense. Any entity defined in subsection (a) shall not be liable for more than 180 days of forfeitures which accrued prior to notification by the Commission of a potential violation. Nothing in this paragraph shall limit the forfeiture imposed on any person as a result of any violation that continues subsequent to such notification. In addition, any person liable for such penalty may also have any license under this Act for cable auxiliary relay service suspended until the Commission determines that the failure involved has been corrected. Whoever knowingly makes any false statement or submits documentation which he knows to be false, pursuant to an application for certification under this section shall be in violation of this section.

"(3) The provisions of paragraphs (3) and (4), and the last 2 sentences of paragraph (2), of section 503(b) shall apply to forfeitures under this subsection.

"(4) The Commission shall provide for notice to the public and appropriate franchising authorities of any penalty imposed under this section.

"(g) Employees or applicants for employment who believe they have been discriminated against in violation of the requirements of this section, or rules under this section, or any other interested person, may file a complaint with the Commission. A complaint by any such person shall be in writing, and shall be signed and sworn to by that person. The regulations under subsection (d)(1) shall specify a program, under authorities otherwise available to the Commission, for the investigation of complaints and violations, and for the enforcement of this section.

"(h)(1) For purposes of this section, the term 'cable operator' includes any operator of any satellite master antenna television system, including a system described in section 602(6)(A).

"(2) Such term does not include any operator of a system which, in the aggregate, serves fewer than 50 subscribers.

"(3) In any case in which a cable operator is the owner of a multiple unit dwelling, the requirements of this section shall only apply to such cable operator with respect to its employees who are primarily engaged in cable telecommunications.

"(i)(1) Nothing in this section shall affect the authority of any State or any franchising authority—

"(A) to establish or enforce any requirement which is consistent with the requirements of this section, including any requirement which affords equal employment opportunity protection for employees;

"(B) to establish or enforce any provision requiring or encouraging any cable operator to conduct business with enterprises which are owned or controlled by members of minority groups (as defined in section 309(i)(3)(C)(ii) or which have their principal operations located within the community served by the cable operator; or

"(C) to enforce any requirement of a franchise in effect on the effective date of this title.

"(2) The remedies and enforcement provisions of this section are in addition to, and not in lieu of, those available under this or any other law.

"(3) The provisions of this section shall apply to any cable operator, whether operating pursuant to a franchise granted before, on, or after the date of the enactment of this section.

"JUDICIAL PROCEEDINGS

"SEC. 635. (a) Any cable operator adversely affected by any final determination made by a franchising authority under section 625 or 626 may commence an action within 120 days after receiving notice of such determination, which may be brought in—

"(1) the district court of the United States for any judicial district in which the cable system is located; or

"(2) in any State court of general jurisdiction having jurisdiction over the parties.

"(b) The court may award any appropriate relief consistent with the provisions of the relevant section described in subsection (a).

"COORDINATION OF FEDERAL, STATE, AND LOCAL AUTHORITY

"SEC. 636. (a) Nothing in this title shall be construed to affect any authority of any State, political subdivision, or agency thereof, or franchising authority, regarding matters of public health, safety, and welfare, to the extent consistent with the express provisions of this title.

"(b) Nothing in this title shall be construed to restrict a State from exercising jurisdiction with regard to cable services consistent with this title.

"(c) Except as provided in section 637, any provision of law of any State, political subdivision, or agency thereof, or franchising authority, or any provision of any franchise granted by such authority, which is inconsistent with this Act shall be deemed to be preempted and superseded.

"(d) For purposes of this section, the term 'State' has the meaning given such term in section 3(v).

"EXISTING FRANCHISES

"SEC. 637. (a) The provisions of—

"(1) any franchise in effect on the effective date of this title, including any such provisions which relate to the designation, use, or support for the use of channel capacity for public, educational, or governmental use, and

"(2) any law of any State (as defined in section 3(v)) in effect on the date of the enactment of this section, or any regulation promulgated pursuant to such law, which relates to such designation, use or support of such channel capacity, shall remain in effect, subject to the express provisions of this title, and for not longer than the then current remaining term of the franchise as such franchise existed on such effective date.

"(b) For purposes of subsection (a) and other provisions of this title, a franchise shall be considered in effect on the effective date of this title if such franchise was granted on or before such effective date.

"CRIMINAL AND CIVIL LIABILITY

"SEC. 638. Nothing in this title shall be deemed to affect the criminal or civil liability of cable programmers or cable operators pursuant to the Federal, State, or local law of libel, slander, obscenity, incitement, invasions of privacy, false or misleading advertising, or other similar laws, except that cable operators shall not incur any such liability for any program carried on any channel designated for public, educational, governmental use or on any other channel obtained under section 612 or under similar arrangements.

"OBSCENE PROGRAMMING

"SEC. 639. Whoever transmits over any cable system any matter which is obscene or otherwise unprotected by the Constitution of the United States shall be fined not more than $10,000 or imprisoned not more than 2 years, or both."

JURISDICTION

SEC. 3. (a)(1) Section 2(a) of the Communications Act of 1934 is amended by adding at the end thereof the following: "The provisions of this Act shall apply with respect to cable service, to all persons engaged within the United States in providing such service, and to the facilities of cable operators which relate to such service, as provided in title VI.".

(2) Section 2(b) of such Act is amended by inserting after "section 301" the following: "and title VI".

(b) The provisions of this Act and amendments made by this Act shall not be construed to affect any jurisdiction the Federal Communications Commission may have under the Communications Act of 1934 with respect to any communication by wire or radio (other than cable service, as defined in section 602(5) of such Act) which is provided through a cable system, or persons or facilities engaged in such communications.

POLE ATTACHMENTS

SEC. 4. Section 224(c) of the Communications Act of 1934 is amended by adding at the end thereof the following new paragraph:

"(3) For purposes of this subsection, a State shall not be considered to regulate the rates, terms, and conditions for pole attachments—

"(A) unless the State has issued and made effective rules and regulations implementing the State's regulatory authority over pole attachments; and

"(B) with respect to any individual matter, unless the State takes final action on a complaint regarding such matter—

"(i) within 180 days after the complaint is filed with the State, or

"(ii) within the applicable period prescribed for such final action in such rules and regulations of the State, if the prescribed period does not extend beyond 360 days after the filing of such complaint.".

UNATHORIZED RECEPTION OF CERTAIN COMMUNICATIONS

SEC. 5. (a) Section 705 of the Communications Act of 1934 (as redesignated by section 6) is amended by inserting "(a)" after the section designation and by adding at the end thereof the following new subsections:

"(b) The provisions of subsection (a) shall not apply to the interception or receipt by any individual, or the assisting (including the manufacture or sale) of such interception or receipt, of any satellite cable programming for private viewing if—

"(1) the programming involved is not encrypted; and

"(2)(A) a marketing system is not established under which—

"(i) an agent or agents have been lawfully designated for the purpose of authorizing private viewing by individuals, and

"(ii) such authorization is available to the individual involved from the appropriate agent or agents; or

"(B) a marketing system described in subparagraph (A) is established and the individuals receiving such programming has obtained authorization for private viewing under that system.

"(c) For purposes of this section—

"(1) the term 'satellite cable programming' means video programming which is transmitted via satellite and which is primarily intended for the direct receipt by cable operators for their retransmission to cable subscribers;

"(2) the term 'agent', with respect to any person, includes an employee of such person;

"(3) the term 'encrypt', when used with respect to satellite cable programming, means to transmit such programming in a form whereby the aural and visual characteristics (or both) are modified or altered for the purpose of preventing the unauthorized receipt of such programming by persons without authorized equipment which is designed to eliminate the effects of such modification or alteration;

"(4) the term 'private viewing' means the viewing for private use in an individual's dwelling unit by means of equipment, owned or operated by such individual, capable of receiving satellite cable programming directly from a satellite; and

"(5) the term 'private financial gain' shall not include the gain resulting to any individual for the private use in such individual's dwelling unit of any programming for which the individual has not obtained authorization for that use.

"(d)(1) Any person who willfully violates subsection (a) shall be fined not more than $1,000 or imprisoned for not more than 6 months, or both.

"(2) Any person who violates subsection (a) willfully and for purposes of direct or indirect commercial advantage or private financial gain shall be fined not more than $25,000 or imprisoned for not more than 1 year, or both, for the first such conviction and shall be fined not more than $50,000 or imprisoned for not more than 2 years or both for any subsequent conviction.

"(3)(A) Any person aggrieved by any violation of subsection (a) may bring a civil action in a United States district court or in any other court of competent jurisdiction.

"(B) The court may—

"(i) grant temporary and final injunctions on such terms as it may deem reasonable to prevent or restrain violations of subsection (a);

"(ii) award damages as described in subparagraph (C); and

"(iii) direct the recovery of full costs, including awarding reasonable attorneys' fees to an aggrieved party who prevails.

"(C)(i) Damages awarded by any court under this section shall be computed, at the election of the aggrieved party, in accordance with either of the following subclauses;

"(I) the party aggrieved may recover the actual damages suffered by him as a result of the violation and any profits of the violator that are attributable to the violation which are not taken into account in computing the actual damages; in determining the violator's profits, the party aggrieved shall be required to prove only the violator's gross revenue, and the violator shall be required to prove his deductible expenses and the elements of profit attributable to factors other than the violation; or

"(II) the party aggrieved may recover an award of statutory damages for each violation involved in the action in a sum of not less than $250 or more than $10,000, as the court considers just.

"(ii) In any case in which the court finds that the violation was committed willfully and for purposes of direct or indirect commercial advantage or private financial gain, the court in its discretion may increase the award of damages, whether actual or statutory, by an amount of not more than $50,000.

"(iii) In any case where the court finds that the violator was not aware and had no reason to believe that his acts constituted a violation of this section, the court in its discretion may reduce the award of damages to a sum of not less than $100.

"(4) The importation, manufacture, sale, or distribution of equipment by any person with the intent of its use to assist in any activity prohibited by subsection (a) shall be subject to penalties and remedies under this subsection to the same extent and in the same manner as a person who has engaged in such prohibited activity.

"(5) The penalties under this subsection shall be in addition to those prescribed under any other provision of this title.

"(6) Nothing in this subsection shall prevent any State, or political subdivision thereof, from enacting or enforcing any laws with respect to the importation, sale, manufacture, or distribution of equipment by any person with the intent of its use to assist in the interception or receipt of radio communications prohibited by subsection (a).

"(e) Nothing in this section shall affect any right, obligation, or liability under title 17, United States Code, any rule, regulation, or order thereunder, or any other applicable Federal, State, or local law.".

"(b) The amendments made by subsection (a) shall take effect on the effective date of this Act.

TECHNICAL AND CONFORMING AMENDMENTS

SEC. 6. (a) Title VI of the Communications Act of 1934 (as in effect before the enactment of this Act) is redesignated as title VII, and sections 601 through 610 are redesignated as sections 701 thorugh 710, respectively.

(b)(1) Section 309(h) of the Communications Act of 1934 is amended by striking out "section 606" and inserting in lieu thereof "section 706".

(2) Section 2511 of title 18, United States Code, is amended—

(A) in subsection (2)(e), by striking out "section 605 or 606" and inserting in lieu thereof "section 705 or 706"; and

(B) in subsection (2)(f), by striking out "section 605" and inserting in lieu thereof "section 705".

(3) Section 105(f)(2)(C) of the Foreign Intelligence Surveillance Act of 1978 (50 U.S.C. 1805(f)(2)(C)) is amended by striking out "section 605" and inserting in lieu thereof "section 705".

SUPPORT OF ACTIVITIES OF THE UNITED STATES TELECOMMUNICATIONS TRAINING INSTITUTE

SEC. 7. Nothing in this Act, the Communications Act of 1934, or any other Act, shall be construed to preclude the Federal Communications Commission or the National Telecommunications and Information Administration within the Department of Commerce from participation (including use of staff and other appropriate resources) in support of any activities of the United States Telecommunications Training Institute.

TELECOMMUNICATIONS POLICY STUDY COMMISSION

SEC. 8. Title VII of the Communications Act of 1934 (as redesignated by section 6 of this Act) is amended by adding at the end thereof the following new section:

"TELECOMMUNICATIONS POLICY STUDY COMMISSION

"SEC. 711. (a) There is hereby established the Telecommunications Policy Study Commission (hereinafter in this section referred to as the 'Commission') which shall—

"(1) compare various domestic telecommunications policies of the United States and other nations, including the impact of all such policies on the regulation of interstate and foreign commerce, and

"(2) prepare and transmit a written report thereon to the Congress, the President, and the Federal Communications Commission.

"(b)(1) Such Commission shall be composed of the chairman and ranking minority members of the Committee on Commerce, Science, and Transportation and the Communications Subcommittee of the Senate and the Committee on Energy and Commerce and the Telecommunications, Consumer Protection and Finance Subcommittee of the House of Representatives (or delegates of such chairmen or members appointed by them from among members of such committees).

"(2) The chairmen of such committees (or their delegates) shall be co-chairmen of the Commission.

"(c)(1) The report under subsection (a)(2) shall be submitted not later than December 1, 1987. Such report shall contain the results of all Commission studies and investigations under this section.

"(2) The Commission shall cease to exist—

"(A) on December 1, 1987, if the report is not submitted in accordance with paragraph (1) on the date specified therein; or

"(B) on such date (but not later than May 1, 1988) as may be determined by the Commission, by order, if the report is submitted in accordance with paragraph (1) on the date specified in such paragraph.

"(d)(1) The members of the Commission who are not officers or employees of the United States, while attending conferences or meetings of the Commission or while otherwise serving at the request of the chairmen, shall be entitled to receive compensation at a rate not in excess of the maximum rate of pay for grade GS-18, as provided in the General Schedule under section 5332 of title 5 of the United States Code, including travel time, and while away from their homes or regular places of business, they may be allowed travel expenses, including per diem in lieu of subsistence as authorized by law (5 U.S.C. 5703) for persons in the Government service employed intermittently.

"(2) The Commission may appoint and fix the pay of such staff as it deems necessary.

"(e)(1) In conducting its activities, the Commission may enter into contracts to the extent it deems necessary to carry out its responsibilities, including contracts with nongovernmental entities that are competent to perform research or investigations in areas within the Comission's responsibilities.

"(2) The Commission is authorized to hold public hearing, forums, and other meetings to enable full public participation.

"(f) The heads of the departments, agencies, and instrumentalities of the executive branch of the Federal Government shall cooperate with the Commission in carrying out this section and shall furnish to the Commission such information as the Commission deems necessary to carry out this section, in accordance with otherwise applicable law.

"(g) There are authorized to be appropriated such sums as may be appropriated to carry out this section for a period of three fiscal years.

"(h) Activities authorized by this section may be carried out only with funds and to the extent approved in appropriation Acts.

"(i) Nothing in this section shall be construed to affect any proceedings by, or activities of, the Federal Communications Commission, except that the Federal Communications Commission shall consider submissions by the Commission submitted pursuant to subsection (a)(2).".

EFFECTIVE DATE

SEC. 9. (a) Except where otherwise expressly provided, the provisions of this Act and the amendments made thereby shall take effect 60 days after the date of enactment of this Act.

(b) Nothing in section 623 or 624 of the Communications Act of 1934, as added by this Act, shall be construed to allow a franchising authority, or a State or any political subdivision of a State, to require a cable operator to restore, retier, or reprice any cable service which was lawfully eliminated, retiered, or repriced as of September 26, 1984.

Approved October 30, 1984.

A Suggested Procedure for Assessment of Communication Needs for New Franchises or Renewal

ASSESSMENT OF COMMUNICATION NEEDS: MEETING WITH INFORMED CITIZENS[1]

This procedure is written from the perspective of a *franchising authority* assessing community needs before seeking a cable franchisee or renewing an existing franchise. The same procedure could be followed by a *cable company* if it is to be asked by the franchising authority to assess community needs prior to applying for a franchise or renewal.

To determine communication needs initially, the cable commission (advisory body) may convene a meeting of representatives of various interest in the community to discuss local needs. One or two meetings with a variety of persons should provide an inventory of communication needs and also serve a community education function.

STEP 1. SELECTION OF PARTICIPANTS

Potential participants should be discussed and selected at an advisory body meeting. Agreement to participate should be secured from at least one community leader or informed citizen in all areas of interest in the community. As a checklist of such areas, the following categories might be helpful:

[1]Adapted from Thomas F. Baldwin, Thomas A. Muth, and Michael O. Wirth, "Assessment of Communication Needs: Meeting with Informed Citizens," *Cable Communications in Minnesota: Suggestions for Systematic Development*, Minnesota Cable Communications Board, Bloomington, Minnesota, August 1975.

- Economic (e.g., business, labor, agriculture)
- Social services
- Government
- Education
- Youth
- Religion
- Recreation
- Health
- Human relations (e.g., minority-majority relations)

The checklist will have to be adapted to the particular community and the range of interests in that community. Backup persons should be identified in case the originally selected persons are not available.

STEP 2. INVITATION OF PARTICIPANTS

The participants selected should be invited by phone, letter, or personal visit, at which time they should be informed of the responsibility of the advisory body to assess communication needs. They should be assured that they are not expected to be communication experts—that the greatest interest is in their knowledge of the needs and problems of the community and the various communication and public information techniques presently being used.

The invited participants should be made to feel that their participation is important and the results of the inquiry will be reflected in the cable franchise agreement.

Two or more meetings may be necessary to cover all the community representatives who should participate.

STEP 3. MEETING

To be most productive, the meeting should be structured somewhat so that the participants know the interests of the advisory body and to ensure that all the potential uses of cable communication are covered.

Introduction

Begin by reminding people of the information supplied in the invitation to participate: (1) the advisory body responsibility to assess communication needs, (2) the importance to potential applicants of knowing the special communication needs of a community, and (3) the importance to the city government of knowing communication needs prior to evaluating franchise or renewal applications.

Background on Cable Communication

Many of the participants may need to know more about cable television. A few minutes should be spent reviewing how cable works and the kinds of services it can provide.

Present Communication Methods

The discussion should be opened by asking the participants about present communication methods and problems in the community. Specific types of communication should be suggested. For example,

- Public relations and public information
  Communication of ideas, plans, advice, news, etc. Seeking feedback or response.
  Citizenship information: political office campaigns and platforms, public meetings, hearings, public safety information, speeches, committee, commission and board meetings, "know your government" programs, news conferences, patriotic and holiday events, etc.
- Specialized news and announcements
  Announcements, news, meetings, special events and other information intended for specialized groups such as trade or professional associations, employees, unions, school parents' associations, civic groups, service clubs, youth groups, church groups, etc.
- Education
  Job training, adult education, cultural enrichment, remedial programs, etc.
- Data transfer
  Moving information from one place to another within a community: business records, real property descriptions, court calendars, school schedules, inventories, medical records, etc.
- Weather and disaster information
  Tornado, flood, hail, and other storm warnings. Civil defense information.
- Communication resources in the community
  Audio and video equipment in private and public organizations (e.g., schools, government agencies, businesses, hospitals, libraries).

General Needs and Problems

Since it may not occur to all participants that specific problems may be addressed by certain communication techniques, the participants should be asked about the most important problems and needs in the community, not necessarily those related to communication. As these problems are expressed, in some cases it may become apparent that there are aspects of the problem that may be approached by communication. The group process of attempting to apply cable communication to these general problems may produce some creative ideas. And, at the same time, some thoughts that may later bear fruit will have been planted.

STEP 4. RECORDING AND SUMMATION

During the meeting(s), one of the advisory body members will have to take notes. At another meeting of the advisory body, an "initial statement of communication needs" should be prepared, with each need described in as much detail as the assessment meeting provided.

STEP 5. PUBLIC COMMENT

The "initial statement" should be read at a city council meeting and given to the local newspaper and other media. It should be available thereafter at city hall. The public should be invited to comment on the statement and make additions. This procedure permits adequate opportunity for public input. After a suitable period has elapsed, on the basis of the public comments, the "initial statement" should be revised if necessary and a final "statement of communication needs" agreed upon.

STEP 6. DISTRIBUTION

Each cable system making inquiry about the franchise or the cable system seeking renewal should receive a copy of the "statement of communication needs." The final statement should also be available at city hall and given to the local newspaper and other local media.

Cable Costs and Revenues

Capital Costs

Headend (35-channel system)	$150,000

Distribution plant	
Aerial	10,000 per mile
Underground	15,000 per mile
Home drop (addressable converter, flexible cable, installation)	140 per drop

Annual Operating Budget for Typical Urban Cable System
(60,000 subscribers)

Revenues	Thousands of Dollars	% of Total	Annual Per Subscriber
Basic	$12,300	59.0	$205
Pay	7,800	37.4	130
Installation	250	1.2	4
Advertising	510	2.4	9
	$20,860	100.0	$348
Expenses			
Franchise fees	$560	4.7	$9
Bad debts	311	2.8	5
Programming expenses	3,557	31.5	59
Technical operations	1,653	14.6	28
Installation	851	7.5	14
Customer service/billing	470	4.2	8
Sales & marketing	1,211	10.7	20
Advertising	164	1.5	3
Local programming	311	2.8	5
Finance & accounting	732	6.5	12
Personnel	361	3.3	6
General & administration	1,119	9.9	19
	$11,300*	100.0	$188

*Operating expenses are usually between 50 and 60 percent of operating revenues.

Sample Advertising Production Rates

Production Rate Card

I.	*Formatted Commercial*	
	—Initial Client Consultation	No Charge
	—Scripting	$25 per 30 seconds
	—One Camera in Studio	$50 per hour*
	—One Camera Remote	$50 per hour**
	—Editing	$50 per hour
	—Voice Over	$20 per hour
	Average commercial package price	$250.00

Each formatted commercial includes music bed and character generator. Talent provided at cost to client.

II.	*Additional Services*			
	Audio/Visual Tag	$20	Chroma Key $20/hr.*
	Revoice Copy	$20	Tape Dubs $20/hr.*
	Reedit Video	$50/hr.	Extras available at cost***
	Slide Transfer	$10/slide	

*One Hour Minimum charge, (client provides tape for duplication)

**Two Hour Minimum charge

***Extras include props, artwork, logos, costumes, actors, etc.

Rates listed above are for information only and do not constitute an offer on behalf of _____. We reserve the right to add charges for additional equipment and contracted labor needed for productions. We will gladly provide an estimate in advance.

30-Spot Contract	*60-Spot Contract*	*300-Spot Contract*
Client receives 1 30 second spot at out-of-pocket cost.	Client receives 2 30 second spots at out-of-pocket cost.	Client receives 4 30 second spots at out-of-pocket cost.
Additional spots at standard rates.	Additional spots at standard rates.	Additional spots at standard rates.

Cable Audience Survey Methods

This is a set of procedures for conducting a cable audience study for programming, marketing, or advertising purposes.

SAMPLING

For surveys, a *sample* of the desired universe is sufficient. The necessary accuracy will determine the number in the sample. Other things being equal, the larger the sample, the lower the sampling error. But this is not a direct relationship. Because the size of the sample must be quadrupled to halve the error, there is a high cost to reducing error. For most purposes, a sample of about 400 is sufficient.

The sample is drawn from a *frame*. The frame is any listing of units in which one is interested. In cable, the desired frames should already be on hand and quite complete. The frames, or universes, that might be surveyed are all homes passed, nonsubscribers, basic subscribers, and pay subscribers (perhaps broken out by the various combinations or packages available). Whatever the frame, it must contain *all* of the households to be surveyed.

DETAILED SAMPLING DEMONSTRATION

For illustration here, we assume that the interest is in surveying nonsubscribers about their knowledge of cable and the appeal of specific new channels.

The cable system computer, or other recordkeeping system, should be capable of producing a complete list of nonsubscribers. The computer may also be programmed to draw a sample, but for these purposes we assume a physical list (printout) of nonsubscribers. If there are 20,000 nonsubscribers, and we wish to sample 400, we divide 400 into 20,000 to determine that we need one household for every 50 households on the list. This number is known as the *skip interval.*

We need to pick a random starting point within that interval. Use a table of random numbers in a statistics text or an almanac. In this case we enter the list blindly, land on a number, and use its last two digits. If the last two digits exceed 50, we keep dropping down the list until a number between 01 and 50 appears. If no table of random numbers is handy, we may use the numbers in a telephone book in the same way. Suppose that the number picked is 28. Thus we start at the head of the list (frame), count to number 28, and select that household as the first of our sample. From that point, 50 households are counted off so that we take households 78, 128, 178, and so on until we get to 19,978, at which point we will have our 400-household sample. To avoid the tedium of counting, a template or rule may be used to measure off the space on the list taken by 50 households.

If the frame is mixed—for example, subscribers and nonsubscribers together, with the subscribers noted by a code of some sort—a similar procedure is used to obtain a sample of nonsubscribers. The total number in the list is known. If there are 44,000 total, we divide 400 into 44,000 to get a skip interval of 110. We select a number from 1 to 110 at random. When we enter the list and get to the randomly determined starting point, if that household is a nonsubscriber, we take it and count off 110 from that point. If any household we land on is not a nonsubscriber, we simply go down the list until we do reach a nonsubscriber and take that household, *but we must go back to the originally selected (but rejected household), to resume the counting.* Suppose that our starting point is household 61 and that it is a nonsubscriber. We take it. The next one would be household number 171. It is a subscriber and so is 172 and 173. Number 174 is not a subscriber, so household 174 becomes the second member of our sample. But, to count off the next 110, we return to 171 and start the counting from that point.

SAMPLING ERROR

What we have just described is a *systematic sample.* It is a *probability sample*, meaning that every household has had an equal chance of being selected. The laws of probability tell us that it is representative of the whole universe from which it is drawn; that is, the sample should have the same characteristics as the universe. In our example, the knowledge of cable and the interest in specific new channels would be about the same in the sample of nonsubscribers as in the whole group of nonsubscribers. They would not be *exactly* the same because of *sampling error.* There would be some difference because, in the luck of the draw, we may have gotten a few too many people who, in our knowledge of cable questions let us say, don't realize there are 24-hour movie channels on cable. Perhaps if we took a *census* of all 20,000 nonsubscribers, we might find 27 percent of the people who do not know about 24-hour movie channels and in our sample of 400 there are 29 percent. The difference can be attributed to sampling error. It may be calculated for a specific sample size and sampling procedure to give a plus or minus range of accuracy with a specific probability. For almost all purposes in marketing and programming research in cable, it can be assumed that the result actually obtained is a good estimate of the universe studied without actually calculating the sampling error.

NONRESPONSE BIAS

In all surveys we can expect a certain amount of *nonresponse*. People are not home when contacted, even after a series of callbacks, or they refuse to cooperate, or in a telephone survey, the number has been disconnected. To have a sample that is representative of the whole, it is absolutely essential to keep the nonresponse to a minimum. The response rate should be 70 percent or better. The danger is in the probability that the nonrespondents will be somehow systematically different from the respondents. For example, if we are trying to judge how many nonsubscribers would be attracted to a particular new cable channel, and we project our estimate from a group of respondents who were home when we called, and 20 percent of the people in the sample were not home after the third callback, then we might have overestimated the likely "lift" that the new channel would provide. Why? Because the people who are not at home are less likely to be interested in the new cable service. They are not home very often to watch television. Our projection would be based on the stay-at-homes who have greater use for television. This is an example of a systematic nonresponse bias which will throw our results off a little. It is the reason for making several callbacks and trying to get absolutely as many people in the sample as possible.

If our original sample were 400 and we got 20 percent who are not home after the third callback and 10 percent refusals and disconnected phones, then we would have 280 people in our final tabulation. It does not help to remove the nonresponse bias if we add additional members to the sample. We will get roughly the same nonresponse rate from the new households sampled, and so the *proportion* of those who are not at home much for television viewing will remain the same. *The point here is that nonresponse error cannot be corrected by increasing the sample, only by maximizing the likelihood of response by callbacks and proper interviewing techniques.*

Because of the very high nonresponse problem in mail surveys, they are not especially useful. A mail questionnaire, sent to subscribers along with the bill, even with an incentive to respond, is not likely to exceed a 30 percent response rate. This is too low to be reliable. There is a high probability that the other 70 percent is systematically different, in a number of unknown ways, from those who do respond. In special circumstances where you are almost certain of a high response rate, mail may be economical.

QUESTIONNAIRES

The purpose of the survey, along with another formal statement about how the information will be used, should be written out clearly before beginning to write the questionnaire. This focuses the questions and keeps the survey on track.

A survey questionnaire is introduced with a brief statement about its purpose. The sponsor is identified. Sometimes the real sponsor is disguised by using a research organization title. This could be an ambiguous research unit of the cable company in an in-house survey. The introduction does not ask cooperation (giving the respondent a chance to say no), but goes immediately to an easy or interesting question to get the respondent started and committed. ("Do you own a television set?")

The rest of the questionnaire should be simple and direct. The questions should not use words that a portion of the study population would not know. The respondent should not have to have a question repeated (a clue to an awkward question). Although it is a great temptation to ask everything that seems interesting, no question, other than perhaps an introductory throwaway question, should be in the questionnaire that does not relate directly to the prestated purpose of the survey and its specific use.

Simple surveys for internal use can be carried out by staff members, and it is important to develop that capability, internally. More complicated surveys might be contracted to one of several research firms, many with experience in cable. Listings are available in the *Broadcasting/Cablecasting Yearbook*.

AUDIENCE COUNT RESEARCH

Audience count research for individual television markets is usually done by telephone or by a combination of telephone and diary, and occasionally by personal interview. The objective of this kind of research is to find out how many people are viewing various channels and programs and how these audiences are demographically composed.

The best technique may be a telephone coincidental method whereby a sample of subscribers is interviewed by phone for each hour of the day in which audience data are desired. Subscribers are simply asked if they are presently viewing and, if so, what channel and program. This might be asked for each member of the household, and it might be necessary to ask if the set being viewed is connected to the cable. Finally, there may be a question or two about household demographics. Because these interviews are very short, quite a few can be conducted. It would require several hundred people for each hour. A variation would be to ask people, rather than what they are viewing coincidentally, to recall their viewing for the day part immediately past (e.g., in the early afternoon for the morning). Or the respondent may be asked to recall viewing for the day before. In this case the day parts should be used as a prompt to the recall. ("What cable programs or channels did you watch before 9 A.M.?" "What programs or channels between 9 A.M. and noon?" "Between noon and 5 P.M.?" "Between 5 P.M. and 8 P.M.?" "Between 8 P.M. and 11 P.M.?" "Between 11 P.M. and 1 A.M.?" "After 1 A M ?"

Another method of audience research is to present the respondent with a list of channels or programs and ask if each had been viewed in some past period (usually the past week). This may be done by phone, if the list is not too long, for example, only the local origination programming or only the automated channels. For a longer list, perhaps covering all services on all channels, a personal interview with a checklist that is actually handed to the respondent and simultaneously read by the interviewer is most appropriate.

In the diary method, a sample of people is called by phone and asked to cooperate. This procedure is called *placement*. Those who agree are mailed a diary in which to record viewing, usually at 15-minute intervals throughout the day for seven days. Although this appears to be the most practical method of collecting audience information, it has some weaknesses. A substantial proportion of the sample is unwilling to keep the diary and, among those who say they will, there are many who do not follow through. Another difficulty is forgetting to keep up. Many people neglect the diary for several days and then try to recall a long period of previous viewing. Because there are so many channels on a cable system the diaries are complex. Recall of viewing is difficult.

If audiences are being counted to determine the levels of interest in the various channel offerings for programming purposes (e.g., to make a decision about which channel to drop in order to add a pay service to a 12-channel system or to determine whether a local origination program is attracting enough viewers to justify the expense), then in-house research conducted by the cable system staff is adequate, probably through one of the telephone methods described earlier.

On the other hand, if the goal is to document audience for selling advertising time, then it is necessary to go outside. In-house staff research would not be credible to advertisers and advertising agencies. Furthermore, advertising agencies are habituated to Nielsen and Arbitron data for television, and it would be worth considering a custom survey by one of those companies. Some agencies will not use data from any other source. This would be most important if the goal is to use the data for selling national advertising. Local agencies might be willing to accept audience studies from a reputable local research agency.

Index